Selected Titles in This Series

(Continued in the back of this publication)

Sign-based Methods in Linear Statistical Models

Translations of
MATHEMATICAL
MONOGRAPHS

Volume 162

Sign-based Methods in Linear Statistical Models

M. V. Boldin
G. I. Simonova
Yu. N. Tyurin

American Mathematical Society
Providence, Rhode Island

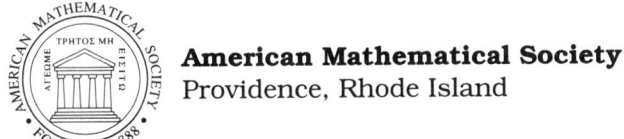

ЗНАКОВЫЕ МЕТОДЫ В ЛИНЕЙНЫХ СТАТИСТИЧЕСКИХ МОДЕЛЯХ

Translated by D. M. Chibisov from an original Russian manuscript.

Supported by the Russian Foundation for Fundamental Research, Grant #95-01-01282

1991 *Mathematics Subject Classification.* Primary 62G; Secondary 62G05, 62G10, 62G20, 62G35, 62M10.

Abstract. In the book a new nonparametric approach to the analysis of statistical data is exposed. It consists in using only the signs of observations or of certain functions of them depending on the data structure. Hence the approach is referred to as sign based. The book treats regression and autoregression models important for applications. For these models the sign-based methods yield the solutions of the principal statistical problems (parameter estimation, hypothesis testing, etc.). Both finite-sample and large-sample properties of the sign procedures are studied. The sign procedures are shown to be robust with respect to gross errors in the data. Numerical algorithms to implement the sign analysis are proposed, and examples of their application to real and simulated data are given. The exposition evolves from elementary to advanced theory to make the book accessible to a broader readership. The book is intended for those studying or applying mathematical statistics.

Library of Congress Cataloging-in-Publication Data

Boldin, M. V.
Sign-based methods in linear statistical models / M. V. Boldin, G. I. Simonova, Yu. N. Tyurin.
p. cm. — (Translations of mathematical monographs, ISSN 0065-9282; v. 162)
Includes bibliographical references.
ISBN 0-8218-0371-9 (alk. paper)
1. Linear models (Statistics) I. Simonova, G. I. (Galina I.) II. Tiurin, IUriĭ Nikolaevich. III. Title. IV. Series.
QA279.B645 1997
519.5′35–dc21 97-3452
CIP

∞ The paper used in this book is acid-free and falls within the guidelines established to ensure permanence and durability.

10 9 8 7 6 5 4 3 2 1 02 01 00 99 98 97

Contents

Preface

For nonparametric statistics, the last half of this century was the time when rank-based methods originated, developed vigorously, reached maturity, and received wide recognition. The rank-based approach in statistics consists in ranking the observed values and using only the ranks rather than the original numerical data. In fitting relationships to observed data, the ranks of residuals from the fitted dependence are used. About a decade ago we began exploring a similar approach based on the signs of observations or residuals. Some sign procedures have been well known for a long time, for instance, the sign test. We pursued this approach in our theoretical research and applied studies. When the authors of this book gathered at Moscow University and when our work received support from the Russian Foundation for Fundamental Research, we decided to set out this subject systematically. That is how this book came into being.

The sign-based approach hinges on the assumption that random errors take positive or negative values with equal probabilities. Under this assumption the sign procedures are distribution free. Another merit of these procedures is that they are very robust to violations of model assumptions, for instance, to the presence even of a considerable amount of gross errors in observations. Surprising as it may seem, sign procedures have fairly high relative asymptotic efficiency, in spite of the obvious loss of information incurred by the use of signs instead of the corresponding numerical values.

Our particular attention was paid to finite samples, where we strived for exact results, i.e., results giving exact significance levels in hypothesis testing or exact confidence levels in confidence estimation. Their numerical computation turns out to be simple and fast enough for implementation on ordinary computers. For this purpose we worked out a package of computer programs, which was used for computations in the examples to be given in the book. We also explored asymptotic properties of sign rules for an increasing number of observations. This asymptotic analysis constitutes the most technical part of the book. The sign rules presented in the book form a set of nonparametric procedures for data processing, which can be applied in all principal statistical problems, like hypothesis testing (including testing linear hypotheses), point estimation, and setting confidence regions for unknown parameters. Regarding their scope and potentialities, the methods based on signs are quite comparable with rank methods.

We develop sign-based methods in the framework of linear models. In the first part of the book these are linear and factor models involving independent observations. In the second part we consider linear models of time series, primarily autoregressive models. In the Introduction we exemplify the capabilities of the sign-based methods and explain the fundamentals of the nonparametric analysis based on signs.

We tried to make our exposition accessible to the broadest possible readership, who are interested in statistical theory and its nonparametric aspects. This particularly concerns the first part. The basic graduate courses in probability theory and mathematical statistics are sufficient background for reading the book. Some experience in reading mathematical literature is also needed, especially for the second part. Knowledge of classical Gaussian analysis of variance and regression analysis is desirable.

The authors made a different contibution into different chapters of the book. The first part is written mostly by Yu. N. Tyurin, the second part mostly by M. V. Boldin. All the computations were done by G. I. Simonova. She wrote the sections dealing with numerical results.

The next lines are the most pleasant for us. We express in them our gratitude to everyone who one way or another supported our research and helped us to present it here. Unfortunately, we can name only few of them. But it does not mean that we have ungratefully forgotten the rest.

We are indebted to the Moscow State University and its Faculty of Mechanics and Mathematics, from which we all graduated and where we work now, for the stimulating professional environment and for the encouragement and support of our research.

We gratefully revere the memory of Academician B. V. Gnedenko who always was so solicitous about our studies.

We are grateful to Prof. N. M. Sotskii for stimulating discussions of theoretical and applied aspects of the statistical analysis of time series.

We are grateful to Prof. A. S. Sharov who consulted with us on some topics in astronomy touched upon in Chapter 1.

We thank Professors A. P. Korostelev, A. A. Makarov, Ya. Yu. Nikitin, and E. V. Khmaladze for their interest to our work and useful comments.

We thank Prof. D. M. Chibisov for translating this book into English. During this job he made a number of valuable comments to make the text clearer and more accurate, for which we are also deeply grateful.

We acknowledge with gratitude the grant from the Russian Foundation of Fundamental Research, which supported our studies.

We are grateful to our families for all possible assistance and infinite patience.

M. V. Boldin
G. I. Simonova
and Yu. N. Tyurin
Moscow, June 1996

Introduction

This book sets out the nonparametric sign-based approach to the statistical analysis of data. We develop this approach as applied to linear models involving independent observations and autoregressive models of time series.

Nonparametric methods, and the sign-based methods among them, extend the scope of statistical methods as compared to the classical parametric ones, which assume the error distribution to be specified up to a finite number of unknown parameters. Among applied statistical methods, the method of least squares is undoubtedly the most important one. This method allows for deep statistical results subject to the assumption that thc random errors have normal (Gaussian) distribution. This model assumption reflects more or less adequately the behavior of real statistical data in many practical situations. A remarkable feature of the normal model is that it provides for a systematic statistical theory solving all principal statistical problems (point and interval estimation, hypothesis testing) for diverse structural data models. Together with computational simplicity this determines the longlasting popularity of the normal theory rules in applied studies.

But we should remember that the normal model, with its ample theoretical opportunities, imposes restrictive mathematical assumptions which may fail in practical applications. Then the normal theory may lead to wrong conclusions. Usually there is no way to verify the validity of the normal model with sufficient certainty. And what is more, it is frequently impossible to find any parametric family at all to contain the particular error distribution. In these circumstances it is preferable to have a statistical methodology not restricted to any specific parametric family and thus applicable to a more general class of random errors. (Such a methodology is said to be nonparametric). Moreover, it is desirable to work out a theory comparable to the normal theory regarding its scope and capabilities, so that it would provide the researcher with a system of statistical rules for data processing solving all principal statistical problems (point and interval estimation, hypothesis testing) for various structural models.

One nonparametric approach of this kind is the approach based on ranks. It was developed extensively during recent decades and is widely used now. Its main underlying assumption is that the random errors are independent and identically distributed with a continuous common distribution.

In this book we develop another nonparametric methodology based on the assumption that the random errors take positive and negative values with equal probabilities. In many cases they need not be identically distributed. This approach consists in using only signs of the observations or residuals rather than their numerical values, so that we refer to it as the sign-based one.

Some sign procedures have been used in statistics for a long time. A well-known example is the sign test for hypothesis testing about the median of the common distribution of independent identically distributed (i.i.d.) observations. Our sign procedures are applicable to more complicated structural data models, in particular, regressive schemes (including factor models) with independent errors and linear models of time series, such as autoregression and moving average schemes and their modifications. In these diverse models the sign-based procedures retain their characteristic features, some of which will be pointed out here.

The most important one, in our view, is that the sign test statistics are distribution free. This enables us not only to construct tests with a fixed significance level for small samples, but also to obtain exact confidence sets for unknown parameters. Such a possibility is an important and rarely encountered property even for the schemes involving i.i.d. observations, to say nothing of more complicated models of time series.

Another appealing property of the sign procedures is their asymptotic normality (as the number of observations unboundedly increases) under minimal restrictions on the underlying probability distribution. This is a convenient property which is by no means common for other procedures widely used in practice. In particular, for the autoregression model, the well-known and commonly used least squares estimators (LSE) and least absolute deviations estimators are $\sqrt{n}$-asymptotically normal only when the observations have a finite variance. Otherwise, if this variance is infinite, these estimators may converge to the true values at an even faster rate than usual $n^{-1/2}$, but unfortunately their asymptotic distributions are not known. Of course, the use of signs instead of the numerical values of observations incurs a loss of information. Remarkably, the asymptotic efficiency of sign procedures relative to even the corresponding optimal procedures is not very low. For a Gaussian regression this asymptotic efficiency is $2/\pi$ (the same as for i.i.d. observations!), while for Gaussian autoregression it is equal to $(2/\pi)^2$. In general, when the random errors have a symmetric density function taking its maximum value at the origin, the asymptotic efficiency of the sign estimator in a linear model relative to the LSE cannot be less than $1/3$, provided that both estimators are asymptotically normal. For underlying distributions with heavy tails the asymptotic efficiency of sign procedures relative to the least squares procedures not only may be greater than one, but may take arbitrarily large values.

The sign estimators are robust to gross errors in the data, in contrast to the LSE (note that in the autoregression model the least absolute deviations estimators are not robust as well).

The sign procedures can also be used both in linear regression and autoregression models when the errors are nonidentically distributed.

Finally, the sign procedures form a system of statistical rules for data processing comparable in its scope with the least squares procedures for normally distributed errors.

Of course, there are other methods of processing statistical data (for example, rank methods), which cover the entire range of problems and possess some of the properties listed above. The advantage of the sign procedures is that these properties hold for them all at once. We will illustrate the performance of the sign methods by several examples. The first two examples are related to the models of independent observations. The third example concerns time series. First of all we illustrate the robustness, though it is a quite expectable property of sign procedures. The examples demonstrate it very clearly.

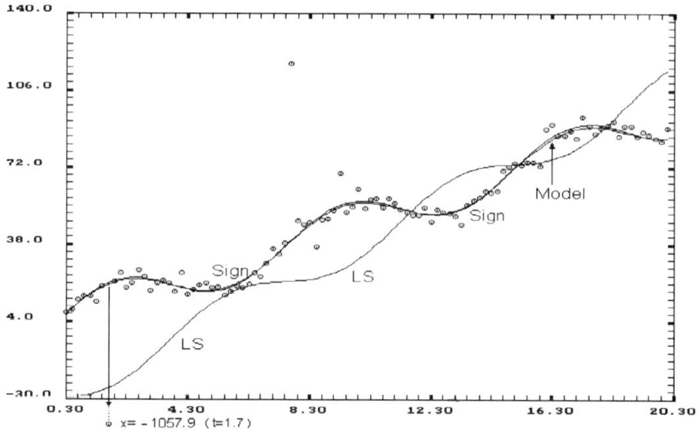

FIGURE I.1. Graphs of estimated functions by the sign method (Sign) and the method of least squares (LS).

EXAMPLE 1. The data in Figure I.1 were generated according to the regression model $x_i = \theta_1 + \theta_2 t_i + \theta_3 \sin(\omega t_i) + \xi_i$, $i = 1, \dots, n$. They simulate the observations over the dependence $x = \theta_1 + \theta_2 t + \theta_3 \sin(\omega t)$ at times $t_1, t_2, \dots, t_n$ with random errors ξ_i, $i = 1, \dots, n$, which are assumed independent. Regarding their distribution we make the following basic assumption.

CONDITION I(i). $\mathbf{P}\{\xi_i < 0\} = \mathbf{P}\{\xi_i > 0\} = 1/2$.

In Example 1 we have taken the following distribution function of the errors ξ_i, $i = 1, \dots, n$:

$$F(x) = (1 - \varepsilon)\Phi(x/\sigma) + \varepsilon K(x/\sigma),$$

where $\Phi(x)$ is the standard normal distribution function and $K(x)$ is the Cauchy distribution function,

$$K(x) = \frac{1}{2} + \frac{1}{\pi} \arctan \sqrt{\frac{\pi}{2}} x,$$

with $\sigma = 3$ and $\varepsilon = 0.2$.

The dots in Figure I.1 show the pairs (t_i, x_i) generated according to the above regression model with θ_1, θ_2, θ_3 given in Table I.1; the number of observations is $n = 101$, $\omega = 4\pi/15$. The figure does not contain an outlying point $x_8 = -1057.9$ for $t_8 = 1.70$.

Suppose that, for a known ω, parameters θ_1, θ_2, θ_3 are unknown and we have to estimate this dependence based on the observations (t_i, x_i), $i = 1, \dots, n$. The curve LS is obtained by the method of least squares. Although the majority of the observations closely follows the true dependence, the estimated curve is completely different from the true one. This is the result of the well-known nonrobustness property of the least squares method. Errors are mostly small and fluctuate similarly to i.i.d. Gaussian random variables. However, about 20 percent of observations contain much larger errors. Such observations are often viewed as outliers contaminating the main body of data. The presence of these outliers in this example is what so strongly affects the LSE.

The curve labelled "Sign" is obtained by the method based on signs. It differs very little from the true dependence of x versus t labelled as "Model". Table I.1

TABLE I.1

Parameter	Model	LSE	Sign
θ_1	3.50	−29.77	3.54
θ_2	4.50	6.87	4.47
θ_3	10.00	−7.38	9.32

contains the numerical values of the estimates obtained by the least squares and sign-based methods.

It is seen from Figure I.1 and Table I.1 that the sign-based method performs well in spite of contamination. Practical experience and theory show that the sign procedures are in general very robust to contamination.

EXAMPLE 2. Our second example concerns the analysis of factors. We consider a two-way layout with several values of the response variable x_{ijk}, $k = 1, \dots, m$, corresponding to each combination of factor levels (i, j) (equal number of observations per cell). The table contains simulated data according to an additive factor model, where

$$x_{ijk} = \mu + \alpha_i + \beta_j + e_{ijk}.$$

Here $\mu, \alpha_1, \alpha_2, \dots, \beta_1, \beta_2, \dots$ are the total mean and the factor effects; the variables e_{ijk} represent the errors by which the response variables x_{ijk} differ from the expected values $\mu + \alpha_i + \beta_j$. The set of errors e_{ijk} in this case is of the same nature as in Example 1, in that it is a homogeneous sample contaminated by a small fraction of outliers. This is modelled by the distribution function of the form

$$F(x) = (1 - \varepsilon)\Phi(x/\sigma_1) + \varepsilon\Phi(x/\sigma_2),$$

where $\sigma_1 = 0.5$, $\sigma_2 = 20$, $\varepsilon = 0.15$. The table has 5 rows and 7 columns, with 12 observations in each cell.

To estimate the unknown parameters μ, α_i, β_j from the observations x_{ijk}, we again apply the least squares and sign-based methods. Unfortunately, in this case we cannot demonstrate the results of estimation as clearly as we could do for regression analysis. Hence we will assess the performance of the estimators by the frequency analysis of the residuals $x_{ijk} - \widehat{\mu} - \widehat{\alpha}_i - \widehat{\beta}_j$, where $\widehat{\mu}$, $\widehat{\alpha}_i$, $\widehat{\beta}_j$ denote the estimates of the corresponding parameters. If the estimates are close to the true values, the residuals vary closely to the (unobservable) errors e_{ijk}. Their histogram gives us an idea of the performance of the corresponding estimator.

Figures I.2a and I.2b show the histograms of the residuals $x_{ijk} - \widehat{\mu} - \widehat{\alpha}_i - \widehat{\beta}_j$ for sign-based (I.2a) and least squares (I.2b) estimates. The figures contain only the parts of the histograms for residuals lying within the interval $[-3, 3]$. The normal density function with $\sigma_1 = 0.5$, which governs about 85% of errors, is also shown in the figures.

It is seen from the historgrams that the residuals corresponding to the sign-based estimator are less scattered. It means that the sign estimates lie near the true parameter values, so that the outliers have little effect on them. The results of estimation of the model parameters by the two methods are presented in Table I.2.

EXAMPLE 3. Our third example illustrates the robustness of the sign-based estimator in a linear model of stationary autoregression. Such models, moving average models as well as more complicated autoregression-moving average models,

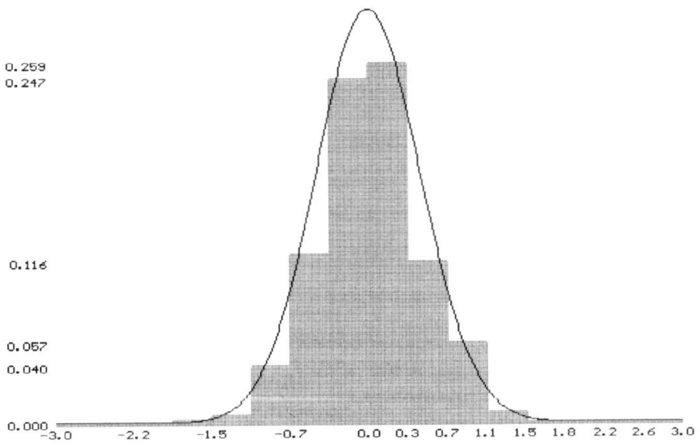

FIGURE I.2A. The histogram of residuals for the sign-based estimator

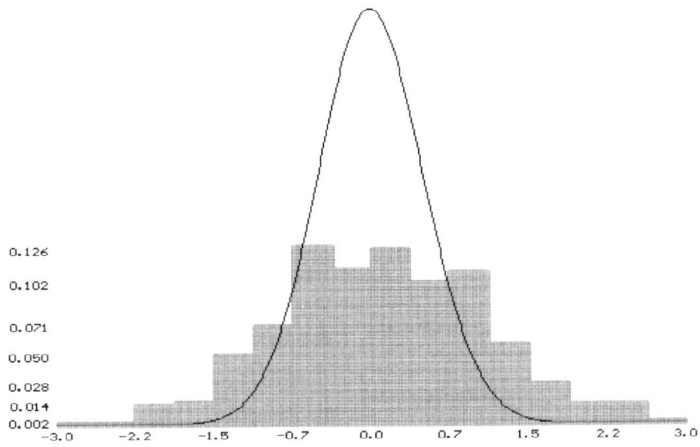

FIGURE I.2B. The histogram of residuals for the LSE

are the most popular and widely used models of time series. Suppose the variables $y_0, \ldots, y_n$ are observed, where

$$y_i = u_i + z_i^\gamma \xi_i, \qquad i \in \mathbb{Z} = \{\ldots, -1, 0, 1, \ldots\}.$$

Here $\{u_i\}$ satisfy the first order autoregression relation

$$u_i = \beta u_{i-1} + \varepsilon_i, \qquad i \in \mathbb{Z}, \quad |\beta| < 1,$$

where $\{\varepsilon_i\}$ are i.i.d. noise variables, $\{z_i^\gamma\}$ are i.i.d. binary random variables taking values 1 and 0 with probabilities γ and $1-\gamma$, respectively; the parameter γ, $0 \leq \gamma \leq 1$, determines the contamination rate; and finally $\{\xi_i\}$ is a sequence of i.i.d. random variables, which model the gross errors contaminating the main body of data. The sequences $\{u_i\}$, $\{z_i^\gamma\}$, $\{\xi_i\}$ are assumed to be mutually independent. Figure I.3 presents the data $y_0, y_1, \ldots, y_n$ generated according to this model for

TABLE I.2

Parameter	Model	LSE	Sign
α_1	0.50	0.99	0.56
α_2	2.00	2.39	2.08
α_3	7.80	7.48	7.66
α_4	−5.30	−4.68	−5.30
α_5	−5.00	−6.19	−5.00
β_1	−6.40	−5.62	−6.36
β_2	7.30	7.32	7.26
β_3	13.50	13.94	13.39
β_4	−7.00	−7.82	−6.87
β_5	5.40	5.47	5.42
β_6	−2.60	−3.27	−2.52
β_7	−10.20	−10.02	−10.31
μ	1.00	0.86	0.96

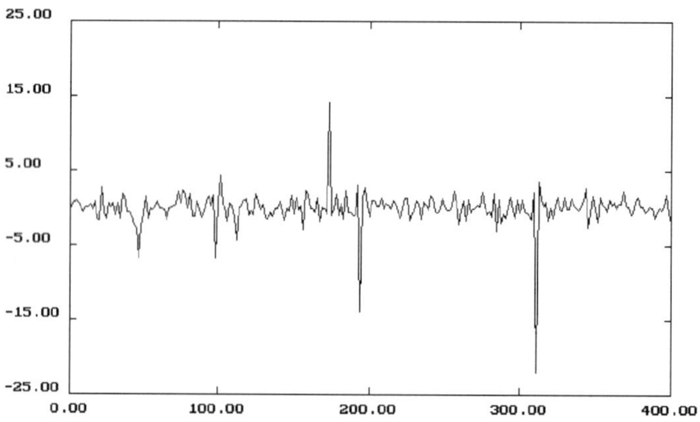

FIGURE I.3. Data generated according to the model $y_i = u_i + z_i^\gamma \xi_i$, $i = 1, \ldots, 400$, $\beta = 0.2$; $\varepsilon_i \sim N(0,1)$; $\gamma = 0.05$; $\xi_i \sim N(0, 100)$.

$n = 400$; $\beta = 0.2$, with $\{\varepsilon_i\}$ being standard normal random variables and $\{\xi_i\}$ normal variables with zero mean and standard deviation $\sigma = 10$; $\gamma = 0.05$.

Like the regression model with independent errors, the simplest and most widely used estimator for the parameters of autoregression models is the LSE, and, like the regression case, the LSE is very sensitive to gross errors. On the contrary, our sign estimator is robust to gross errors. This is illustrated by the results of estimation based on the above data. Table I.3 presents the results of three independent simulations.

Of course, there are many other methods of robust estimation (see, for example, Huber [**44**] and Hampel et al. [**36**]). However, an advantage of the sign-based method is that it yields exact rules (tests and confidence sets) valid for finite samples. Moreover the sign-based inference is valid irrespective of any specific form of the error distribution, so that in this sense the sign procedures are nonparametric.

TABLE I.3.
$N = 400; \beta = 0.2; \varepsilon_i \sim N(0,1); \gamma = 0.05; \xi_i \sim N(0,100)$.

Sample	LSE	Sign estimator
1	0.04	0.17
2	0.02	0.23
3	0.03	0.21

Let us describe the main steps of the sign analysis in the framework of the models treated in this book. In these models the statistical data $\mathbf{X}$, say, have the following structure

$$\mathbf{X} = \mathbf{f}(\boldsymbol{\theta}, \boldsymbol{\xi}),$$

where $\mathbf{f}(\cdot,\cdot)$ is a specified function, $\boldsymbol{\theta}$ is an unknown (finite-dimensional) parameter, and $\boldsymbol{\xi}$ are random errors.

For regression models (and other linear models of independent observations) this formula simplifies to

$$\mathbf{X} = \mathbf{f}(\boldsymbol{\theta}) + \boldsymbol{\xi}. \tag{I.1}$$

In Example 1 we have $\boldsymbol{\theta} = (\theta_1, \theta_2, \theta_3)$. In Example 2 parameter $\boldsymbol{\theta}$ is formed by the total mean μ and main effects $\alpha_1, \alpha_2, \dots$ and $\beta_1, \beta_2, \dots$. Finally, in Example 3 it is the univariate parameter β.

While the statistical analysis in the Gaussian case usually begins with estimation of $\boldsymbol{\theta}$, the nonparametric analysis of the model (I.1) begins by constructing a sensible (or, preferably, optimal in some sense) test for the hypothesis

$$H_0\colon \boldsymbol{\theta} = \boldsymbol{\theta}_0, \tag{I.2}$$

where $\boldsymbol{\theta}_0$ is some specified parameter value. For this purpose we need a statistic $\mathbf{Y}$ (depending on $\mathbf{X}$ and $\boldsymbol{\theta}_0$) which would be *distribution-free* under H_0 given by (I.2). This means that $\mathbf{Y}$ has the same distribution for any error distribution compatible with model assumptions. For the sign analysis this is condition (I.i).

In the above examples $\mathbf{Y}$ is formed by

$$\begin{gathered} \operatorname{sign}\left(x_i - \theta_1^0 - \theta_2^0 t_i - \theta_3^0 \sin(\omega t_i)\right), \qquad i = 1, \dots, n, \\ \|\operatorname{sign}(x_{ij} - \mu^0 - \alpha_i^0 - \beta_j^0)\|, \\ \operatorname{sign}(u_k - \beta_0 u_{k-1}), \qquad k = 1, 2, \dots, n, \end{gathered}$$

respectively. In each case, when $\boldsymbol{\theta}_0$ is the true parameter value, the random variables involved, under Condition I(i), are mutually independent and take on the values ± 1 with probabilities $1/2$.

Note that in the rank approach, random errors are assumed to be i.i.d. with a continuous distribution function. Then the ranks of residuals form a distribution-free statistic.

When dealing with a distribution-free statistic $\mathbf{Y}$, the composite hypothesis $H_0\colon \boldsymbol{\theta} = \boldsymbol{\theta}_0$ about the distribution of $\mathbf{X}$ becomes a simple hypothesis about the distribution of $\mathbf{Y}$. This fact enables one to construct valid tests for H_0 (based on $\mathbf{Y}$), i.e., tests with a preassigned significance level, which is kept constant for all admissible distributions of random errors. In many cases such tests (using signs, ranks, or a combination of them) can be found to possess some optimality properties thus increasing their power. This will be discussed in the book. Here we state the resulting tests. In Example 1 the locally optimal sign test statistic for $H_0\colon \theta_1 = 0,\ \theta_2 = 0,\ \theta_3 = 0$ is

$$\Big(\sum_{i=1}^{n} \operatorname{sign} x_i\Big)^2 + \Big(\sum_{i=1}^{n} t_i \operatorname{sign} x_i\Big)^2 + \Big(\sum_{i=1}^{n} \sin(\omega t_i) \operatorname{sign} x_i\Big)^2 .$$

In Example 3 the locally optimal sign test statistic for $H_0\colon \beta = \beta_0$ against one-sided alternatives $H^+\colon \beta > \beta_0$ has the form

$$\sum_{m=1}^{n-1} {\beta_0}^{m-1} \Big(\sum_{k=m+1}^{n} \operatorname{sign}(u_k - \beta_0 u_{k-1})(u_{k+m} - \beta_0 u_{k+m-1}) \Big). \tag{I.3}$$

For testing H_0 against two-sided alternatives $H\colon \beta \neq \beta_0$, one can use the absolute value of (I.3) as the test statistic.

The distributions of test statistics may depend on a particular value of $\boldsymbol{\theta}_0$. Nevertheless these distributions and their quantiles can be computed with a desirable accuracy for any $\boldsymbol{\theta}_0$. For example, one can use the Monte Carlo method. The null hypothesis is rejected when the observed value of the test statistic lies beyond the corresponding critical value.

By "inverting" sign tests one can construct confidence sets and even point estimates for parameter $\boldsymbol{\theta}$. Namely, the set of those $\boldsymbol{\beta}$, for which the hypothesis $H\colon \boldsymbol{\theta} = \boldsymbol{\beta}$ is accepted by the sign test, can be taken as a confidence set for $\boldsymbol{\theta}$. If we use a test of level α, the corresponding coverage probability (confidence level) equals $1-\alpha$. For instance, let $c_{n,\alpha}(\beta)$ and $c_{n,1-\alpha}(\beta)$ be the α- and $(1-\alpha)$-quantiles of the random variable (I.3). Then the set

$$\Big\{\beta\colon c_{n,\alpha}(\beta) \le \sum_{m=1}^{n-1} \beta^{m-1} \sum_{k=m+1}^{n} \operatorname{sign}(u_k - \beta u_{k-1})(u_{k+m} - \beta u_{k+m-1}) \le c_{n,1-\alpha}(\beta)\Big\}$$

is a confidence set for β of confidence level at least $1 - 2\alpha$. We will show in this book how such tests can be constructed in an explicit form. For large n one can use approximate quantiles derived from the asymptotic normality of the test statistic.

For regression models, confidence sets have a simpler form. Namely, if a test for $H_0\colon \boldsymbol{\theta} = \mathbf{0}$ accepts the hypothesis when $\mathbf{X} - \mathbf{f}(0) \in \mathbb{A}$, then

$$\{\,\boldsymbol{\theta}\colon \mathbf{X} - \mathbf{f}(\boldsymbol{\theta}) \in \mathbb{A}\,\} \tag{I.4}$$

is a confidence set for $\boldsymbol{\theta}$.

The next step is to derive a point estimate from confidence sets. It is seen that confidence sets generated by a certain test statistic for different confidence levels form a nested family of sets. Hence the common part of these sets can be taken for the point estimate. For the test statistic (I.3) this is the point where this statistic as a function of β_0 changes its sign (if this step function takes on 0, such points may form an interval). Thus in the autoregression problem of Example 3 the sign estimate is taken to be the solution of

$$\sum_{m=1}^{n-1} \beta^{m-1} \sum_{k=m+1}^{n} \operatorname{sign}(u_k - \beta u_{k-1})(u_{k+m} - \beta u_{k+m-1}) \div 0,$$

where $\div 0$ denotes the point where the function crosses the zero level. For other statistics the point estimate is taken to be the parameter value minimizing the test statistic.

Again, the problem simplifies for the regression models, where we can restrict ourselves to hypotheses of the form $H_0\colon \boldsymbol{\theta} = \mathbf{0}$. Typically, a sign test rejects this hypothesis for large values of a test statistic $q(\mathbf{Y})$, say. Therefore,

$$\widehat{\boldsymbol{\theta}}_n = \arg\min_{\boldsymbol{\theta}} \, q\left(\mathbf{X} - \mathbf{f}(\boldsymbol{\theta})\right) \tag{I.5}$$

can be taken for an estimate of $\boldsymbol{\theta}$. In particular, in Example 1 the point estimator is obtained as a solution of the problem

$$\begin{aligned}&\left[\sum_{i=1}^{n} \operatorname{sign}(x_i - \theta_1 - \theta_2\, t_i - \theta_3 \sin\omega t_i)\right]^2 \\ &\quad + \left[\sum_{i=1}^{n} t_i \operatorname{sign}(x_i - \theta_1 - \theta_2\, t_i - \theta_3 \sin\omega t_i)\right]^2 \\ &\quad + \left[\sum_{i=1}^{n} (\sin\omega t_i) \operatorname{sign}(x_i - \theta_1 - \theta_2\, t_i - \theta_3 \sin\omega t_i)\right]^2 \Longrightarrow \min_{\theta_1,\theta_2,\theta_3} .\end{aligned}$$

We study the asymptotic properties of the sign estimators and show that they are constistent and $\sqrt{n}$-asymptotically normal. Their asymptotic efficiency relative to, say, LSE, may be arbitrarily large for heavy-tailed error distributions. The sign estimators are robust to contamination, as was illustrated above.

Let us list the principal steps of the statistical analysis based on signs. These steps are similar for many statistical models.

1. Transformation of the data into a suitable set of signs.
2. Construction of a locally optimal nonparametric sign test for a specific parameter value.
3. Construction of finite-sample confidence sets.
4. Derivation of the corresponding nonparametric sign-based point estimators.
5. Investigation of asymptotic properties of tests and estimators thus obtained.
6. Testing linear hypotheses (for multiparameter schemes).
7. The study of the robustness properties of the procedures.

It is to be noted that the same scheme of dealing with statistical data (new set of observations, hypothesis testing, confidence estimation, point estimation, linear hypotheses) is also applicable when using ranks rather than signs. In this book we discuss the rank approach only briefly, mostly for its comparison with the rank-based approach.

The methodology just described was first proposed by Hodges and Lehmann [**41**] in the framework of the two-sample location problem when the two underlying distributions differ only by an unknown location parameter. They used Wilcoxon's rank sum test statistic. Since then this nonparametric program for rank methods was realized due to the efforts of many authors. In this book we carry out the same program for sign-based methods.

Part 1

Linear Models of Independent Observations

CHAPTER 1

Sign-based Analysis of One-parameter Linear Regression

In this chapter we demonstrate what the sign-based method is and how it works in a mathematically simple model of one-parameter regression. For the sake of illustration we apply this method to a set of real data which are adequately described by this model. For that purpose we have chosen the data related to Hubble's Law in astronomy to show how the Hubble constant could be determined from these data by a sign-based method. The idea to exemplify the one-parameter linear model by Hubble's Law is picked up from Hettmansperger [**39**].

1.1. Hubble's Law: A historical overview

In 1929 the American astronomer Edwin P. Hubble published a paper which was to play a great role in contemporary science [**43**]. It dealt with extragalactic nebulae, the stellar systems similar to the galaxy containing our sun and planets. The paper entitled "A relation between distance and radial velocity among extra-galactic nebulae" established a striking phenomenon that the galaxies move away from us and from one another with velocities increasing proportionally to distance. This proportionality is now referred to as "Hubble's Law" and the proportionality coefficient as "the Hubble constant". Its numerical value is still under discussion, though the relationship itself is firmly established. It is interpreted as the evidence that our universe is expanding.

The linear relation in Hubble's Law inevitably has a statistical nature. Apart from their common involvement in the expansion of the universe, the galaxies have proper movements relative to one another, so that the resulting radial velocities deviate from strictly linear dependence on distance. In our statistical analysis we will treat these deviations as independent random errors.

In 1917 a new 100-inch reflector telescope was brought into operation at the Mount Wilson Observatory in California. It was then world's largest telescope, much more powerful than any previously available instrument. Hubble started working at Mount Wilson soon after World War I. He joined a group of astronomers studying nebulae. The true nature of nebulae was unknown at that time, and the term "nebula" was applied to various objects that could not be resolved into individual stars. In particular, it was debatable whether they belonged to our galaxy or if certain kinds of nebulae could be galaxies comparable with ours and lying far away from it. Thus especially significant for understanding the nature of nebulae would be the determination of their distances.

In 1923, using the 100-inch reflector, Hubble began to observe the Andromeda nebula, a very large nebula that is presumably close to us, known to mankind for centuries. Soon he succeeded in finding a Cepheid variable in this nebula.

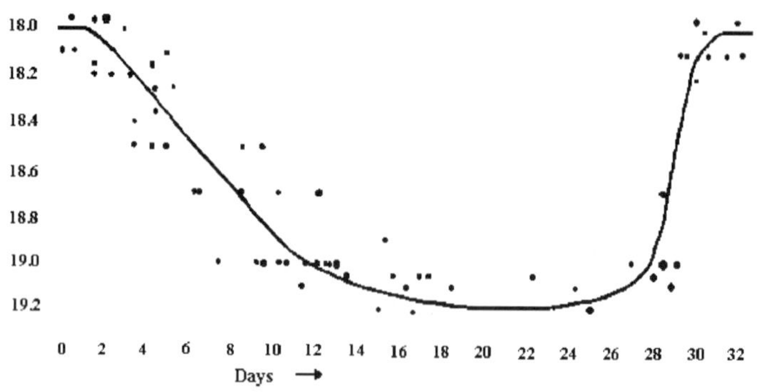

FIGURE 1.1.1. The light curve of the first Cepheid in the Andromeda nebula discovered by Hubble. Reproduced from Sharov and Novikov [**77**] by courtesy of the authors.

This discovery gave a clue to measuring the distance to the Andromeda nebula. The point is that the Cepheids, which are specific stars with periodically varying brightness, admit a reliable determination of distance due to a certain relationship (established by Shapley in 1918) between the period of brightness variation of a Cepheid and its true emission of light at maximum luminosity. By observing a Cepheid, it is not very difficult to find its period and then determine the absolute luminosity using the period-luminosity dependence. Its comparison with apparent magnitude of the star gives an estimate of the distance to the Cepheid. If this Cepheid belongs to a nebula, this provides the distance to the nebula itself. In Figure 1.1.1 you see the light curve plotted by Hubble of the first extragalactic Cepheid discovered by him.

Somewhat later Hubble identified Cepheids in another interesting and large object, the Triangulum nebula. Thus for the first time it became possible to find the distances to two nebulae based on reliable distance indicators. The distances were very large, almost a million light years, implying, in particular, that these nebulae were located far beyond our galaxy and that they were galaxies themselves, comparable with ours.

At the same time Hubble conducted the study of other nebulae. Investigation of the Andromeda and Triangulum nebulae showed that the distribution of stars over luminosity in these galaxies was similar to that in our galaxy, which confirmed the hypothesis that the brightest stars in different galaxies must have about the same absolute luminosity. Hence the distance to more remote galaxies could be inferred from the apparent magnitude of their brightest stars, though with less accuracy than using the method of Cepheids. By 1929, based on rich observational material, Hubble was able to determine the distances to more than 20 galaxies. These data are given in Table 1.1.1, taken from Vaucouleurs et al. [**91**]. The nebulae are numbered according to NGC (New General Catalog).

Now we recall that measurement of the radial velocities of stellar objects is based on the shifts in their spectra. Owing to the Doppler effect, the spectral frequencies decrease or increase depending on whether the luminous body recedes from or approaches the observer; hence, the spectral lines in the optical range shift toward the red or blue end of the spectrum, respectively. From these shifts the radial velocities of the objects can be determined with high accuracy.

TABLE 1.1.1

Nebula	Distance (Mpc)	Velocity (km/s)
SMC	0.032	170
LMC	0.034	290
NGC 6822	0.214	−130
598	0.263	−70
221	0.275	−185
224	0.275	−220
5457	0.450	200
4736	0.500	290
5194	0.500	270
4449	0.630	200
4214	0.800	300
3031	0.900	−30
3627	0.900	650
4826	0.900	150
5236	0.900	500
1068	1.000	920
5055	1.100	450
7331	1.100	500
4258	1.400	500
4151	1.700	960
4382	2.000	500
4472	2.000	850
4486	2.000	800
4649	2.000	1090

The first measurement of the radial velocity for a nebula was made in 1912 by V. M. Slipher. This was for the Andromeda nebula, which was found to approach our galaxy at a velocity of 300 km/s. By 1917 Slipher obtained 25 radial velocities of nebulae; by 1925 this number reached 45. This was a laborious task. Photographing the spectrum of faint nebulae required exposures of tens of hours, taking many nights.

Already in 1914, having obtained about 15 measurements of radial velocities, Slipher noticed that most of them were positive, i.e., with a few exceptions, the nebulae were moving away from our solar system with very high velocities. The subsequent measurements confirmed this phenomenon, which became known as "red-shift". In 1916–17 the possibility of the linear dependence between the radial velocities and the distances for remote objects was inferred by W. de Sitter from Einstein's general relativity theory.

In his 1929 paper, combining the data on radial velocities (both Slipher's and those he had obtained at Mount Wilson) with his estimates of distances for 24 nebulae, Hubble demonstrated with certainty that the linear dependence does exist (see Table 1.1.1 and Figure 1.1.2).

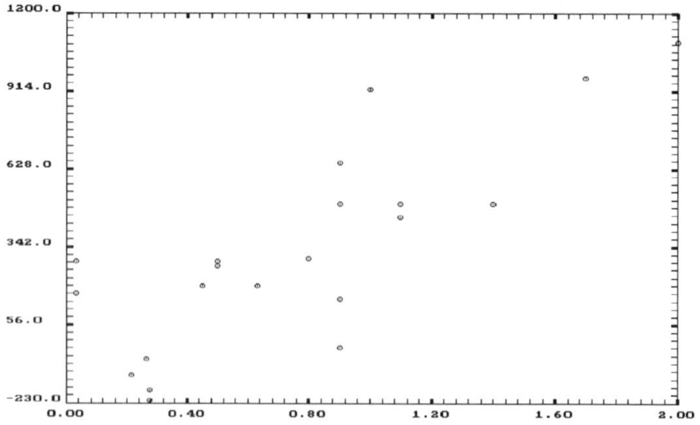

FIGURE 1.1.2. (Data from Hubble [**43**]). The x-coordinate: distances (Mpc), the y-coordinate: velocities (km/s)

The concept of an expanding universe based on Hubble's Law is now firmly established, playing a fundamental role in the contemporary cosmological theories. However, Hubble's evaluations of distances have been reconsidered due to subsequent developments in astronomy, which led to appreciable reduction of the Hubble constant (which, however, does not affect the linear relationship itself). From 500 km/s per megaparsec as given by Hubble, it dropped to 50–100 km/s per megaparsec. Its value is still under discussion, the opinions of most experts being split between two alternative estimates, 50 ± 10 km/(s·Mpc) and 80 ± 10 km/(s·Mpc). This problem is of particular importance because different estimates of the constant imply different estimates for the age of the universe, which is derived by extrapolation back its present expansion. The larger of the two figures leads to a duration shorter than the estimated age of some nebular objects (globular clusters)!

The Hubble Space Telescope put in orbit around the Earth in 1990, was designed, in particular, to determine the Hubble constant with higher accuracy. For that it was necessary to detect and explore Cepheids in galaxies significantly more distant than the Andromeda galaxy. Recently, Freedman et al. [**29**] reported such observations, from which the Hubble constant was derived as 80 ± 17 km/(s·Mpc). This value agrees with the so-called "short chronology". Thus the problem of the age discrepancies became even more acute.

1.2. Determination of the Hubble constant by the sign-based method

Denote by r the distance of a nebular object from the Earth, and by v its radial velocity, i.e., the component of its velocity taken along the direction from the observer to the object. Suppose we have measurements r_i, v_i, $i = 1, \ldots, n$, of these quantities for n objects. In view of the measurement errors and the statistical nature of Hubble's Law mentioned above, v_i and r_i are not strictly proportional. The relationship between them should be written as

$$v_i = r_i\theta + \xi_i, \qquad i = 1, \ldots, n. \tag{1.2.1}$$

Here, the proportionality coefficient θ is the Hubble constant to be determined, and $\xi_1, \ldots, \xi_n$ are random deviations from the linear dependence. They comprise both measurement errors of r_i, v_i, and deviations from the proportionality due to the proper movements of the nebulae. We will assume these deviations to be independent random variables. This assumption is fundamental for the statistical model. Moreover, we assume that these deviations "randomly fluctuate" about the linear relationship, which for the purpose of the sign analysis is expressed by the basic assumption

$$\mathbf{P}\{\xi_i > 0\} = \mathbf{P}\{\xi_i < 0\} = 1/2. \tag{1.2.2}$$

In statistical theory it is often assumed that the random errors are identically distributed. If this assumption appears inadequate, one has to specify a certain rule of variation of their distribution. For instance, one could assume that the distributions of ξ_i in (1.2.1) differ only by a scale parameter varying as a certain function of r_i. Unfortunately, it is difficult to make a justified choice of such a dependence. At the same time it is clear that any error in setting a statistical model inevitably leads to wrong conclusions from it. For the sign-based analysis it is not necessary to assume that the random errors $\xi_1, \dots, \xi_n$ are identically distributed.

Let us list the assumptions which constitute our statistical model for the analysis of data in Table 1.1.1. The variables v_i and r_i, $i = 1, \dots, n$ $(n = 24)$ satisfy the relationship (1.2.1). The deviations $\xi_1, \dots, \xi_n$ are independent random variables fulfilling (1.2.2). The coefficient θ in (1.2.1) is unknown and is to be estimated from the observations. Denote the true value of θ (the Hubble constant) by θ^0. As an estimate for θ^0 we will take the value of θ which provides the best fit of the linear dependence (1.2.1) to the observations (r_i, v_i), $i = 1, \dots, n$. Our assessment of this fit will be based not on the residuals $v_i - \theta r_i$ themselves, but on their signs

$$\big(\operatorname{sign}(v_1 - \theta r_1), \operatorname{sign}(v_2 - \theta r_2), \dots, \operatorname{sign}(v_n - \theta r_n)\big).$$

According to the "nonparametric program" exposed in the introduction, the statistical analysis of (1.2.1) begins with the hypothesis testing of $H_0\colon \theta = 0$. For testing this hypothesis against the alternative $H\colon \theta \neq 0$ we need a sensible statistical test.

We will use the sign test (2.2.9) from Chapter 2, or, rather, its two-sided version (2.2.11). In §2.2 we show that for identically distributed random errors the test (2.2.9) with test statistic

$$\sum_{i=1}^{n} r_i \operatorname{sign} v_i \tag{1.2.3}$$

is a locally most powerful sign test against one-sided alternatives. According to (2.2.9) the hypothesis $H_0\colon \theta = 0$ is rejected in favor of $H^+\colon \theta > 0$ if

$$\sum_{i=1}^{n} r_i \operatorname{sign} v_i \geq z, \tag{1.2.4}$$

where the critical constant z is chosen in such a way that, under H_0,

$$\mathbf{P}\Big\{\sum_{i=1}^{n} r_i \operatorname{sign} v_i \geq z\Big\} \tag{1.2.5}$$

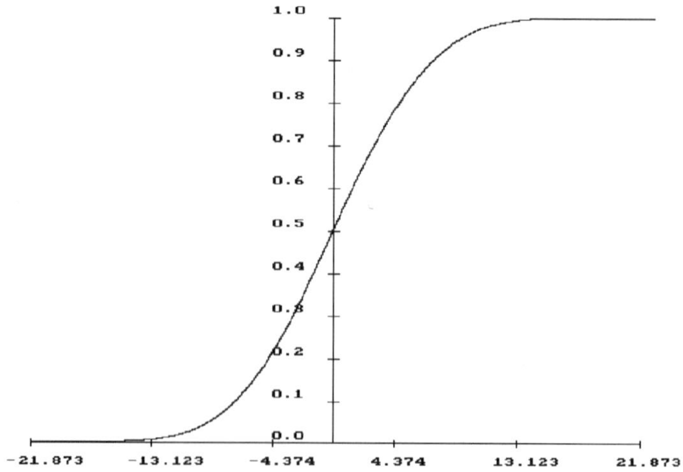

FIGURE 1.2.1. The distribution function of T_n (1.2.7) for $n = 24$.

would have a prescribed (small) value (significance level). For two-sided alternatives we propose using the two-sided version of this test, i.e., to reject the hypothesis $H_0\colon \theta = 0$ in favor of $H\colon \theta \neq 0$ when

$$\left| \sum_{i=1}^{n} r_i \operatorname{sign} v_i \right| \geq \text{const}\,. \tag{1.2.6}$$

For nonidentically distributed random errors, which is apparently the case in the problem under consideration, the test (1.2.4) is not optimal. However, it retains the prescribed significance level provided (1.2.2) is fulfilled. Hence, it can still be used in these circumstances.

To determine the critical values in (1.2.4) or (1.2.6) we need the distribution of the test statistic (1.2.3) under H_0. Consider the random variable

$$T_n = \sum_{i=1}^{n} r_i \zeta_i, \tag{1.2.7}$$

where $\zeta_1, \ldots, \zeta_n$ are mutually independent and take the values $+1$ and -1 with probability $1/2$. The random variable (1.2.7) has the same distribution as (1.2.5) under H_0. Its distribution for $n = 24$ and r_i from Table 1.1.1 was computed by the Monte Carlo method with $50,000$ replications. (All computations in the book are made with the aid of the software package "SIGN" worked out by the authors). The distribution function of T_n thus obtained is plotted in Figure 1.2.1.

It is seen that the distribution of T_n (1.2.7) is approximately normal. This becomes particularly clear when the distribution function is plotted on the probability normal paper, which is shown in Figure 1.2.2.

Asymptotically, as $n \to \infty$, T_n is distributed as $N(0, \sum_{i=1}^{n} r_i^2)$ provided the sequence $r_1, \ldots, r_n$ satisfies the condition

$$\frac{\max_{1 \leq i \leq n} r_i^2}{\sum_{i=1}^{n} r_i^2} \to 0. \tag{1.2.8}$$

TABLE 1.2.1

ε	$T_{n,\varepsilon}$	$T_{n,1-\varepsilon}$
0.075	−7.9	7.9
0.05	−9.0	9.0
0.025	−10.6	10.6
0.0125	−12.0	12.0

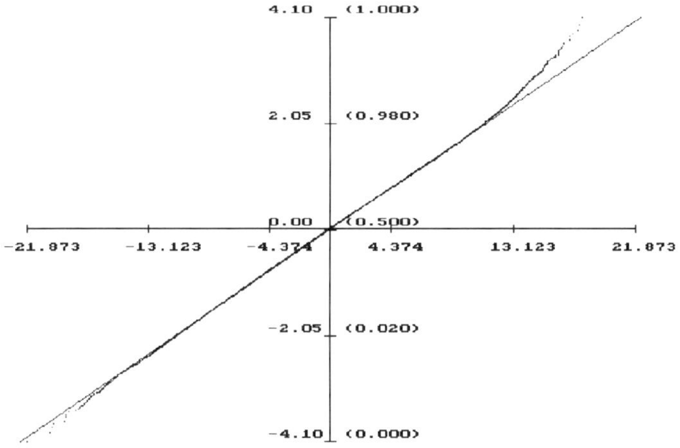

FIGURE 1.2.2. The distribution function of T_n: The y-axis scaled by the quantiles of the standard normal law.

The left-hand side of (1.2.8) for the data in Table 1.1.1 is approximately 0.14. Thus in our case $n = 24$ turned out to be "sufficiently large" and the left-hand side of (1.2.8) "sufficiently small" for the normal approximation to provide a satisfactory accuracy in the central part of the distribution.

For testing $H_0: \theta = 0$ and for setting confidence intervals for θ according to (I.4) (if H_0 is rejected) we need the quantiles of T_n. Since T_n has a discrete distribution, the equation

$$\mathbf{P}\{T_n \leq x\} = \varepsilon$$

for its quantiles either

(a) has no solutions,

or

(b) its solutions form a half-open interval of the real line.

In the case (a) we define the ε-quantile of $T_{n,\varepsilon}$ to be

$$T_{n,\varepsilon} = \sup\big\{x\colon \mathbf{P}\{T_n \leq x\} \leq \varepsilon\big\}.$$

In the case (b) we set $T_{n,\varepsilon}$ to be the midpoint of the interval. Since T_n is symmetrically distributed about zero, $T_{n,1-\varepsilon} = -T_{n,\varepsilon}$. In our case the distribution function of T_n with satisfactory accuracy can be treated as continuous (see Figures 1.2.1 and 1.2.2). Therefore, we will use approximate values of $T_{n,\varepsilon}$ obtainable, for example, from Figure 1.2.1, without special reservations. Some ε-quantiles of T_n computed by the Monte Carlo method are given in Table 1.2.1.

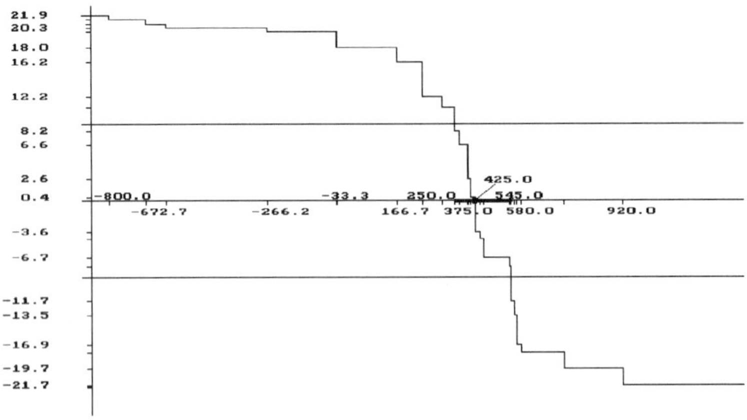

FIGURE 1.2.3. The function $T_n(\theta)$ and 90%-confidence interval (357, 540) for the Hubble constant.

The test statistic (1.2.4) for the data in Table 1.1.1 is equal to $\sum_{i=1}^n r_i \operatorname{sign} v_i \approx 18.02$, which exceeds the two-sided 2.5%-critical value. Thus $H_0 \colon \theta = 0$ is rejected with high certainty. This means that the statistical relationship between radial velocities and distances contains a nontrivial linear dependence.

According to (I.4), for interval estimation of the true value θ^0 we should consider the function

$$T_n(\theta) = \sum_{i=1}^n r_i \operatorname{sign}(v_i - \theta r_i). \tag{1.2.9}$$

Then a confidence set containing θ^0 with probability at least $1 - 2\alpha$ has the form

$$\big\{\theta \colon |T_n(\theta)| \le T_{n,1-\alpha}\big\}. \tag{1.2.10}$$

The function (1.2.9) can be rewritten as

$$T_n(\theta) = \sum_{i=1}^n |r_i| \operatorname{sign}\Big(\frac{v_i}{r_i} - \theta\Big). \tag{1.2.11}$$

It is seen from this formula that $T_n(\theta)$ is a nonincreasing step function varying from $\sum_{i=1}^n |r_i|$ to $-\sum_{i=1}^n |r_i|$ by jumps $2|r_i|$ at points v_i/r_i, $i = 1, \dots, n$. These properties show that (1.2.12) is an interval whose endpoints are some values of v_i/r_i. The function $T_n(\theta)$ for the data in Table 1.1.1 is plotted in Figure 1.2.3 along with the 90%-confidence interval for θ^0.

The left, θ_l, and right, θ_u, endpoints of confidence intervals for other confidence probabilities are given in Table 1.2.2.

It remains to discuss the point estimate. According to (I.5) it is obtained as a solution of the problem

$$\Big|\sum_{i=1}^n r_i \operatorname{sign}(v_i - r_i\theta)\Big| \Longrightarrow \min_\theta . \tag{1.2.12}$$

By the properties of $T_n(\theta)$ stated above, the θ's minimizing the left-hand side of (1.2.12) form an interval. In this case the minimal value is 0.359, and it is attained

TABLE 1.2.2

Confidence level $1-2\alpha$	Left endpoint θ_l	Right endpoint θ_u
0.85	375.00	540.00
0.90	357.14	540.00
0.95	357.14	545.00
0.975	314.46	555.60

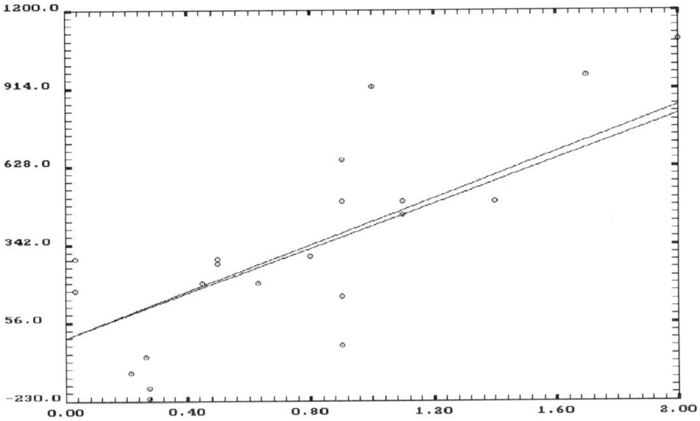

FIGURE 1.2.4. Estimation of the Hubble constant: function $v = 425r$; x-axis: distances (Mpc); y-axis: velocities (km/s).

on the interval $(\bar{\theta}_1, \bar{\theta}_2) = (409.1, 425.0)$. Any number from this interval can be taken as an estimate $\widehat{\theta}_n$ for θ^0. Note that the indeterminacy in the choice of the numerical value of $\widehat{\theta}_n$ in (1.2.12) is much less than the statistical indeterminacy in the estimation of θ^0, which is characterized by the length of confidence intervals for θ^0 given in Table 1.2.2. The essential difference in size between these two types of intervals allows us to speak about "point" estimation even when the solutions of the extremal problem (I.5) form an interval. Figure 1.2.4 shows the estimate for the slope of the line around which the 24 measurements of distance and velocity fluctuate. The two lines $v = \bar{\theta}_1 r$ and $v = \bar{\theta}_2 r$ whose slopes are the endpoints of the minimization interval of $|T_n(\theta)|$ differ very little.

The sign estimators are closely related to the least absolute deviations (LAD) estimators. The LAD estimator in the model (1.2.1) is defined as the solution of the extremal problem

$$\sum_{i=1}^{n} |v_i - \theta r_i| \Longrightarrow \min_{\theta}.$$

The objective function in this extremal problem is a convex function of θ, so that this problem always has a solution. It is easily seen that it can be found from the equation

$$\sum_{i=1}^{n} \operatorname{sign}(v_i - \theta r_i) \div 0. \tag{1.2.13}$$

The sign $\div$ means crossing of zero level, i.e., we define the solution of this equation to be the point $\widehat{\theta}_{\mathrm{LD}}$ at which the monotone function $T_n(\theta)$ changes its sign. In our example the sign of $T_n(\theta)$ changes when the argument equals 425.0. Therefore $\widehat{\theta}_{\mathrm{LD}} = 425.0$. The properties of $T_n(\theta)$ as in (1.2.9) mentioned above allow us to give the following description of the LAD estimator: $\widehat{\theta}_{\mathrm{LD}}$ is the median of the discrete probability distribution which assigns probabilities

$$p_i = \frac{|r_i|}{\sum_{i=1}^n |r_i|} \quad \text{to the points} \quad \frac{v_i}{r_i}, \qquad i = 1, \dots, n,$$

i.e.,

$$\widehat{\theta}_{LD} = \operatorname{med}\left\{ \left(\frac{v_i}{r_i},\ p_i \right),\ i = 1, \dots, n \right\}. \tag{1.2.14}$$

Treatment of contemporary data. As was mentioned above, the data in Table 1.1.1, which were used by Hubble, were reconsidered afterward, particularly, due to more precise determination of distances. In Table 1.2.3 we present their contemporary values, as given by Vaucouleurs et al. in [**91**]. The distances are obtained by using LEDA (Lyon–Mendon Extragalactic Database, 1995). Here we apply to the contemporary data in Table 1.2.3 the same computations as in the previous section. Notice that the estimate obtained from such a small sample (only 24 observations) is very close to the contemporary values of the Hubble constant.

The corresponding 24 points on the plane and the graphs of the distribution function of T_n as in (1.2.5) are shown in Figures 1.2.5, 1.2.6, and 1.2.7.

Table 1.2.4 gives some ε-quantiles of T_n computed by the Monte Carlo method with 50,000 replications.

The step function $T_n(\theta)$ defined by (1.2.11) varies from 150.92 to -150.92. Figure 1.2.8 shows its graph and the 90%-confidence interval for the Hubble constant. This interval is (53.02, 85.73). The left and right endpoints of confidence intervals for other confidence levels are given in Table 1.2.5.

The extremal problem (1.2.10) has a unique solution 66.76. The corresponding minimal value is ≈ 2.58. The line with slope 66.76, which is the sign estimate for the Hubble constant, is shown in Figure 1.2.9.

TABLE 1.2.3

Nebula	Distance (Mpc)	Velocity (km/s)
SM	0.083	190
LM	0.047	277
NGC 6822	1.60	−26
598	1.44	−204
221	1.57	−205
224	0.98	−295
5457	15.30	221
4736	8.05	297
5194	4.30	463
4449	3.68	211
4214	2.75	298
3031	5.10	−49
3627	9.70	703
4826	7.10	474
5236	7.28	503
1068	12.75	1093
5055	9.55	516
7331	15.75	835
4258	7.60	480
4151	6.28	956
4382	6.34	722
4472	6.30	983
4486	8.12	1282
4649	9.25	1114

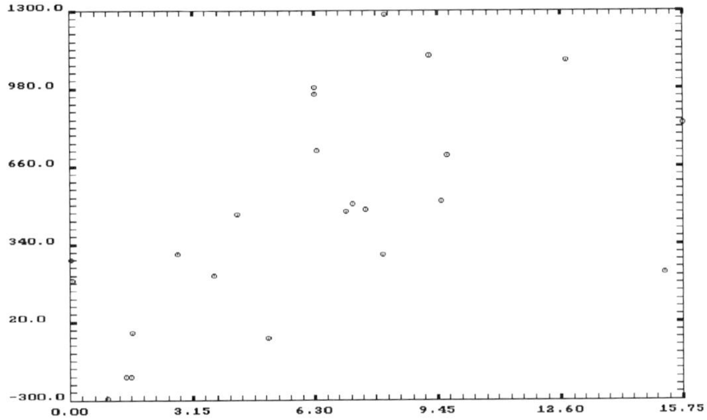

FIGURE 1.2.5. Contemporary data. x-axis: distances (Mpc), y-axis: velocities (km/s).

TABLE 1.2.4

ε	$T_{n,\varepsilon}$	$T_{n,1-\varepsilon}$
0.075	−54	54
0.05	−62	62
0.025	−73	73
0.0125	−82	82

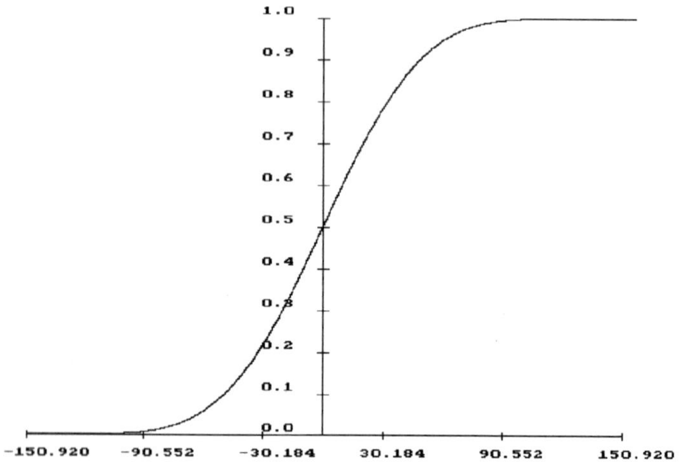

FIGURE 1.2.6. The distribution function of T_n (1.2.7), $n = 24$.

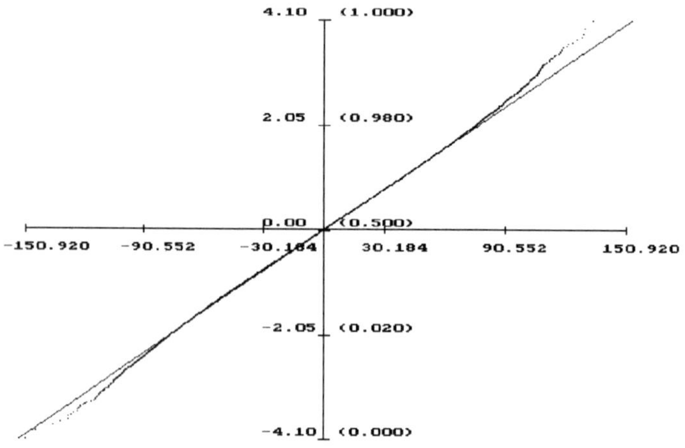

FIGURE 1.2.7. The distribution function of T_n: The y-axis scaled by the quantiles of the standard normal law.

TABLE 1.2.5

Confidence level $1-2\alpha$	Left endpoint θ_l	Right endpoint θ_u
0.85	53.02	85.73
0.90	53.02	85.73
0.95	53.02	108.36
0.975	53.02	113.88

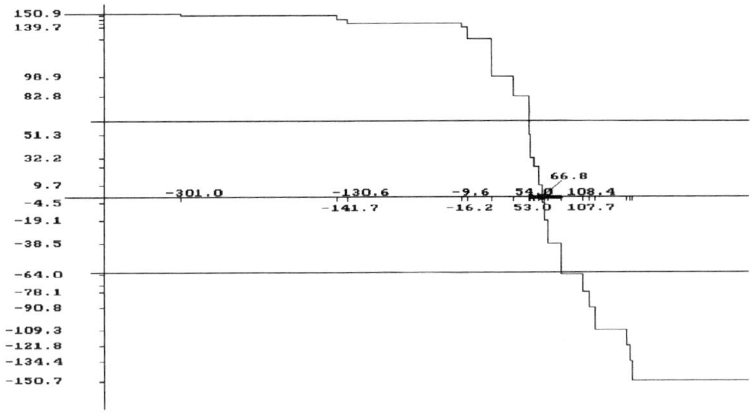

FIGURE 1.2.8. The Function $T_n(\theta)$ and 90%-confidence interval (53.02,85.73) for the Hubble constant.

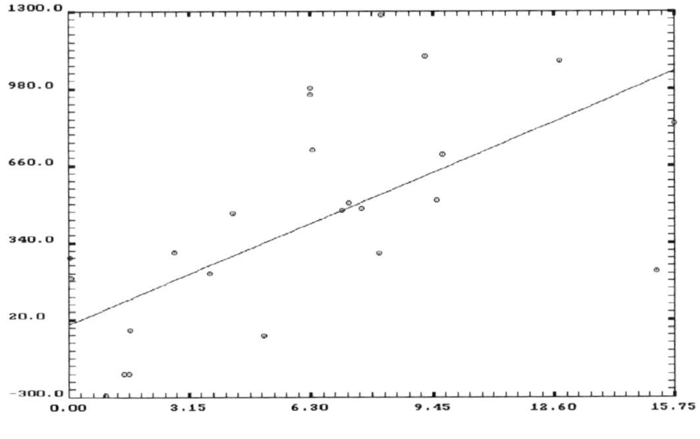

FIGURE 1.2.9. Estimation of the Hubble constant: function $v = 66.76\,r$.

1.3. Asymptotic results

In this section we study the large sample properties of sign procedures. In this case the distributions of statistics, critical values, confidence bounds, etc., take a simpler form than for finite n. This is due to the Central Limit Theorem and the corresponding normal approximation. We have already formulated a theorem that gives sufficient conditions for the normal approximation for statistic (1.2.5) or random variable (1.2.7). This theorem also implies the normal approximation for the quantiles $T_{n,\varepsilon}$. There is no theoretical solution, though, as to how many observations one needs for the reliable use of asymptotic results. Rather, in various applications this problem is solved empirically, by numerical comparisons of approximate formulas provided by the asymptotic theory with computational results. For that reason in our practical recommendations we prefer, whenever possible, the procedures admitting exact solutions, or we use computational methods with guaranteed accuracy (for example, Monte Carlo). Nevertheless asymptotic methods constitute an important and useful part of statistical theory. The main asymptotic results related to the sign estimator $\widehat{\theta}_n$ given by (1.2.12) or (1.2.13) concern its consistency and asymptotic normality.

Consistency. Recall the general definition of consistency: an estimator $\widehat{\theta}_n$ of parameter θ is called *consistent* if $\widehat{\theta} \to \theta$ in probability as $n \to \infty$ in the entire range of possible values of θ. In practical applications this property is interpreted as an approximate equality $\widehat{\theta}_n \approx \theta^0$ for large n, where θ^0 is the value of θ in the particular experiment (the "true value" of θ).

The sign estimator $\widehat{\theta}_n$ in the model (1.2.1) was defined as the solution of the extremal problem (1.2.13). We will show that it is consistent under some natural regularity conditions. Put

$$B_n^2 = \sum_{i=1}^{n} r_i^2$$

and consider the following equivalent problem

$$\left| \frac{1}{B_n^2} T_n(\theta) \right| \Longrightarrow \min_{\theta} .$$

As was pointed out, $T_n(\theta)$ is a nonincreasing function. Therefore, in order to prove the consistency of $\widehat{\theta}_n$ it suffices to show that for a fixed θ, as $n \to \infty$,

(a) $\mathbf{P}\left\{ \frac{1}{B_n^2} T_n(\theta) < 0 \right\} \to 1, \quad \text{if } \theta > \theta^0;$

(b) $\mathbf{P}\left\{ \frac{1}{B_n^2} T_n(\theta) < 0 \right\} \to 0, \quad \text{if } \theta < \theta^0;$

(c) $\frac{1}{B_n^2} T_n(\theta^0) \to 0 \quad$ in probability.

It is readily shown that under the condition (1.2.8) $T_n(\theta)$ defined by (1.2.11) is asymptotically normal $N\big(\mathbf{E}T_n(\theta), \operatorname{Var} T_n(\theta)\big)$ for any fixed θ as $n \to \infty$. Here

$$\mathbf{E}T_n(\theta) = \sum_{k=1}^{n} r_k p_k, \qquad \operatorname{Var} T_n(\theta) = \sum_{k=1}^{n} r_k^2 (1 - p_k^2)$$

with

$$p_k := \mathbf{E} \operatorname{sign}(v_k - \theta r_k).$$

Hence,

$$\mathbf{P}\{T_n(\theta) < 0\} \sim \Phi\Big(-\frac{\mathbf{E}T_n(\theta)}{\sqrt{\operatorname{Var} T_n(\theta)}}\Big).$$

It remains to check that

$$\frac{-\sum_{k=1}^n r_k p_k}{\sqrt{\sum_{k=1}^n r_k^2(1-p_k^2)}} \to \begin{cases} +\infty & \text{for } \theta < \theta^0, \\ -\infty & \text{for } \theta > \theta^0. \end{cases} \tag{1.3.1}$$

Consider the case $\theta > \theta^0$; the other one is treated similarly.

To that end we need additional assumptions about the design of experiment and the properties of random errors. Concerning the design of experiment, which is specified by the regression constants $r_1, r_2, \dots, r_n$, let us note that in our example

$$r_k \geq r_0 \qquad \text{for all} \quad k = 1, 2, \dots \tag{1.3.2}$$

for some $r_0 > 0$.

In asymptotic analysis the random errors are most commonly assumed to be identically distributed. This assumption is adequate if all the measurements are made in about the same conditions. In regression model (1.2.1) it appears natural to assume that the magnitude of the random error ξ_k increases with distance r_k. However, this increase should not be too fast; otherwise, the measurements at large distances will be practically noninformative. A natural assumption is that at large distances the scale of the error increases not faster than the distance itself. Under the condition (1.3.2) this assumption can be expressed by the inequalities

$$\begin{aligned} \mathbf{P}\{\xi_k/r_k < t\} &\geq G(t) \quad \text{for } t > 0 \\ \mathbf{P}\{\xi_k/r_k < t\} &\leq G(t) \quad \text{for } t < 0, \end{aligned} \tag{1.3.3}$$

where $G(t)$ is some monotone increasing continuous function with $G(0) = 1/2$. For identically distributed errors condition (1.3.3) along with (1.3.2) means that their common distribution function monotonically increases in a neighborhood of zero.

Under (1.3.3) we obtain that for $\theta > \theta^0$

$$p_k \geq 2G(\theta - \theta^0) - 1 > 0;$$

hence,

$$\frac{\sum_{k=1}^n r_k p_k}{\sqrt{\sum_{k=1}^n r_k^2(1-p_k^2)}} \geq \frac{2G(\theta-\theta^0)-1}{\sqrt{1-(2G(\theta-\theta^0)-1)^2}} \, \frac{\sum_{k=1}^n r_k}{\sqrt{\sum_{k=1}^n r_k^2}}.$$

Now it is sufficient for (1.3.1) to hold that

$$\frac{\sum_{k=1}^n r_k}{\sqrt{\sum_{k=1}^n r_k^2}} \to \infty \qquad \text{as} \quad n \to \infty. \tag{1.3.4}$$

It is natural to assume in our example that the number of nebulae in a ball of radius R is approximately proportional to its volume, i.e., to R^3. If the observations are numbered in the ascending distance order, we obtain that $r_n = O(n^{1/3})$. It is easy to see that (1.3.4) is fulfilled for such a rate of increase of r_n (as well as for any other polynomial rate $r_n = O(n^\alpha)$, $\alpha > 0$). In the case of a yet faster distance increase rate, one would have to jointly consider the properties of sequences $\{r_n\}$ and $\{p_n\}$ for the proof of (1.3.1). Finally, the assertion (c) follows by Chebyshev's inequality since $B_n^2 \to \infty$ as $n \to \infty$ by (1.2.8).

The following theorem summarizes the above considerations.

THEOREM 1.3.1 (consistency of the sign estimator (1.2.11)). *Suppose that the design of experiment* $(r_1, r_2, \ldots, r_n)$ *in the problem* (1.2.1) *is such that*

$$r_k \geq r_0 > 0, \qquad \frac{\max_{1\leq k\leq n} r_k^2}{\sum_{k=1}^n r_k^2} \to 0, \qquad \frac{\sum_{k=1}^n r_k}{\sqrt{\sum_{k=1}^n r_k^2}} \to \infty$$

as $n \to \infty$. *Regarding the distribution of independent random errors* ξ_k, $k = 1, 2, \ldots, n$, *we assume that there exists a continuous monotone increasing function* $G(t)$ *with* $G(0) = 1/2$ *such that for* $r_k \geq r_0 > 0$

$$\mathbf{P}\{\xi_k/c_k < t\} \geq G(t) \quad \textit{for } t > 0,$$
$$\mathbf{P}\{\xi_k/c_k < t\} \leq G(t) \quad \textit{for } t < 0.$$

Then the sign estimator (1.2.12) *is consistent.*

Asymptotic normality. Having established that the sign estimator is consistent under some natural assumptions on the design of experiment and the distribution of random errors, i.e., that it converges to the true parameter value, it is natural to inquire about the rate of this convergence.

It is well known that estimators based on samples from distributions "regularly" depending on parameter, converge at a rate of $n^{-1/2}$. For regression problems the number of observations n does not tell us the amount of information about unknown parameters contained in observations. Rather, in the problem (1.2.1) one has to consider the quantity

$$B_n^2 = \sum_{k=1}^n r_k^2. \tag{1.3.5}$$

As will be shown, under some natural conditions the rate of convergence is B_n^{-1}. By saying that an estimator $\widehat{\theta}_n$ converges to the true parameter value θ^0 at a rate of B_n^{-1}, we mean that the random variable

$$B_n(\widehat{\theta}_n - \theta^0) \tag{1.3.6}$$

has a nondegenerate limiting distribution as $n \to \infty$. Usually we obtain the normal distribution as the limiting one. We will give sufficient conditions for asymptotic normality restricting ourselves to the case of identically distributed random errors.

In order to prove asymptotic normality of (1.3.6), we need a more detailed study of the function T_n given by (1.2.7) in a neighborhood of $\theta = \theta^0$. Introduce a new variable t by equality $\theta = \theta^0 + t/B_n$. Then

$$T_n(\theta) = \sum_{k=1}^n r_k \operatorname{sign}\Big(-t\,\frac{r_k}{B_n} + \xi_k\Big). \tag{1.3.7}$$

Moreover, let

$$N_n := \frac{\sum_{i=1}^n r_i^2}{\max_{1\leq i\leq n} r_i^2}. \tag{1.3.8}$$

Under (1.2.8), $N_n \to \infty$ as $n \to \infty$. Consider the problem

$$\left[\sum_{k=1}^n \frac{r_k}{B_n} \operatorname{sign}\Big(\xi_k - t\,\frac{r_k}{B_n}\Big)\right]^2 \Longrightarrow \min_{t\in\mathbb{R}^1}, \tag{1.3.9}$$

equivalent to (1.2.13).

Our derivations will be based on the following version of stochastic differentiation of $T_n(\theta)$. Recall that this function, being discontinuous (piecewise constant), is not differentiable in the ordinary sense.

THEOREM 1.3.2. *Suppose that the experiment design* $(r_1, r_2, \dots, r_n)$ *in the problem* (1.2.1) *satisfies the condition* (1.2.8) *as* $n \to \infty$. *Moreover, suppose that the random errors* $\xi_1, \xi_2, \dots, \xi_n$ *are independent and identically distributed with common distribution function* $F(x)$ *having a continuous derivative* $F'(x)$ *in a neighborhood of zero, and* $F(0) = 1/2$. *Then*

$$\frac{1}{B_n}\sum_{k=1}^{n} r_k \operatorname{sign}\Big(-t\,\frac{r_k}{B_n} + \xi_k\Big) = \frac{1}{B_n}\sum_{k=1}^{n} r_k \operatorname{sign}\xi_k - 2\,t\,F'(0) + X_n(t), \tag{1.3.10}$$

as $n \to \infty$, *where* $X_n(t) \to 0$ *in probability. This convergence is uniform in* t *over* $|t| < N_n^\alpha$ *for any* $0 < \alpha < 1/4$, *i.e., for any* $\varepsilon > 0$

$$\mathbf{P}\Big\{\sup_{|t|<N_n^\alpha} |X_n(t)| > \varepsilon\Big\} \to 0. \tag{1.3.11}$$

PROOF. By (1.3.10)

$$X_n(t) = \sum_{k=1}^{n} \frac{r_k}{B_n}\Big(\operatorname{sign}\Big(\xi_k - t\,\frac{r_k}{B_n}\Big) - \operatorname{sign}\xi_k + 2F'(0)\,t\,\frac{r_k}{B_n}\Big). \tag{1.3.12}$$

Put

$$Y_n(t) = \sum_{k=1}^{n} \frac{r_k}{B_n}\Big(\operatorname{sign}\Big(\xi_k - t\,\frac{r_k}{B_n}\Big) - \operatorname{sign}\xi_k\Big). \tag{1.3.13}$$

Divide the interval $\{t\colon |t| \le N_n^\alpha\}$ into intervals of length $h(n) = N_n^{-\beta}$, $0 < \beta < 1/4$. Denote by Ω the set of these intervals. We will write ω for an arbitrary element of Ω and τ for the midpoint of the interval ω. For the proof of (1.3.11) it suffices to show that

$$\sum_{\omega\in\Omega} \mathbf{P}\Big\{\sup_{t\in\omega} |X_n(t)| > \varepsilon\Big\} \to 0. \tag{1.3.14}$$

The sum in (1.3.14) contains $2N_n^\alpha / N_n^{-\beta} = 2N_n^{\alpha+\beta}$ terms. Hence (1.3.11) will be shown if we obtain a uniform upper bound of order $o\,(N_n^{\alpha+\beta})$ for the probabilities

$$\mathbf{P}\Big\{\sup_{t\in\omega} |X_n(t)| > \varepsilon\Big\}.$$

Recall that the oscillation of a function $f(x)$ on a set $\mathbb{M}$ is

$$W(f, \mathbb{M}) := \sup_{x\in\mathbb{M}} f(x) - \inf_{x\in\mathbb{M}} f(x)$$

and that

$$W(f+g, \mathbb{M}) \le W(f, \mathbb{M}) + W(g, \mathbb{M}).$$

Now

$$\begin{aligned}(1.3.15)\qquad \mathbf{P}\{\sup_{t\in\omega}|X_n(t)|>\varepsilon\} &\le \mathbf{P}\Big\{W(X_n,\omega)+|X_n(\tau)|>\varepsilon\Big\}\\ &\le \mathbf{P}\Big\{W(X_n,\omega)>\frac{\varepsilon}{2}\Big\}+\mathbf{P}\Big\{|X_n(\tau)|>\frac{\varepsilon}{2}\Big\}\\ &\le \mathbf{P}\Big\{W(Y_n,\omega)+2F'(0)h(n)>\frac{\varepsilon}{2}\Big\}\\ &\qquad+\mathbf{P}\Big\{|X_n(\tau)|>\frac{\varepsilon}{2}\Big\}\\ &\le \mathbf{P}\Big\{W(Y_n,\omega)>\frac{\varepsilon}{3}\Big\}+\mathbf{P}\Big\{|X_n(\tau)|>\frac{\varepsilon}{2}\Big\},\end{aligned}$$

since $h(n)\to 0$.

The last two probabilities will be estimated by Chebyshev's inequality. To that end, we need upper bounds for the first two moments of $Y_n(t)$ and $X_n(t)$. We begin with the second term.

Let Δ, $\Delta\in\mathbb{R}^1$, belong to the neighborhood of zero where, by the assumption, $F'(x)$ exists and is continuous. Note that

$$\operatorname{sign}(\xi_k-\Delta)-\operatorname{sign}\xi_k=\begin{cases}0, & \text{for } \xi_k>0,\ \xi_k>\Delta;\\ 2, & \text{for } \xi_k<0,\ \xi_k>\Delta;\\ -2, & \text{for } \xi_k>0,\ \xi_k<\Delta;\\ 0, & \text{for } \xi_k<0,\ \xi_k<\Delta,\end{cases}$$

whereas any other possibility has probability zero. Therefore,

$$\begin{aligned}\mathbf{E}\{\operatorname{sign}(\xi_k-\Delta)-\operatorname{sign}\xi_k\}&=2\mathbf{P}\{0<\xi_k<\Delta\}-2\mathbf{P}\{\Delta<\xi_k<0\}\\ &=-2F'(\kappa)|\Delta|,\end{aligned}$$

where κ is a point between 0 and Δ. Note also that for $|t|<N_n^\alpha$

$$\max_{1\le k\le n}\Big|t\frac{r_k}{B_n}\Big|\le N_n^{-(1-\alpha)}\to 0.$$

Therefore, for $|t|<N_n^\alpha$

$$|\mathbf{E}X_n(t)|\le 2N_n^\alpha\sum_{k=1}^n\frac{r_k^2}{B_n^2}\big|F'(0)-F'(\kappa_k)\big|,$$

where κ_k is a point between 0 and $t\frac{r_k}{B_n}$, $k=1,\dots,n$. Hence,

$$\max_{1\le k\le n}|\kappa_k|\le N_n^{-(1-\alpha)}\to 0,$$

so that $|F'(0)-F'(\kappa_k)|\le \text{const}\,|\kappa_k|$ for any $k=1,\dots,n$. Thus,

$$(1.3.16)\qquad \sup_{|t|<N_n^\alpha}|\mathbf{E}X_n(t)|\le \text{const}\,N_n^{-(1-2\alpha)}.$$

In a similar way we prove that

$$\sup_{|t|<N_n^\alpha}\operatorname{Var}X_n(t)\le \text{const}\,N_n^{-(1-\alpha)}.$$

Hence, using (1.3.16) we obtain for sufficiently large n

$$
\begin{aligned}
(1.3.17)\qquad & \mathbf{P}\Big\{|X_n(\tau)| > \frac{\varepsilon}{2}\Big\} \\
& \le \mathbf{P}\Big\{|X_n(\tau) - \mathbf{E}X_n(\tau)| + |\mathbf{E}X_n(\tau)| > \frac{\varepsilon}{2}\Big\} \\
& \le \mathbf{P}\Big\{|X_n(\tau) - \mathbf{E}X_n(\tau)| > \frac{\varepsilon}{3}\Big\} \\
& \le \Big(\frac{\varepsilon}{3}\Big)^{-2} \operatorname{Var} X_n(\tau) \le \operatorname{const} N_n^{-(1-\alpha)}.
\end{aligned}
$$

Next consider $\mathbf{E}W(Y_n,\omega)$ and $\operatorname{Var} W(Y_n,\omega)$. Since $\frac{r_k}{B_n}\operatorname{sign}\big(\xi_k - t\,\frac{r_k}{B_n}\big)$ are monotone nonincreasing functions of t, for $\omega = [\lambda,\mu]$ we have

$$
W(Y_n,\omega) = \sum_{k=1}^{n} \frac{r_k}{B_n}\Big(\operatorname{sign}\Big(\xi_k - \mu\,\frac{r_k}{B_n}\Big) - \operatorname{sign}\Big(\xi_k - \lambda\frac{r_k}{B_n}\Big)\Big).
$$

Arguing as before, we obtain that

$$
\mathbf{E}(Y_n,\omega) \le \operatorname{const} h(n), \qquad \operatorname{Var}(Y_n,\omega) \le \operatorname{const} N_n^{-1} h(n)
$$

for any interval ω of length h belonging to the set $\{t\colon |t| < N_n^{\alpha}\}$. Hence,

$$
\begin{aligned}
(1.3.18)\qquad & \mathbf{P}\Big\{W(Y_n,\omega) > \frac{\varepsilon}{3}\Big\} \\
& \le \mathbf{P}\Big\{|W(Y_n,\omega) - \mathbf{E}W(Y_n,\omega)| + \mathbf{E}W(Y_n,\omega) > \frac{\varepsilon}{3}\Big\} \\
& \le \mathbf{P}\Big\{|W(Y_n,\omega) - \mathbf{E}W(Y_n,\omega)| > \frac{\varepsilon}{4}\Big\} \\
& \le \Big(\frac{\varepsilon}{4}\Big)^{-2} \operatorname{Var} W(Y_n,\omega) \le \operatorname{const} N_n^{-1} h(n).
\end{aligned}
$$

Taking into account that there are $2N_n^{\alpha+\beta}$ terms in (1.3.14), we obtain by (1.3.15), (1.3.17), and (1.3.18) that

$$
\begin{aligned}
\mathbf{P}\{\sup_{|t|<N_n^{\alpha}} |X_n(t)| > \varepsilon\} & \le \operatorname{const} N_n^{\alpha+\beta}(N_n^{-1}N_n^{-\beta} + N_n^{-1+\alpha}) \\
& \le \operatorname{const} N_n^{-1+2\alpha+\beta} \to 0,
\end{aligned}
$$

if $0 < \alpha < 1/4,\ 0 < \beta < 1/4$. This completes the proof. □

Now we can obtain the asymptotic distribution of the sign estimator $\widehat{\theta}_n$ given by (1.3.9) in the model (1.2.1).

THEOREM 1.3.3. *Assume that the conditions of Theorem* 1.3.2 *are fulfilled and* $F'(0) > 0$. *Then*

$$
\widehat{\theta}_n = \theta^0 + \frac{1}{2\,F'(0)\,B_n^2} \sum_{k=1}^{n} r_k \operatorname{sign}\xi_k + o_p\Big(\frac{1}{B_n}\Big);
$$

hence, asymptotically,

$$
\widehat{\theta}_n \sim N\Big(\theta^0, \frac{1}{(2F'(0)B_n)^2}\Big).
$$

PROOF. We will use Theorem 1.3.2. Take an arbitrary α, $0 < \alpha < 1/4$, and divide the range of t in (1.3.9) into three parts:

$$A = \{t \leq -N_n^\alpha\}, \qquad B = \{t \geq N_n^\alpha\}, \qquad C = \{|t| < N_n^\alpha\}.$$

We will show that in the domains A and B the objective function in (1.3.9) unboundedly increases in probability as $n \to \infty$, whereas it is bounded in probability for $t = 0$, which lies in C. Therefore, the point of global minimum of (1.3.9) belongs to the set C with probability arbitrarily close to 1 for sufficiently large n. Hence, we can restrict t in the problem (1.3.9) to the set $|t| < N_n^\alpha$.

Consider the set A. Recall that the left-hand side of (1.3.10) is monotone nonincreasing in t. Hence for $t \leq -N_n^\alpha$

$$\sum_{k=1}^n \frac{r_k}{B_n} \operatorname{sign}\Big(\xi_k - t\,\frac{r_k}{B_n}\Big) \geq \sum_{k=1}^n \frac{r_k}{B_n} \operatorname{sign}\Big(\xi_k - N_n^\alpha\,\frac{r_k}{B_n}\Big).$$

Denote the right-hand side by ε_n. It is easily seen that

$$\mathbf{E}\varepsilon_n = -2\,N_n^\alpha \sum_{k=1}^n \frac{r_k^2}{B_n^2} F'(\kappa_k),$$

where $\kappa_k \in (-N_n^\alpha \frac{r_k}{B_n}, 0)$, so that $\max_{1\leq k\leq n} |\kappa_k| \to 0$. Hence $F'(\kappa_k) \to F'(0)$ uniformly in k, where $F'(0) > 0$ by assumption. Hence $\mathbf{E}\varepsilon_n \to -\infty$. Obviously, the variance of ε_n is bounded. Therefore, $\varepsilon_n \xrightarrow{\mathbf{P}} -\infty$. Thus the absolute value of (1.3.10) as well as the left-hand side of (1.3.9) tend to infinity in probability in the set A.

A similar result holds for the set B; namely, the minimal value of the left-hand side of (1.3.9) goes to $+\infty$ in probability.

Consider now (1.3.9) with restriction $|t| < N_n^\alpha$. By Theorem 1.3.2 this problem can be restated as

$$\Big(\frac{1}{B_n}\sum_{k=1}^n r_k \operatorname{sign}\xi_k - 2F'(0)\,t + X_n(t)\Big)^2 \Longrightarrow \min_{t:\ |t|<N_n^\alpha}, \tag{1.3.19}$$

where $X_n(t) \xrightarrow{\mathbf{P}} 0$ uniformly over this set. Hence (1.3.19) is equivalent to

$$\Big(\frac{1}{B_n}\sum_{k=1}^n r_k \operatorname{sign}\xi_k - 2F'(0)\,t\Big)^2 \Longrightarrow \min_{t:\ |t|<N_n^\alpha} \tag{1.3.20}$$

in the sense that the difference between the solutions of (1.3.19) and (1.3.20) converges to zero in probability. If we neglect the restriction on t, the minimum in (1.3.20) equals zero and is attained for

$$\widehat{t}_n = \frac{1}{2F'(0)} \sum_{k=1}^n \frac{r_k}{B_n} \operatorname{sign}\xi_k.$$

Note that $\widehat{t}_n$ has asymptotically the normal distribution $N\big(0, \frac{1}{(2\,F'(0))^2}\big)$. Hence

$$\mathbf{P}\{|\widehat{t}_n| < N_n^\alpha\} \to 1.$$

Therefore, for sufficiently large n $\widehat{t}_n$ also solves (1.3.20) with probability arbitrarily close to 1. Thus the difference between $\widehat{t}_n$ and the solution of (1.3.19) tends to zero in probability. This proves the theorem. □

1.4. The influence function

The concept of the influence function provides a quantitative measure of the effect which a single observation can have on the estimator. It is one of the characteristics, along with, say, bias or variance, which determines the properties of an estimator. Similarly to these characteristics, the influence function has its finite-sample and asymptotic forms. The influence functions and their role in the asymptotic statistical theory are discussed for i.i.d. samples and linear models with random regressors in the excellent book by Hampel et al. [**36**]. We are unaware of corresponding results for linear models with nonrandom factors, although Hampel et al. [**36**] point out (p. 47) that by a suitable modification, a similar theory can be worked out for general linear models.

Instead of the influence function we will consider a closely related concept of the sensitivity curve proposed by Tukey [**82**] (see also Hampel et al. [**36**], §2.1e). The sensitivity curve is defined for finite samples. As pointed out by Hampel et al. [**36**], for large samples the sensitivity curve usually converges to the influence function. We will determine the sensitivity curve and find its limit.

Let an observation (r, v), where $v = r\theta^0 + \varepsilon$, be added to observations (r_i, v_i) in the model $v_i = r_i\theta^0 + \xi_i$, $i = 1, \dots, n$. Let $\widehat{\theta}_n$ be the estimator based on the observations (r_i, v_i), $i = 1, \dots, n$, and let $\widehat{\theta}^*_{n+1}$ be the estimator based on the same observations with addition of the artificial nonrandom observation (r, v). Somewhat modifying Tukey's definition (due to the substitution of n by B_n^2 in our case) we define the sensitivity curve at the point (r, v) as

$$\mathrm{SC}_n(r, \varepsilon) = B_n^2\big(\widehat{\theta}^*_{n+1} - \widehat{\theta}_n\big).$$

We will show that under the conditions of Theorem 1.3.3, as $n \to \infty$,

$$\mathrm{SC}_n(r, \varepsilon) = \frac{r \operatorname{sign} \varepsilon}{2F'(0)} + o_p(1). \tag{1.4.1}$$

It is seen that the limit of this function in probability is bounded with respect to ε. Therefore, the effect of a single gross error in the measurement of the response variable on the estimate remains bounded, however large the error be. Note, though, that a large error in the measurement of the regressor v may have a large effect on the estimator. This property is unpleasant, but not surprising, since we study the problem under the classical assumption made in linear analysis that the regressors are specified precisely.

For the sake of comparison, let us state the sensitivity curve of the LSE in the same problem. Assuming that the errors ξ_i are i.i.d. with $\mathbf{E}\xi_i = 0$, $\operatorname{Var} \xi_i < \infty$, it is equal to $r\varepsilon + o_p(1)$ as $n \to \infty$, which is obtained by a direct calculation. It is seen that this sensitivity curve is an unbounded function of the error ε, which reflects the well-known fact that even a single gross error can drastically distort the result of estimation by the least squares method.

PROOF OF (1.4.1). We assume that the conditions of Theorem 1.3.3 are fulfilled. The sign estimators $\widehat{\theta}_n$ and $\widehat{\theta}^*_{n+1}$ are defined as the solutions of the extremal problems

$$\left|\sum_{i=1}^{n} r_i \operatorname{sign}\left(v_i - r_i\theta\right)\right| \Longrightarrow \min_{\theta} \tag{1.4.2}$$

and

$$\left|\sum_{i=1}^{n} r_i \operatorname{sign}\left(v_i - r_i\theta\right) + r \operatorname{sign}(v - r\theta)\right| \Longrightarrow \min_{\theta}. \tag{1.4.3}$$

The asymptotic analysis of (1.4.3) as $n \to \infty$ is carried out in the same way as for the estimator $\widehat{\theta}_n$ given by (1.2.12). We will keep the same notation. Denote by $\widehat{t}^*_{n+1}$ the solution of (1.4.3) in terms of the new variable $t = B_n(\theta - \theta^0)$, so that $\widehat{t}^*_{n+1} = B_n(\widehat{\theta}^*_{n+1} - \theta^0)$. Rewrite (1.4.3) in the form

$$\left|\frac{1}{B_n}\sum_{i=1}^{n} r_i \operatorname{sign}\left(\xi_i - t\frac{r_i}{B_n}\right) + \frac{r}{B_n}\operatorname{sign}\left(\varepsilon - t\frac{r}{B_n}\right)\right| \Longrightarrow \min_{t}. \tag{1.4.4}$$

As in the proof of Theorem 1.3.3, we show that under the conditions of this theorem the solution of (1.4.4) with probability arbitrarily close to 1 lies in the interval $C = \{t\colon |t| \le \operatorname{const} N_n^{\alpha}\}$ for sufficiently large n. In this interval the expansion (1.3.10) is valid. Moreover, in the interval C for sufficiently large n

$$\operatorname{sign}\left(\varepsilon - t\frac{r}{B_n}\right) = \operatorname{sign}\varepsilon.$$

Therefore, in the interval C the problem (1.4.4) is equivalent to

$$\left|\frac{1}{B_n}\sum_{i=1}^{n} r_i \operatorname{sign}\xi_i + \frac{r}{B_n}\operatorname{sign}\varepsilon - 2F'(0)\,t + X_n(t)\right| \Longrightarrow \min_{t\colon |t|\le \operatorname{const} N_n^{\alpha}}. \tag{1.4.5}$$

The same arguments as in the proof of Theorem 1.3.3 show that the difference between the solutions of the problems (1.4.4) and (1.4.5) tends to zero in probability, and consequently

$$\widehat{t}^*_{n+1} = \frac{1}{2\,F'(0)}\left(\frac{1}{B_n}\sum_{i=1}^{n} r_i \operatorname{sign}\xi_i + \frac{r}{B_n}\operatorname{sign}\varepsilon\right) + o_P(1),$$

or

$$\widehat{t}^*_{n+1} = \widehat{t}_n + \frac{1}{2\,F'(0)}\frac{r\operatorname{sign}\varepsilon}{B_n} + o_P(1).$$

This implies (1.4.1). □

CHAPTER 2

Sign Tests

2.1. General linear model

Linear models arise in various fields of mathematical statistics. This term is used when a mathematical model of observations is constructed by means of linear operations. Of course, the linear models in regression and factor analysis are most generally known, and one often means these very types of model when using this term. However, linear models play an important role in the analysis of time series as well. These are models of autoregression and moving average. They differ from the previously mentioned ones, for example, by the fact that in those models observations are stochastically independent, whereas the characteristic property of time series is dependence between the observed random variables. In the first part of the book we apply nonparametric (mostly sign-based) methods to linear models of independent observations. In the second part these ideas and methods will be developed for the linear models of time series, specifically, for autoregression models. We will treat only models with a scalar response. This means that for any fixed values of explanatory variables (factors, treatments) only one dependent variable (response) is recorded and analyzed. Of course, this restricts the applicability of the results because in many situations one has to simultaneously study several characteristics and their interaction. However, multiresponse (multivariate) linear analysis is well developed only for Gaussian models, and nonparametric methods in multivariate analysis have not been sufficiently advanced so far.

Let us formulate the linear model in geometric terms. Let $\mathbf{X}$ be the vector of observations. These observations are said to form a linear model if $\mathbf{X}$ is representable as

$$\mathbf{X} = \mathbf{l} + \boldsymbol{\xi}, \tag{2.1.1}$$

where $\boldsymbol{\xi}$ is a random vector with independent components and $\mathbf{l}$ is a nonrandom vector. The vectors $\mathbf{l}$ and $\boldsymbol{\xi}$ are unobservable and unknown to the statistician, but $\mathbf{l}$ is assumed to lie in a given linear subspace $\mathbb{L}$, i.e., $\mathbf{l} \in \mathbb{L}$. Diverse linear models arise from different representations for $\mathbf{X}$, various descriptions of $\mathbb{L}$, and various assumptions on the random errors $\boldsymbol{\xi}$. In theoretical studies it is convenient to consider $\mathbf{X}$ as an n-variate (according to the number of observations) column-vector, $\mathbf{X} = (x_1, \ldots, x_n)^T$, and specify $\mathbb{L}$ by a basis. Let $r = \dim \mathbb{L}$, with $r < n$, and (column) vectors $\mathbf{C}_1, \ldots, \mathbf{C}_r$ form a basis of $\mathbb{L}$. Let

$$\mathbf{C}_\alpha = (c_{1\alpha}, c_{2\alpha}, \ldots, c_{n\alpha})^T, \qquad \alpha = 1, \ldots, r,$$

and let $\mathbf{C}$ be the matrix with columns $\mathbf{C}_\alpha$, i.e.,

$$\mathbf{C} = \|c_{i\alpha}\|, \qquad i = 1, \ldots, n; \quad \alpha = 1, \ldots r.$$

We write

$$\boldsymbol{\theta} = (\theta_1, \ldots, \theta_r)^T \tag{2.1.2}$$

for the vector of unknown parameters and

$$\boldsymbol{\xi} = (\xi_1, \ldots, \xi_n)^T \tag{2.1.3}$$

for the vector of random errors. We will assume these random errors to be independent.

For the sign-based methods, the basic assumption on the distribution of random errors, apart from their independence, is stated in terms of their medians. Namely, we will assume that the errors have zero medians. For example, in the model (2.1.1) this means that

$$\mathbf{P}\{\xi_i < 0\} = \mathbf{P}\{\xi_i > 0\} = 1/2 \tag{2.1.4}$$

for any $i = 1, \ldots, n$. This assumption allows for the analysis based on signs. We do not require the errors to be indentically distributed, though in some instances we will need this property, for example, for deriving optimal rules and for their asymptotic analysis. Some other, more traditional assumptions on (2.1.3) will be stated later on.

In the above matrix notation (2.1.1) can be rewritten in the form commonly used in regression analysis:

$$\mathbf{X} = \mathbf{C}\boldsymbol{\theta} + \boldsymbol{\xi}. \tag{2.1.5}$$

In the one- and two-way layout problems it is natural to represent the set of data in the form of a one- or two-way table. In the one-way model (several sample problem) there is a single factor taking on k different values (levels). For each level j, $j = 1, \ldots, k$, one has n_j independent identically distributed observations x_{ij}, $i = 1, \ldots, n_j$, so that $\mathbf{X}$ can be represented by a table consisting of k columns of size $n_1, n_2, \ldots, n_k$. The linear model can be conveniently written here in a coordinate form by assuming that

$$x_{ij} = \mu + \tau_j + \xi_{ij} \tag{2.1.6}$$

for $i = 1, \ldots, n_j$ and $j = 1, \ldots, k$, where ξ_{ij} are independent random variables describing the part of the response variability which is not explained by the effect of the corresponding factor. The quantity μ is the average value from which the observed values x_{ij} deviate (by τ_j due to the effect of the jth level of factor and by ξ_{ij} due to random fluctuation). For the identifiability of the model (2.1.6) and by the very meaning of the variables τ_j we should impose a linear constraint on them. For example, we can require that

$$\sum_{j=1}^{k} \tau_j = 0. \tag{2.1.7}$$

Geometrically, the subspace L of dimension k is specified here by $k + 1$ vectors, according to the number of parameters μ, $\tau_1, \ldots, \tau_k$. The parameter τ_j is associated with the table of the form described above whose jth column consists of ones and all other entries are zeros. The table corresponding to μ consists entirely of ones, so that it is the sum of vectors (tables) corresponding to $\tau_1, \ldots, \tau_k$. These $k + 1$ vectors in the k-dimensional subspace do not form its basis, but the constraint

(2.1.7) ensures a unique representation of any vector $\mathbf{l} \in \mathbb{L}$ as a linear combination of these vectors. In what follows we will also encounter other specifications of $\mathbb{L}$.

It remains to discuss various assumptions which are usually imposed on the random errors represented by the vector $\boldsymbol{\xi}$. We have already stated the most important assumption that the random variables forming $\boldsymbol{\xi}$ are independent. In the classical (Gaussian) analysis they are assumed to have a normal distribution with mean 0 and variance σ^2 (or, briefly, $N(0, \sigma^2)$). Under these assumptions the well-known method of least squares is appropriate. We will suppose the reader to be familiar with the recommendations of the Gaussian theory of errors, which will serve us as a standard for comparison. The requirements in nonparametric setup are less restrictive: the errors are assumed to be identically distributed, with their common probability law unknown to the statistician and subject only to the condition that its distribution function is continuous. These properties allow for an application of rank-based methods, which will be discussed below briefly. For methods based on signs, only continuity of the distribution at zero point is needed, and it is not necessary to assume that the random errors are identically distributed.

We conclude with the list of assumptions on the distribution of random errors and the design of experiment to be used in what follows. Let us stress that we do not require that all of them are fulfilled simultaneously. Rather, only those needed in each particular instance will be invoked.

Assumptions on the error distribution function $F(x) = \mathbf{P}\{\xi_i < x\}$.

CONDITION 2.1(i). $F(0) = 1/2$.

CONDITION 2.1(ii). *$F(x)$ has a continuous density $f(x)$ in a neighborhood of zero, and $f(0) > 0$.*

CONDITION 2.1(iii). *$f(x)$ is absolutely continuous in a neighborhood of zero and $f'(0) = 0$.*

CONDITION 2.1(iv). *$F(x)$ satisfies the Lipschitz condition at zero, i.e., there exists a constant $L > 0$ such that*

$$|F(u_1) - F(u_2)| < L|u_1 - u_2|.$$

CONDITION 2.1(v). *There exist a constant $M > 0$ and a neighborhood of zero $\{u : |u| < d\}$ such that*

$$|F(u_1) - F(u_2)| > M|u_1 - u_2|, \text{ if } |u_1| < d, \ |u_2| < d.$$

CONDITION 2.1(vi). *In a neighborhood of zero F has a density f satisfying the Lipschitz condition, i.e., for some $N > 0$, $d > 0$,*

$$|f(u_1) - f(u_2)| < N|u_1 - u_2|, \qquad \text{if } \ |u_1| < d, \quad |u_2| < d.$$

Assumptions on the design matrix $\mathbf{C} = \|c_{i\alpha}\|$:

CONDITION 2.1(vii). $\max_{1\le i\le n} \max_{1\le \alpha\le r} |c_{i\alpha}| < K < \infty$.

CONDITION 2.1(viii). $\frac{1}{n}\mathbf{C}^T\mathbf{C} \to \mathbf{\Sigma} > 0$.

2.2. Locally optimal sign tests in the regression problem

In this section we consider the testing problem of

$$H : \mathbf{l} = \mathbf{0} \tag{2.2.1}$$

in linear models (2.1.1). We have already pointed out in the introduction that this hypothesis is rarely of importance for practical applications, but the test statistics to be derived for the testing of this hypothesis will be needed throughout the sequel, including estimation and testing linear hypotheses. To derive reasonable (and even optimal in a certain sense) tests for (2.2.1), we make some assumptions on the distribution of the random errors in the model (2.1.1), additionally to the basic assumption (2.1.4). A detailed treatment will be given to the model in the regression form (2.1.5). Corresponding results for factor models will be obtained by a suitable modification of these results.

So, let observations $\mathbf{X} = (x_1, \dots, x_n)^T$ follow the regression model (2.1.5), where $\mathbf{C}$ is a given design matrix, $\boldsymbol{\theta}$ is the vector of unknown parameters (see (2.1.2)), and $\boldsymbol{\xi}$ is the vector of mutually independent random errors (2.1.3). In what follows we assume that the columns of the matrix $\mathbf{C}$ are linearly independent, i.e., that $\mathbf{C}$ has a full rank. Denoting the ith row of $\mathbf{C}$ by $\mathbf{c}_i = (c_{i1}, \dots, c_{ir})$, we can write x_i as $x_i = \mathbf{c}_i\boldsymbol{\theta} + \xi_i$, $i = 1, \dots n$, where $\mathbf{c}_i\boldsymbol{\theta}$ denotes the scalar product, $\mathbf{c}_i\boldsymbol{\theta} = \sum_{\alpha=1}^{r} c_{i\alpha}\theta_\alpha$. The hypothesis (2.2.1) in this model becomes

$$H_0 \colon \boldsymbol{\theta} = \mathbf{0}, \tag{2.2.2}$$

to be tested against alternatives $\boldsymbol{\theta} \neq \mathbf{0}$. We will construct *distribution-free* tests for H_0 which remain valid (i.e., retain the prescribed level) for any error distributions subject to the condition (2.1.4). These test procedures will be based only on the signs of observations $x_1, \dots, x_n$.

We will construct sign tests for (2.2.1) optimal in a certain local sense to be explained below. Consider the vector of signs

$$\mathbf{S}(\mathbf{X}) = (\operatorname{sign} x_1, \operatorname{sign} x_2, \dots, \operatorname{sign} x_n)^T. \tag{2.2.3}$$

The possible values of the random vector $\mathbf{S}(\mathbf{X})$ are vectors consisting of $+1$ and -1. Let

$$\mathbf{S} = (s_1, s_2, \dots, s_n)^T \tag{2.2.4}$$

be an arbitraty vector of this form. Let Q be a set of vectors of the form (2.2.4) to be used as a critical region for testing H_0 as in (2.2.2). In other words, H_0 is rejected whenever the following event occurs:

$$\{\mathbf{X} : \mathbf{S}(\mathbf{X}) \in Q\}. \tag{2.2.5}$$

Consider the power of this test in the model (2.1.5) as a function of $\boldsymbol{\theta}$:

$$\mathbf{P}\{\mathbf{S}(\mathbf{X}) \in Q \mid \boldsymbol{\theta}\} \tag{2.2.6}$$

or, briefly, $\mathbf{P}\{Q \mid \boldsymbol{\theta}\}$. Notice that

$$\mathbf{P}\{Q|\boldsymbol{\theta}\} = \sum_{\mathrm{s} \in Q} \mathbf{P}\{\mathrm{s} \mid \boldsymbol{\theta}\},$$

where

$$\mathbf{P}\{\mathrm{s} \mid \boldsymbol{\theta}\} = \mathbf{P}\{\mathbf{S}(\mathbf{X}) = \mathrm{s} \mid \boldsymbol{\theta}\}.$$

The requirements of the choice of the "best" sign test (2.2.5) will be stated in terms of its power function. First, we fix an arbitrary significance level ε, $\varepsilon > 0$, and restrict our choice by the tests with this level. Hence, the first requirement on Q is

$$\mathbf{P}\{Q \mid \boldsymbol{\theta}\}\big|_{\boldsymbol{\theta}=\mathbf{0}} = \varepsilon. \tag{2.2.7}$$

Since, subject to condition (2.1.4), the probability of any vector (2.2.4) under H_0 is 2^{-n}, only multiples of 2^{-n} can be taken for the level ε in (2.1.5). For such ε any sign test Q contains $K = \varepsilon 2^n$ points of the form (2.2.4). Among sign tests of a given level, the best one would be the test with power function exceeding the power functions of any other test. However, for alternatives $H\colon \boldsymbol{\theta} \neq \mathbf{0}$ there exists no test with this property. For this reason we will focus our attention on the local behavior of the power function in a neighborhood of $\boldsymbol{\theta} = \mathbf{0}$.

We begin with the one-parameter model

$$x_i = c_i\theta + \xi_i, \qquad i = 1, \dots, n,$$

with parameter $\theta \in \mathbb{R}^1$. Such a model appeared in Chapter 1 as model (1.2.1). In the one-parameter model one can construct locally most powerful sign tests for testing the hypothesis (2.2.2) against one-sided alternatives $H^+\colon \theta > 0$ or $H^-\colon \theta < 0$. Then one can use their two-sided versions for testing (2.2.2) against two-sided alternatives. One acts in a similar way when using two-sided versions of Student's test in the normal theory and when dealing with rank tests.

Assume that the random errors are independent and identically distributed. Their common distribution function

$$F(u) = \mathbf{P}\{\, \xi_i < u \,\}$$

will be assumed to satisfy Conditions 2.1(i, ii).

Consider, for definiteness, the alternative $H^+\colon \theta > 0$. Among the tests with a given significance level we will look for the one whose power has the highest rate of increase in a vicinity of the point $\theta = 0$. In other words, we look for the test Q which maximizes

$$\frac{d}{d\theta}\mathbf{P}\{\mathbf{S}(\mathbf{X}) \in Q \mid 0\}$$

subject to condition (2.2.7). Since

$$\frac{d}{d\theta}\mathbf{P}\{\mathbf{S}(\mathbf{X}) \in Q \mid \theta\} = \sum_{\mathbf{s} \in Q} \frac{d}{d\theta}\mathbf{P}\{\mathbf{S}(\mathbf{X}) = \mathbf{s} \mid \theta\},$$

we should include into Q the vectors $\mathbf{s}$ for which $\frac{d}{d\theta}\mathbf{P}\{\mathbf{S}(\mathbf{X}) \in Q \mid 0\}$ are as large as possible. Therefore

$$Q = \left\{\mathbf{s} : \frac{d}{d\theta}\mathbf{P}\{\mathbf{S}(\mathbf{X}) = \mathbf{s} \mid 0\} \geq \text{const}\right\},$$

where the constant is chosen such that (2.2.7) be satisfied. Note that the distribution of the test statistic may have larger atoms than 2^{-n}, so that the given ε, even a multiple of 2^{-n} may be unattainable for any critical constant. To achieve the preassigned level, one can use randomization or include into the critical region the necessary number of points from the set

$$\left\{\mathbf{s} : \frac{d}{d\theta}\mathbf{P}\left\{\mathbf{S}(\mathbf{X}) = \mathbf{s} \mid 0\right\} = \text{const}\right\}.$$

The choice of this subset does not affect the local power properties of the test.

It remains to evaluate $\frac{d}{d\theta}\mathbf{P}\{\mathbf{S}(\mathbf{X}) = \mathbf{s} \mid 0\}$. The likelihood $\mathbf{P}\{\mathbf{S}(\mathbf{X}) = \mathbf{s} \mid \boldsymbol{\theta}\}$ with $\mathbf{S}(\mathbf{X})$ and $\mathbf{s}$ as in (2.2.3) and (2.2.4), is given by the formula

$$\mathbf{P}\{\mathbf{S}(\mathbf{X}) = \mathbf{s} \mid \boldsymbol{\theta}\} = \prod_{i=1}^{n} \left(\mathbf{P}\{x_i > 0\}\right)^{(1+s_i)/2} \left(\mathbf{P}\{x_i < 0\}\right)^{(1-s_i)/2}. \tag{2.2.8}$$

It is easily seen that under the conditions on $F(u)$ stated above,

$$\begin{aligned} \mathbf{P}\{\mathbf{S}(\mathbf{X}) = \mathbf{s} \mid \theta\} &= \prod_{i=1}^{n} \left(\mathbf{P}\{x_i > 0\}\right)^{(1+s_i)/2} \left(\mathbf{P}\{x_i < 0\}\right)^{(1-s_i)/2} \\ &= 2^{-n} \prod_{i=1}^{n} [1 + 2f(0)c_i s_i \theta + o(\theta)] \\ &= 2^{-n} \left[1 + 2f(0)\left(\sum_{i=1}^{n} c_i s_i\right)\theta + o(\theta)\right]. \end{aligned}$$

Hence

$$\frac{d}{d\theta}\mathbf{P}\{\mathbf{S}(\mathbf{X}) = \mathbf{s} \mid 0\} = 2^{-(n-1)} f(0) \sum_{i=1}^{n} c_i s_i.$$

Therefore, for any significance level, the required test has the form

$$Q = \left\{\mathbf{s} : \sum_{i=1}^{n} c_i s_i \geq \text{const}\right\}, \tag{2.2.9}$$

where the constant is chosen so that condition (2.2.7) is satisfied. Clearly, the locally most powerful sign test against $H^- : \theta < 0$ has the form

$$Q = \left\{\mathbf{s} : \sum_{i=1}^{n} c_i s_i \leq \text{const}\right\}.$$

In both cases the test statistic is

$$\sum_{i=1}^{n} c_i \operatorname{sign} x_i. \tag{2.2.10}$$

For testing $H_0 \colon \theta = 0$ against $H \colon \theta \neq 0$, we will use the two-sided version of the test (2.2.9), which rejects H_0 in favor of $H \colon \theta \neq 0$ when

$$\left|\sum_{i=1}^{n} c_i \operatorname{sign} x_i\right| \geq \text{const}. \tag{2.2.11}$$

As will be shown in Theorem 2.2.1, under certain additional conditions this test is locally optimal.

It is noteworthy that under the basic condition (2.1.4) the statistic (2.2.10) is distribution free under H_0 even when $\xi_1, \dots, \xi_n$ are not identically distributed.

Hence, in this case the significance levels of tests (2.2.9), (2.2.11) still retain their values.

Now we turn to the multiparameter model (2.1.5), where $\boldsymbol{\theta} \in \mathbb{R}^r$.

Assume that the random errors are i.i.d. with their common distribution satisfying Conditions 2.1(i–iii). The conditions $F(0) = 1/2$ and $f'(0) = 0$ necessarily hold for symmetric about zero distributions with a differentiable density function. We do not require this symmetry, assuming only some of its consequences. It will be shown that under these conditions the derivatives of the power function

$$\frac{d\mathbf{P}\{Q \mid \boldsymbol{\theta}\}}{d\boldsymbol{\theta}} = \left(\frac{\partial \mathbf{P}\{Q \mid \boldsymbol{\theta}\}}{\partial \theta_1}, \dots, \frac{\partial \mathbf{P}\{Q \mid \boldsymbol{\theta}\}}{\partial \theta_r}\right),$$
$$\frac{d^2\mathbf{P}\{Q \mid \boldsymbol{\theta}\}}{d\boldsymbol{\theta}^2} = \left\| \frac{\partial^2 \mathbf{P}\{Q \mid \boldsymbol{\theta}\}}{\partial \theta_i \partial \theta_j} \right\|_{i,j=1}^{r}$$

exist in a neighborhood of $\boldsymbol{\theta} = \mathbf{0}$. As before, our choice of the test for the hypothesis (2.2.2) will be based on the local properties of the power function (2.2.6), which are determined by its behavior for $\boldsymbol{\theta}$ close to the hypothesized value $\boldsymbol{\theta} = \mathbf{0}$. We restrict the class of competing sign tests to that of the so-called locally unbiased tests satisfying

$$\left. \frac{d\mathbf{P}\{Q \mid \boldsymbol{\theta}\}}{d\boldsymbol{\theta}} \right|_{\boldsymbol{\theta}=0} = \mathbf{0}. \tag{2.2.12}$$

The behavior of the power function (2.2.6) in a neighborhood of $\boldsymbol{\theta} = \mathbf{0}$ for a locally unbiased test Q is determined by the quadratic term of its Taylor expansion, i.e., by the quadratic form

$$\frac{1}{2}\boldsymbol{\theta}^T \left(\frac{d^2\mathbf{P}\{Q \mid \mathbf{0}\}}{d\boldsymbol{\theta}^2} \right) \boldsymbol{\theta}. \tag{2.2.13}$$

The performance of Q as a critical region for testing H_0 will be the better, the faster $\mathbf{P}\{Q \mid \boldsymbol{\theta}\}$ increases for $\boldsymbol{\theta}$ moving away from $\mathbf{0}$. Since in the multiparameter case there is no uniformly most powerful test even in the local sense, we have to optimize some reasonably chosen scalar characteristic of Q. The above considerations suggest that we may characterize the locally unbiased tests by the curvature of the surface in the $(r+1)$-dimensional space determined by the equation (2.2.6) at the point $\boldsymbol{\theta} = \mathbf{0}$ (where the condition (2.2.12) holds). There are several definitions of curvature for functions of a multivariate argument. Usually in geometry the Gaussian curvature is considered. But we suggest using the mean curvature, which is proportional to the trace of the matrix of the quadratic form (2.1.13). The definitions of various curvature coefficients can be found, for example, in Dubrovin, Novikov, and Fomenko [**22**], Chapter 2, §8.

Thus we get the following extremal problem: select the set Q such that

$$\operatorname{tr} \left. \frac{d^2\mathbf{P}\{Q \mid \boldsymbol{\theta}\}}{d\boldsymbol{\theta}^2} \right|_{\boldsymbol{\theta}=\mathbf{0}} \Longrightarrow \max \tag{2.2.14}$$

under the restrictions (2.2.7) and (2.2.12). The region Q solving this problem will determine the *locally optimal sign test* of level ε. Notice that by (2.2.6)

$$\operatorname{tr} \frac{d^2\mathbf{P}\{Q \mid \mathbf{0}\}}{d\boldsymbol{\theta}^2} = \sum_{\mathbf{s} \in Q} \operatorname{tr} \frac{d^2\mathbf{P}\{\mathbf{s} \mid \mathbf{0}\}}{d\boldsymbol{\theta}^2}.$$

Under Conditions 2.1(i–iii) we can obtain explicit expressions for the derivatives $\frac{d}{d\boldsymbol{\theta}}\mathbf{P}\{\mathbf{s} \mid \boldsymbol{\theta}\}$ and $\frac{d^2}{d\boldsymbol{\theta}^2}\mathbf{P}\{\mathbf{s} \mid \boldsymbol{\theta}\}$. For the likelihood (2.2.8), one easily derives that under Conditions 2.1(i–iii)

$$\big(\mathbf{P}\{x_i > 0\}\big)^{(1+s_i)/2}\big(\mathbf{P}\{x_i < 0\}\big)^{(1-s_i)/2} = \frac{1}{2}\big(1 + 2s_i f(0)\mathbf{c}_i\boldsymbol{\theta} + o(|\boldsymbol{\theta}|^2)\big).$$

Hence,

(2.2.15)

$$\mathbf{P}\{\mathbf{S}(\mathbf{X}) = \mathbf{s} \mid \boldsymbol{\theta}\} = \Big(\frac{1}{2}\Big)^n \Big(1 + 2f(0)\sum_{i=1}^n s_i\mathbf{c}_i\boldsymbol{\theta} + 4f^2(0)\sum\sum_{1\le i\ne j\le n}(s_i\mathbf{c}_i\boldsymbol{\theta})(s_j\mathbf{c}_j\boldsymbol{\theta}) + o(|\boldsymbol{\theta}|^2)\Big).$$

Therefore, $\operatorname{tr}\frac{d^2}{d\boldsymbol{\theta}^2}\mathbf{P}\{\mathbf{s} \mid \boldsymbol{\theta}\}\big|_{\boldsymbol{\theta}=\mathbf{0}}$ is proportional to

$$\sum\sum_{1\le i\ne j\le n} s_i s_j \sum_{\alpha=1}^r c_{i\alpha}c_{j\alpha}. \tag{2.2.16}$$

Note that (2.2.16) takes the same value for $\mathbf{s}$ and $-\mathbf{s}$. Hence the test of maximum mean curvature simultaneously contains or does not contain both $\mathbf{S}$ and $-\mathbf{S}$. As follows from (2.1.15),

$$\frac{d\mathbf{P}\{\mathbf{s} \mid \boldsymbol{\theta}\}}{d\boldsymbol{\theta}}\bigg|_{\boldsymbol{\theta}=\mathbf{0}} = -\frac{d\mathbf{P}\{-\mathbf{s} \mid \boldsymbol{\theta}\}}{d\boldsymbol{\theta}}\bigg|_{\boldsymbol{\theta}=\mathbf{0}},$$

so that for any pair $\mathbf{s}\cup(-\mathbf{s})$ we have

$$\frac{d\mathbf{P}\{\mathbf{s}\cup(-\mathbf{s})|\boldsymbol{\theta}\}}{d\boldsymbol{\theta}}\bigg|_{\boldsymbol{\theta}=\mathbf{0}} = \mathbf{0}.$$

Therefore the test of maximum mean curvature is necessarily locally unbiased.

Thus we obtain that the optimal sign test in the sense (2.2.14) has the form

$$Q = \Big\{\mathbf{s}\colon \sum\sum_{1\le i\ne j\le n} s_i\, s_j \sum_{\alpha=1}^r c_{i\alpha}c_{j\alpha} > \text{const}\Big\}. \tag{2.2.17}$$

By adding the independent of $\mathbf{s}$ quantity

$$\sum\sum_{1\le i=j\le n} s_i\, s_j \sum_{\alpha=1}^r c_{i\alpha}c_{j\alpha} = \sum_{i=1}^n\sum_{\alpha=1}^r c_{i\alpha}^2,$$

to both sides of the inequality in (2.2.17), the locally optimal sign test becomes

$$Q = \Big\{\mathbf{s}\colon \sum_{\alpha=1}^r\Big(\sum_{i=1}^n c_{i\alpha}\, s_i\Big)^2 > \text{const}\Big\}.$$

We summarize the results just obtained in the following theorem.

THEOREM 2.2.1. *Under Conditions* 2.1(i–iii) *the test with critical region*

$$Q = \Big\{ \mathbf{X} : \sum_{\alpha=1}^{r} \Big(\sum_{i=1}^{n} c_{i\alpha} \operatorname{sign} x_i \Big)^2 > \text{const} \Big\} \tag{2.2.18}$$

is a locally optimal sign test (i.e., a locally unbiased sign test of maximal mean curvature) for the hypothesis (2.2.2) *in the model* (2.1.5). *Under the hypothesis* (2.2.2) *its test statistic is distribution free provided only the condition* (2.1.4) *is fulfilled.*

The sign statistic in (2.2.18) can be represented in the matrix form as

$$\sum_{\alpha=1}^{r} \Big(\sum_{i=1}^{n} c_{i\alpha} \operatorname{sign} x_i \Big)^2 = |\mathbf{C}^T \mathbf{S}(\mathbf{X})|^2 = \big(\mathbf{S}(\mathbf{X})\big)^T \big(\mathbf{C}\mathbf{C}^T\big) \big(\mathbf{S}(\mathbf{X})\big). \tag{2.2.19}$$

The determination of the critical constant in (2.2.18) to satisfy (2.2.7) for a given level ε will be discussed in §2.3. The sign test (2.2.11) is a one-parameter form of (2.2.18).

REMARK. The test (2.2.18) is optimal also in another sense, though closely related to (2.2.14). Namely, among tests of a given level, it has maximum average power over a sphere of an infinitesimal radius. To state it more precisely, consider an arbitrary sign test for (2.2.2) with level no greater than ε. Let $\beta(\boldsymbol{\theta})$ denote its power function. Then the limit

$$\lim_{\rho \to 0} \rho^{-r-1} \int_{\boldsymbol{\theta} : |\boldsymbol{\theta}| = \rho} \big(\beta(\boldsymbol{\theta}) - \beta(\mathbf{0})\big)\, d\boldsymbol{\theta} \tag{2.2.20}$$

under condition (2.2.7) is maximized by the sign test (2.2.18) of level ε.

Along with (2.1.18), one can also consider other quadratic forms of $\mathbf{C}^T\mathbf{S}(\mathbf{X})$. The corresponding tests maximize the average power defined similarly to (2.1.20) with integration over suitable ellipsoids shrinking to $\boldsymbol{\theta} = \mathbf{0}$. The quadratic sign tests are admissible within the class of all sign tests. Recall that the test A is said to be *admissible* within some class of tests if this class contains no test of the same or smaller level with power no less, and somewhere greater, than that of A. For (large-sample) evaluation of critical values, the most convenient quadratic test is

$$Q = \big\{ \mathbf{S}(\mathbf{X}) \colon \big(\mathbf{S}(\mathbf{X})\big)^T \mathbf{C} \big(\mathbf{C}^T\mathbf{C}\big)^{-1} \mathbf{C}^T \mathbf{S}(\mathbf{X}) > \text{const} \big\}. \tag{2.2.21}$$

It will be shown in §2.3 that under certain assumptions on the design matrix $\mathbf{C}$ its test statistic

$$\big(\mathbf{S}(\mathbf{X})\big)^T \mathbf{C} \big(\mathbf{C}^T\mathbf{C}\big)^{-1} \mathbf{C}^T \mathbf{S}(\mathbf{X}) \tag{2.2.22}$$

has asymptotically the chi-squared distribution as $n \to \infty$. Therefore, the corresponding approximation can be used for the determination of the critical constant in (2.2.21). Finally, note that in the one-parameter case all quadratic sign tests are equivalent to the test (2.2.11).

2.3. Evaluation of critical values: Asymptotic theory

Here we discuss how the critical values of statistics (2.2.17), (2.2.21), or other sign test statistics can be evaluated. Define independent random variables $\zeta_i = \operatorname{sign}\xi_i$, $i = 1, \ldots, n$, and let $\boldsymbol{\zeta} = (\zeta_1, \ldots, \zeta_n)^T$. Under the condition (2.1.4)

$$\mathbf{P}\{\zeta_i = 1\} = \mathbf{P}\{\zeta_i = -1\} = 1/2. \tag{2.3.1}$$

Under the hypothesis (2.2.1) the random variables $\operatorname{sign} x_i$ in the model (2.1.1) subject to condition (2.1.4) have the same distribution as ζ_i, $i = 1, \ldots, n$. Hence the sign statistic (2.2.19) is distributed under H_0 as

$$\sum_{\alpha=1}^{r}\Big(\sum_{i=1}^{n} c_{i\alpha}\zeta_i\Big)^2. \tag{2.3.2}$$

The distribution of this random variable depends on the design matrix $\mathbf{C}$. Although there is no analytic expression for this distribution, it can be numerically computed with any desirable accuracy. (The computation of this distribution and its quantiles will be dicussed in §2.5.) Note that the random variable (2.3.2) has a discrete distribution. Let us recall what we mean by the quantile of a discrete random variable.

DEFINITION 2.3.1. The α-quantile ξ_α of a discrete random variable ξ is:

(a) $\xi_\alpha = \sup\{x : \mathbf{P}\{\xi \le x\} \le \alpha\}$ if the equation $\mathbf{P}\{\xi \le x\} = \alpha$ has no solution, and

(b) $\xi_\alpha = \frac{1}{2}\big(\inf\{x : \mathbf{P}\{\xi \le x\} = \alpha\} + \sup\{x : \mathbf{P}\{\xi \le x\} = \alpha\}\big)$, when this equation has a solution (i.e., ξ_α is taken to be the midpoint of the interval formed by the solutions).

Let $q_{1-\varepsilon}$ denote the $(1-\varepsilon)$-quantile of the random variable (2.3.2). Then

$$\mathbf{P}\Big\{\sum_{\alpha=1}^{r}\Big(\sum_{i=1}^{n} c_{i\alpha}\zeta_i\Big)^2 > q_{1-\varepsilon}^n\Big\} \le \varepsilon. \tag{2.3.3}$$

Similarly, let $\widetilde{q}_{1-\varepsilon}^{\,n}$ be the $(1-\varepsilon)$-quantile of the random variable

$$\boldsymbol{\zeta}^T\mathbf{C}\big(\mathbf{C}^T\mathbf{C}\big)^{-1}\mathbf{C}^T\boldsymbol{\zeta}. \tag{2.3.4}$$

Then the rules for testing the hypothesis $H_0\colon \boldsymbol{\theta} = \boldsymbol{\theta}^0$, for a specified $\boldsymbol{\theta}^0$, by means of sign tests (2.2.18) and (2.2.21) can be formulated as follows.

RULE 2.3.1. *Given ε, $0 < \varepsilon < 1$, reject the hypothesis $H_0\colon \boldsymbol{\theta} = \boldsymbol{\theta}^0$ in the model (2.1.5) when*

$$\big[\mathbf{S}(\mathbf{X} - \mathbf{C}\boldsymbol{\theta}^0)\big]^T\mathbf{C}\mathbf{C}^T\big[S(\mathbf{X} - \mathbf{C}\boldsymbol{\theta}^0)\big] > q_{1-\varepsilon}^n \tag{2.3.5}$$

(by the test (2.2.18)), or when

$$\big[\mathbf{S}(\mathbf{X} - \mathbf{C}\boldsymbol{\theta}^0)\big]^T\mathbf{C}\big(\mathbf{C}^T\mathbf{C}\big)^{-1}\mathbf{C}^T\big[\mathbf{S}(\mathbf{X} - \mathbf{C}\boldsymbol{\theta}^0)\big] > \widetilde{q}_{1-\varepsilon}^{\,n} \tag{2.3.6}$$

(by the test (2.2.21)). Both tests have the significance level at most ε.

Let us emphasize that the tests (2.3.5) and (2.3.6) are valid for nonidentically distributed errors as well, provided (2.1.4) is fulfilled.

Under certain assumptions on the matrix $\mathbf{C}$ one can obtain asymptotic null distributions of quadratic sign statistics as $n \to \infty$. Henceforth we consider bounded and asymptotically nondegenerate designs, i.e., we assume Conditions 2.1(vii, viii) stated above: there exists a number K (independent of n) such that

$$\max_{1\le i\le n}\max_{1\le\alpha\le r}|c_{i\alpha}|<K \qquad \text{for all} \quad n$$

and

$$\frac{1}{n}\mathbf{C}^T\mathbf{C}\to\mathbf{\Sigma},$$

where the matrix $\mathbf{\Sigma}$ is nondegenerate (and hence positive definite).

Under Conditions 2.1(vii, viii) the random vector $\frac{1}{\sqrt{n}}\mathbf{C}^T\boldsymbol{\zeta}$ is asymptotically distributed as $N(\mathbf{0},\mathbf{\Sigma})$. This implies the following conclusions about the asymptotic distributions of test statistics (2.2.19) and (2.2.22).

THEOREM 2.3.2. *Under the hypothesis* (2.2.2) *about the model* (2.1.5), *one has, subject to* (2.1.4) *and Conditions* 2.1(vii, viii) *as* $n\to\infty$,

$$\frac{1}{n}\sum_{\alpha=1}^{r}\left(\sum_{i=1}^{n}c_{i\alpha}\operatorname{sign}x_i\right)^2\to\sum_{\alpha=1}^{r}\lambda_\alpha\eta_\alpha^2 \quad (\textit{in distribution}), \tag{2.3.7}$$

where $\lambda_1,\dots,\lambda_r$ *are the eigenvalues of the matrix* $\mathbf{\Sigma}$, $\eta_1,\dots,\eta_r$ *are i.i.d. standard normal random variables, and*

$$(\mathbf{S}^T(\mathbf{X})\mathbf{C})\left(\mathbf{C}^T\mathbf{C}\right)^{-1}(\mathbf{C}^T\mathbf{S}(\mathbf{X}))\to\chi^2(r) \quad (\textit{in distribution}), \tag{2.3.8}$$

with $\chi^2(r)$ *denoting a chi-squared random variable with* r *degrees of freedom.*

PROOF. Let $\mathbf{Y}$ be an r-variate random vector with distribution $N(0,\mathbf{\Sigma})$. By the multidimensional central limit theorem, the vector $\frac{1}{\sqrt{n}}\mathbf{C}^T\boldsymbol{\zeta}$ is asymptotically distributed as $\mathbf{Y}$ as $n\to\infty$. Hence, the statistics (2.3.7) and (2.3.8) are asymptotically distributed as $|\mathbf{Y}|^2$ and $\mathbf{Y}^T\mathbf{\Sigma}^{-1}\mathbf{Y}$, respectively. Let $\mathbf{Q}$ be the orthogonal matrix such that $\mathbf{Q}\mathbf{\Sigma}\mathbf{Q}^T=\mathbf{\Lambda}$, where $\mathbf{\Lambda}=\operatorname{diag}(\lambda_1,\dots,\lambda_r)$. Here $\lambda_1>0,\dots,\lambda_r>0$, since the matrix $\mathbf{\Sigma}$ is positive definite. Let $\mathbf{Z}=\mathbf{QY}$, $\mathbf{Z}=(z_1,\dots,z_r)^T$. Clearly, $\mathbf{Z}=N(0,\mathbf{\Lambda})$; hence, z_α, $\alpha=1,\dots,r$, can be written as $z_\alpha=\sqrt{\lambda_\alpha}\eta_\alpha$, where $\eta_1,\dots,\eta_r$ are independent standard normal random variables. Now $|\mathbf{Y}|^2=\sum_{\alpha=1}^r z_\alpha^2=\sum_{\alpha=1}^r\lambda_\alpha\eta_\alpha^2$,

$$\mathbf{Y}^T\mathbf{\Sigma}^{-1}\mathbf{Y}=\mathbf{Z}^T(\mathbf{Q}^T\Sigma^{-1}\mathbf{Q})\mathbf{Z}=\mathbf{Z}^T\mathbf{\Lambda}^{-1}\mathbf{Z}=\sum_{\alpha=1}^{r}\eta_\alpha^2,$$

which was to be proved. □

The simple form of the limiting distribution of the test statistic in (2.2.17) is one of its advantages over other quadratic sign statistics.

There is one more asymptotic justification for the statistic (2.2.18), which is based on a principle due to Roy [**76**].

Consider the model (2.1.5) and take some unit vector $\mathbf{l}=(l_1,\dots,l_n)^T$, $|\mathbf{l}|=1$, in the subspace $\mathbb{L}=\{\mathbf{C}\boldsymbol{\theta},\ \boldsymbol{\theta}\in\mathbb{R}^r\}$. Consider the one-dimensional alternative $H^{(1)}\colon \mathbf{C}\boldsymbol{\theta}=t\mathbf{l}$, $t\in\mathbb{R}^1$, to the hypothesis $H_0\colon\boldsymbol{\theta}=\mathbf{0}$. The results of §2.2 imply that the locally optimal sign test for testing H_0 against $H^{(1)}$ has the form

$$\left|\sum_{i=1}^{n}l_i\operatorname{sign}x_i\right|>\text{const.} \tag{2.3.9}$$

Of course, for finite n the constant in the right-hand side of (2.3.9) depends on the vector $\mathbf{l}$. However, as $n \to \infty$, the statistic in (2.3.9) is asymptotically normal $N(0,1)$ under H_0. Indeed, a sufficient condition for this asymptotic normality is

$$\max_{1\le i\le n} |l_i|^2 \to 0, \tag{2.3.10}$$

which is fulfilled when the matrix $\mathbf{C}$ satisfies Conditions 2.1(vii, viii). Therefore the critical values in (2.3.9) are practically independent of $\mathbf{l}$.

The hypothesis H_0 is rejected against the alternative $H^{(\mathbf{l})}$ for large values of the statistic (2.3.9). The alternative $H\colon \boldsymbol{\theta} \neq \mathbf{0}$ is the union of alternatives $H^{(\mathbf{l})}$. Hence, following Roy [**76**], it is reasonable to reject H_0 if at least one of the statistics (2.3.9) is large as $\mathbf{l}$ takes all possible values $\mathbf{l} \in \mathbb{L}$, $|\mathbf{l}| = 1$. This means that the test statistic for H_0 is taken to be

$$\max_{\mathbf{l}\in\mathbb{L}} \Bigl(\sum_{i=1}^{n} l_i \operatorname{sign} x_i\Bigr)^2. \tag{2.3.11}$$

It is seen that (2.2.11) equals $|\operatorname{proj}_{\mathbb{L}} \mathbf{S}(\mathbf{X})|^2$, where $\operatorname{proj}_{\mathbb{L}} \mathbf{S}(\mathbf{X})$ is the Euclidean projection of the vector $\mathbf{S}(\mathbf{X})$ onto the subspace $\mathbb{L}$. It is well known in the least squares theory (see, for example, Bickel and Doksum [**5**], §7.2) that the matrix of the orthogonal projection operator onto the subspace $\mathbb{L}$ is $\mathbf{C}(\mathbf{C}^T\mathbf{C})^{-1}\mathbf{C}^T$. Therefore,

$$\operatorname{proj}_{\mathbb{L}} \mathbf{S}(\mathbf{X}) = \mathbf{C}\bigl(\mathbf{C}^T\mathbf{C}\bigr)^{-1}\mathbf{C}^T\mathbf{S}(\mathbf{X}),$$

which implies

$$\max_{\mathbf{l}\in\mathbb{L}} \Bigl(\sum_{i=1}^{n} l_i \operatorname{sign} x_i\Bigr)^2 = \mathbf{S}^T(\mathbf{X})\mathbf{C}\bigl(\mathbf{C}^T\mathbf{C}\bigr)^{-1}\mathbf{C}^T\mathbf{S}(\mathbf{X}).$$

Thus (2.3.11) coincides with the statistic in (2.2.22).

2.4. Example: Two-way layout

In some problems the parameter $\boldsymbol{\theta}$ in the model (2.1.5) is subject to constraints in the form of linear equations. In this case $\boldsymbol{\theta}$ is restricted to a subspace of smaller dimension than the number of its components. Then the criterion for the choice of locally optimal tests for H_0 as in (2.2.2) is the mean curvature of the power function on this subspace. It can often be evaluated without reparametrization. We will demonstrate this possibility in the setup of a two-way layout.

Consider the two-factor experiment with an equal number of observations per cell. We assume the additive model of factor effects. Thus we observe the variables x_{ijk} representable by the model

$$x_{ijk} = \mu + \alpha_i + \beta_j + \xi_{ijk}, \qquad (i = 1,\dots,r, \quad j = 1,\dots,t, \quad k = 1,\dots,m). \tag{2.4.1}$$

Here r and t are the number of the levels of the first and the second factors, respectively, and m is the number of observations for each combination of factor levels; $\mu, \alpha_1, \dots, \alpha_r, \beta_1, \dots, \beta_t$ are unknown parameters. As usual (see, for example, Bickel and Doksum [**5**], **7.3.C**) we assume that

$$\sum_{i=1}^{r} \alpha_i = 0, \qquad \sum_{j=1}^{t} \beta_j = 0. \tag{2.4.2}$$

In Chapter 4 we will consider this model in the framework of testing linear hypotheses. Here we note that if we are interested in simultaneous estimation of all unknown parameters we need a test for the hypothesis

$$H_0 : \mu = 0, \qquad \alpha_1 = \cdots = \alpha_r = 0, \qquad \beta_1 = \cdots = \beta_t = 0. \tag{2.4.3}$$

In the construction of the locally optimal test we have to consider the mean curvature of the power function on the subspace determined by the equations (2.4.2). In order to achive equal accuracy of our statistical inference for all parameters, it is natural to normalize them by putting

$$\mu' = \mu\sqrt{rt}, \qquad \alpha_i' = \alpha_i\sqrt{t}, \qquad \beta_j' = \beta_j\sqrt{r}. \tag{2.4.4}$$

With this parametrization, dropping the prime in the notation (2.4.4), the model (2.4.1) becomes

$$x_{ijk} = \frac{\mu}{\sqrt{rt}} + \frac{\alpha_i}{\sqrt{t}} + \frac{\beta_j}{\sqrt{r}} + \xi_{ijk} \qquad (i = 1, \dots, r; j = 1, \dots, t; k = 1, \dots, m). \tag{2.4.5}$$

The new parameters satisfy the same equations (2.4.2). Define

$$Z_{ij} = \frac{1}{\sqrt{m}} \sum_{k=1}^{m} \operatorname{sign} x_{ijk}. \tag{2.4.6}$$

It can be shown that in this case the principal part of $\mathbf{P}\{\mathbf{S}(X) \mid \boldsymbol{\theta}\} + \mathbf{P}\{-\mathbf{S}(X) \mid \boldsymbol{\theta}\}$ according to (2.2.15) is equal, up to a nonrandom linear transform, to

$$\left(\mu Z_{..} + \frac{1}{\sqrt{r}} \sum_{i=1}^{r} Z_{i.}\alpha_i + \frac{1}{\sqrt{t}} \sum_{j=1}^{t} Z_{.j}\beta_j\right)^2. \tag{2.4.7}$$

As is customary in the analysis of variance, a subscript replaced by a dot means averaging over this subscript,

$$Z_{i\cdot} = \frac{1}{t}\sum_{j=1}^{t} Z_{ij}, \qquad Z_{\cdot j} = \frac{1}{r}\sum_{i=1}^{r} Z_{ij}, \qquad Z_{\cdot\cdot} = \frac{1}{rt}\sum_{i=1}^{r}\sum_{j=1}^{t} Z_{ij}. \tag{2.4.8}$$

The mean curvature of (2.4.7) as a function of μ, α, β at the point

$$\mu = 0, \qquad \alpha_1 = \cdots = \alpha_r = 0, \qquad \beta_1 = \cdots = \beta_t = 0$$

is equal to the sum of the coefficients of the squared variables, i.e.,

$$Z_{\cdot\cdot}^2 + \frac{1}{r}\sum_{i=1}^{r} Z_{i\cdot}^2 + \frac{1}{t}\sum_{j=1}^{t} Z_{\cdot j}^2.$$

Recall now that the mean curvature equals the sum of the normal curvatures over an arbitrary set of orthogonal directions. In the $(r + t + 1)$-dimensional space $\mathbb{R}^{r+t+1}$, where the parameter $\boldsymbol{\theta} = (\mu, \alpha_1, \dots, \alpha_r, \beta_1, \dots, \beta_t)^T$ lies, consider two one-dimensional mutually ortogonal subspaces $\mathbb{M}_1$ and $\mathbb{M}_2$ spanned by the unit vectors

$$\mathbf{e} = \left(0, \frac{1}{\sqrt{r}}, \dots, \frac{1}{\sqrt{r}}, 0, \dots, 0\right)^T, \quad \mathbf{f} = \left(0, 0, \dots, 0, \frac{1}{\sqrt{t}}, \dots, \frac{1}{\sqrt{t}}\right)^T. \tag{2.4.9}$$

Taking into account the constraints (2.4.2) we see that the parameter $\boldsymbol{\theta}$ in the problem (2.4.5) varies over the orthogonal complement of $\mathbb{M}_1 \oplus \mathbb{M}_2$ to $\mathbb{R}^{r+t+1}$. It is easily seen that the curvature of the function (2.4.7) (at the origin) on each of the subspaces $\mathbb{M}_1$, $\mathbb{M}_2$ equals $(Z_{..})^2$. Hence the required statistic is

$$\frac{1}{r}\sum_{i=1}^{r} Z_{i\cdot}^2 + \frac{1}{t}\sum_{j=1}^{t} Z_{\cdot j}^2 - Z_{..}^2 .$$

It is easy to see that this expression is proportional to

$$\sum_{i=1}^{r}\sum_{j=1}^{t} (Z_{i\cdot} + Z_{\cdot j} - Z_{..})^2 . \tag{2.4.10}$$

Therefore, the locally optimal sign test in the additive two-factor model (2.4.5) with constraints (2.4.2) has the form (in the notation (2.4.6))

$$\left\{\mathbf{X} \colon \sum_{i=1}^{r}\sum_{j=1}^{t} (Z_{i\cdot} + Z_{\cdot j} - Z_{..})^2 > \text{const} \right\}. \tag{2.4.11}$$

It can be shown that for fixed r, t and $m \to \infty$ the statistic (2.4.10) under the null hypothesis (2.4.3) is asymptotically distributed as $\chi^2(r+t-1)$. It is noteworthy that (2.4.10) has the same form as the nominator of the F-test for the null hypothesis (2.4.3) in the normal model (up to division by degrees of freedom).

2.5. Computation of critical values: Finite samples

In order to implement the tests derived in §2.2 we have to compute the quantiles of the random variables (2.3.2) or (2.3.5). (Recall Definition 2.3.1 of the quantile of a discrete random variable.)

For small samples the distributions of (2.3.2) and (2.3.5), and hence their quantiles, can be found by enumeration, trying all the 2^n equiprobable combinations of signs. The corresponding values of the statistic depend on the design matrix $\mathbf{C}$. Usually we apply this method for $n \leq 16$. For other moderate n we compute the distribution by the Monte Carlo method. For large n one can use the asymptotic distributions given in §2.3.

In order to simulate (2.3.2) or (2.3.5) we have to generate the sequence $\{\zeta_i\}$ with properties (2.3.1). It is more convenient to deal with variables taking the values 0 and 1. In this case the computation of the statistic reduces to generating a sequence of Bernoulli random variables of given length with subsequent summation of the corresponding elements of the design matrix.

These methods allow us to obtain the exact significance level of the sign test, closest or equal to the preassigned value.

Examples of computation of critical values in the simplest linear regression model were given in Chapter 1 when we discussed the estimation of Hubble's constant. The critical values of test statistics in linear regression problems depend on the design matrix $\mathbf{C}$, so that they have to be computed in each particular problem.

In the two-factor problem discussed above, the design is specified by the number of levels of each factor and the number of observations in each cell of the table.

Consider the test statistic (2.4.10) for the null hypothesis (2.4.3) in the two-factor problem (2.4.10). The distribution of this statistic coincides with that of the random variable

$$\zeta = \frac{4}{m}\sum_{i=1}^{r}\sum_{j=1}^{t}\left(B_{i\cdot} + B_{\cdot j} - B_{\cdot\cdot} - \frac{m}{2}\right)^2, \tag{2.5.1}$$

where B_{ij}, $i = 1, \dots, r$, $j = 1, \dots, t$, denote independent random variables having the binomial distribution with parameters $(m, 1/2)$. As before, the dot denotes averaging over the corresponding subscript. The exact distribution of the discrete random variable ζ can be found by computing all its values and corresponding probabilities. Note that the number of possible values of this variable is not large. For example, it takes 3 values for $r = t = 2$, $m = 1$, 81 values for $r = 2$, $t = 4$, $m = 4$, etc. Table 2.5.1 contains quantiles q and corresponding probabilities $p := \mathbf{P}\{\zeta \le q\}$ for some combinations of parameters r, t, m. The 0.90- and 0.95-quantiles of the limiting $\chi^2(r + t - 1)$ distribution for ζ as stated in §§2.3 and 2.4 are presented in the column $m = \infty$ for comparison.

TABLE 2.5.1. Quantiles of the random variable ζ as in (2.5.1)

	$m=1$	$m=2$	$m=3$	$m=4$	$m=\infty$
	q p	q p	q p	q p	q p
$r=2$ $t=2$	0.000 0.125	4.000 0.758	5.333 0.860	5.000 0.874	
	3.000 0.625	5.500 0.945	6.333 0.909	6.000 0.915	6.251 0.900
	4.000 1.000	6.000 0.977	6.667 0.961	6.750 0.956	7.815 0.950
$t=3$	2.000 0.281	6.000 0.854	6.000 0.818	6.667 0.864	
	4.667 0.844	7.333 0.924	6.889 0.901	7.667 0.918	7.779 0.900
	6.000 1.000	8.000 0.959	8.667 0.951	9.000 0.954	9.488 0.950
$t=4$	6.000 0.617	7.250 0.868	7.500 0.856	7.625 0.857	
	6.500 0.930	9.250 0.945	8.833 0.907	9.000 0.908	9.236 0.900
	8.000 1.000	10.000 0.968	10.167 0.963	10.625 0.957	11.070 0.950
$t=5$	6.000 0.557	9.200 0.887	9.467 0.880	9.400 0.858	
	7.600 0.791	10.000 0.911	10.533 0.921	10.400 0.903	10.645 0.900
	8.400 0.967	11.200 0.966	11.867 0.951	12.000 0.951	12.592 0.950
$r=3$ $t=3$	6.333 0.691	7.778 0.859	7.741 0.865	7.889 0.853	
	7.222 0.973	9.111 0.924	9.222 0.913	8.889 0.902	9.236 0.900
	9.000 1.000	9.778 0.953	10.407 0.960	10.556 0.951	11.070 0.950
$t=4$	8.667 0.880	9.000 0.855	9.333 0.854	9.333 0.852	
	9.333 0.924	10.333 0.922	10.222 0.904	10.500 0.908	10.645 0.900
	10.000 0.995	11.667 0.957	12.000 0.952	12.167 0.954	12.592 0.950
$t=5$	10.200 0.856	10.400 0.852	10.511 0.851	10.600 0.851	
	10.733 0.911	11.467 0.903	11.756 0.908	11.800 0.901	12.017 0.900
	11.267 0.955	12.933 0.953	13.356 0.951	13.533 0.950	14.067 0.950
$r=4$ $t=4$	10.000 0.879	10.500 0.854	10.583 0.857	10.500 0.851	
	11.000 0.912	11.500 0.903	11.667 0.901	11.750 0.901	12.017 0.900
	11.750 0.956	13.000 0.953	13.333 0.953	13.500 0.951	14.067 0.950

CHAPTER 3

Sign Estimators

3.1. Sign estimators and their computation

We have already demonstrated in the introduction how test statistics can be used for point and interval estimation (see (I.4), (I.5)). The sign tests discussed in Chapter 2 enable us to introduce sign estimators in the model (2.1.5) by using the test statistics (2.2.19), (2.2.22). For example, we can consider the point estimators

$$\widehat{\boldsymbol{\theta}}_n = \arg\min_{\boldsymbol{\theta}\in\mathbb{R}^r} (\mathbf{S}(\mathbf{X}-\mathbf{C}\boldsymbol{\theta}))^T\mathbf{C}\mathbf{C}^T\mathbf{S}(\mathbf{X}-\mathbf{C}\boldsymbol{\theta}), \tag{3.1.1}$$

or

$$\widehat{\boldsymbol{\theta}}_n = \arg\min_{\boldsymbol{\theta}\in\mathbb{R}^r} (\mathbf{S}(\mathbf{X}-\mathbf{C}\boldsymbol{\theta}))^T\mathbf{C}(\mathbf{C}^T\mathbf{C})^{-1}\mathbf{C}^T\mathbf{S}(\mathbf{X}-\mathbf{C}\boldsymbol{\theta}), \tag{3.1.2}$$

or similar estimators based on other quadratic sign tests statistics for testing $H_0\colon \boldsymbol{\theta}=\mathbf{0}$.

According to (I.4) the corresponding confidence sets for unknown parameters in the model (2.1.5) of confidence level $1-\varepsilon$ have the form

$$\big\{(\mathbf{S}(\mathbf{X}-\mathbf{C}\boldsymbol{\theta}))^T\mathbf{C}\mathbf{C}^T\mathbf{S}(\mathbf{X}-\mathbf{C}\boldsymbol{\theta}) \le q_{1-\varepsilon}\big\}, \tag{3.1.3}$$

$$\big\{(\mathbf{S}(\mathbf{X}-\mathbf{C}\boldsymbol{\theta}))^T\mathbf{C}(\mathbf{C}^T\mathbf{C})^{-1}\mathbf{C}^T\mathbf{S}(\mathbf{X}-\mathbf{C}\boldsymbol{\theta}) \le \widetilde{q}_{1-\varepsilon}\big\}, \tag{3.1.4}$$

where $q_{1-\varepsilon}$, $\widetilde{q}_{1-\varepsilon}$ are defined by (2.3.3) and (2.3.4), respectively.

In this section we describe computational algorithms based on the results of Chapter 2 and demonstrate their functioning in examples with simulated data. The computer programs based on these algorithms constitute an interactive system SIGN, which works on an IBM PC and compatible computers. This system was used for computations in the examples throughout this book.

First we consider the point estimation in the model (2.1.5). We restrict ourselves to the computation of estimators (3.1.1). Rewrite the problem in (3.1.1) in the coordinate form:

$$\sum_{\alpha=1}^{r}\Big(\sum_{i=1}^{n} c_{i\alpha}\,\mathrm{sign}\Big(x_i-\sum_{\beta=1}^{r} c_{i\beta}\theta_\beta\Big)\Big)^2 \Longrightarrow \min_{\boldsymbol{\theta}\in\mathbb{R}^r}. \tag{3.1.5}$$

This problem always has a solution, since the objective function is piecewise constant, taking on a finite number of values when the vector-valued parameter $\boldsymbol{\theta}$ runs over the parameter space $\mathbb{R}^r$. Clearly, this solution may not be unique, so that in general the solutions form a set in the r-dimensional parameter space. This fact is illustrated by the example of determination of Hubble's constant in §1.2. The parameter space in this example is the real line. Processing the data from Hubble's paper of 1929 we obtained the sign estimate in the form of a nondegenerate interval $(\bar{\theta}_1, \bar{\theta}_2)$, whereas for another set of observations (contemporary data) the sign estimate had a single value.

In SIGN the following two algorithms for the computation of the sign estimator (3.1.1) for $\boldsymbol{\theta} \in \mathbb{R}^r$ are realized:

1. For $r = 2$, a search algorithm trying all values of the function in (3.1.5). In this case confidence sets of a given confidence level are also constructed.
2. For an arbitrary dimension r, an iterative algorithm for solution of the problem (3.1.5).

Here we describe these algorithms.

(1). Search algorithm for $r = 2$. Each term of the sum in (3.1.5) is a function of two parameters, θ_1 and θ_2. Draw the lines $x_i - c_{i1}\theta_1 - c_{i2}\theta_2 = 0$, $i = 1, \ldots, n$, in the plane (θ_1, θ_2). Let us call the intersection points of these lines the nodal points, or nodes. These lines partition the plane into (bounded or unbounded) polygons, in each of which the function in (3.1.5) is constant. Suppose that the design matrix $\mathbf{C}$ and the data vector $\mathbf{X}$ are such that every two lines intersect and each node is the intersection point of only two lines. (It is not difficult to extend the algorithm to the case where some nodes are intersection points of several lines.) Let the point M_{ij} with coordinates $(\theta_1^{ij}, \theta_2^{ij})$ be the intersection point of the ith and jth lines. These lines split the plane into 4 parts, in each of which the differences $x_i - c_{i1}\theta_1 - c_{i2}\theta_2$ and $x_j - c_{j1}\theta_1 - c_{j2}\theta_2$ have definite signs. For any other line $x_k - c_{k1}\theta_1 - c_{k2}\theta_2 = 0$, $k \neq i, j$, the sign of its left-hand side in each domain adjacent to M_{ij} is determined by the sign of $x_k - c_{k1}\theta_1^{ij} - c_{k2}\theta_2^{ij}$. Hence the function in (3.1.5) can be easily evaluated at each nodal point M_{ij}, on the adjacent intervals of the lines $x_i - c_{i1}\theta_1 - c_{i2}\theta_2 = 0$, and $x_j - c_{j1}\theta_1 - c_{j2}\theta_2 = 0$, and inside the adjacent polygons. By enumeration of all the nodes one determines the minimal value of the objective function in (3.1.5) and the set in the plane where this minimum is attained. This set may be a polygon, an interval, a point, or a union of such sets. Some of these possibilities will be realized in the examples to be given below. For moderate n, the search algorithm can be used to construct confidence regions for (θ_1, θ_2), which we will also demonstrate.

(2). Iterative algorithm. As we pointed out, the minimum set of (3.1.5) can be found by enumeration of all values of the target function for $r > 2$ as well. However, the number of computations needed for the search algorithm rapidly increases with the number of observations. Hence, for the minimization problem (3.1.5) we propose an iterative algorithm which consists in consecutive minimization of each term in (3.1.5). We have no proof of its convergence, but it gave correct results in all numerous cases of its implementation.

The algorithm consists in consecutive computation of r-dimensional vectors $\boldsymbol{\theta}(k) = (\theta_1(k), \ldots, \theta_r(k))^T$ for $k = 1, 2, \ldots$ by the following rule:

1. Take an initial point $\boldsymbol{\theta}(0) = (\theta_1(0), \ldots, \theta_r(0))^T$.
2. At the kth step, starting with $\boldsymbol{\theta}(k-1) = (\theta_1(k-1), \ldots, \theta_r(k-1))^T$, $k = 1, 2, \ldots$, compute $\boldsymbol{\theta}(k) = (\theta_1(k), \ldots, \theta_r(k))^T$, where each $\theta_\alpha(k)$, $\alpha = 1, \ldots, r$, is obtained as

$$\theta_\alpha(k) = \arg \min_{\theta_\alpha \in \mathbb{R}^1} \left(\sum_{i=1}^{n} c_{i\alpha} \operatorname{sign}\Big(x_i - c_{i\alpha}\theta_\alpha - \sum_{\beta<\alpha} c_{i\beta}\theta_\beta(k) - \sum_{\beta>\alpha} c_{i\beta}\theta_\beta(k-1) \Big) \right)^2.$$

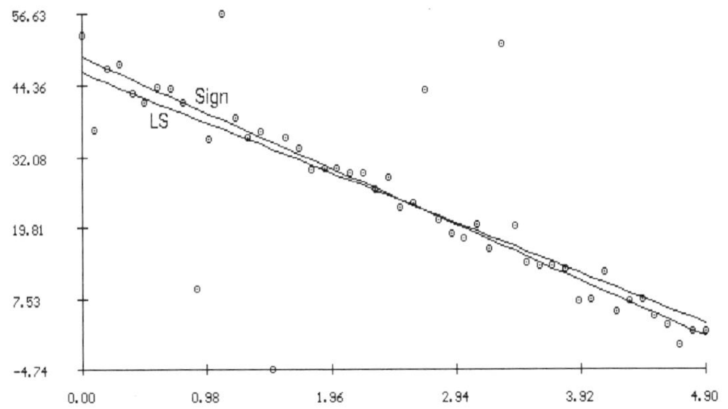

FIGURE 3.1.1. Graphs of $x(t) = \widehat{\theta}_1 + \widehat{\theta}_2 t$ for least squares estimates (LS) and sign estimates (SIGN).

3. The algorithm is terminated at step k if, for a given accuracy ε,

$$\sqrt{\sum_{\alpha=1}^{r} (\theta_\alpha(k) - \theta_\alpha(k-1))^2} < \varepsilon$$

or if the function in (3.1.5) vanishes.

REMARK 1. Our iterative procedure is of Seidel type in that for the computation of $\theta_\alpha(k)$ at the kth step of iteration it uses the values $\theta_1(k), \dots, \theta_{\alpha-1}(k)$ previously computed at this step. This accelerates the procedure and reduces the number of steps needed to reach the desired accuracy.

REMARK 2. The implementation of the iterative procedure to various simulated data shows that the required number of steps depends on the choice of the initial point. In many examples we started with the LSE and performed just a few steps to obtain the sign estimate.

EXAMPLE 1. Consider the regression model

$$x_i = \theta_1 + c_i\theta_2 + \xi_i, \qquad i = 1, \dots, n, \tag{3.1.6}$$

where i.i.d. errors ξ_i have Tukey's contaminated distribution function $F(x) = (1-\delta)\Phi(x/\sigma_0) + \delta\Phi(x/\sigma_1)$, for $\delta = 0.15$, $\sigma_0 = 2$, $\sigma_1 = 30$, with $\Phi(x)$ denoting the standard normal distribution function. Figure 3.1.1 shows simulated data according to this model for $\theta_1 = 50$, $\theta_2 = -10$, and $n = 50$, and estimated linear relationships obtained by the method of least squares (LS) and by the iterative sign procedure (SIGN).

Figure 3.1.2 (on the next page) shows the minimum set of the function (3.1.5) obtained by the search algorithm. This set is the union of two intervals with endpoints M_1, M_2, and M_2, M_3, respectively, where $M_1 = (49.092, -9.886)$, $M_2 = (49.085, -9.852)$, $M_3 = (49.928, -10.028)$. The iterative sign procedure gave the value $(49.09, -9.852)$, which coincides with M_2 within the accuracy of computation. The procedure made 7 iterations with the LSE taken for the initial point. The least abslolute deviation estimate also equals M_2. The LSE is $(46.37, -8.827)$, which deviates much more from the true value $(50, -10)$.

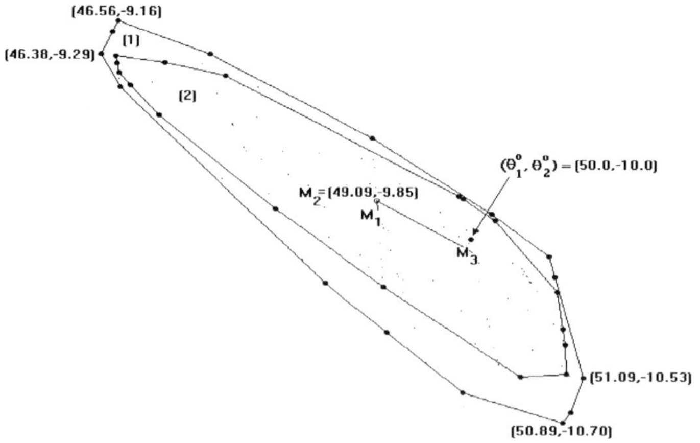

FIGURE 3.1.2. Confidence sets of level 0.95 (1) and 0.90 (2) in Example 1.

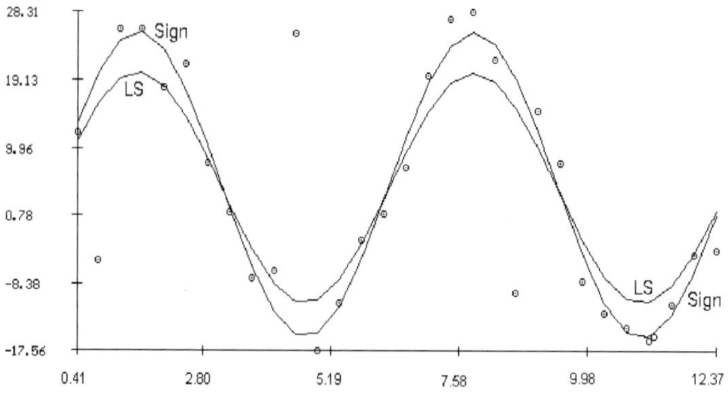

FIGURE 3.1.3. Graphs of functions $x(t) = \widehat{\theta}_0 + \widehat{\theta}_1 \sin(t)$ for least squares estimates (LS) and sign estimates (SIGN). The true curve is indistinguishable from the curve SIGN.

In order to construct confidence sets, we used the Monte Carlo method to compute the distribution function and the 0.90- and 0.95-quantiles of

$$\sum_{\alpha=1}^{r}\Big(\sum_{i=1}^{n} c_{i\alpha}\zeta_i\Big)^2,$$

where ζ_i's are i.i.d. with $\mathbf{P}\{\zeta_i = 1\} = \mathbf{P}\{\zeta_i = -1\} = 1/2$. The sets of points where the test statistic from (3.1.3) does not exceed the corresponding quantile were found by search. These sets are not convex. Their convex hulls are shown in Figure 3.1.2. The small dots indicate the intersection points of the lines $x_i = \theta_1 + c_i\theta_2$, $i = 1, \ldots, n$. Note that the LSE does not fall into these confidence sets, while both these sets cover the true point $(\theta_1, \theta_2) = (50.0, -10.0)$.

EXAMPLE 2. Data corresponding to the relationship $x_i = \theta_0 + \theta_1 \sin(t_i) + \xi_i$, $t_i \in [0.41, 12.37]$, $i = 1, \ldots, n = 30$, were simulated for $\theta_0 = 5$, $\theta_1 = 20$ and ξ_i i.i.d.

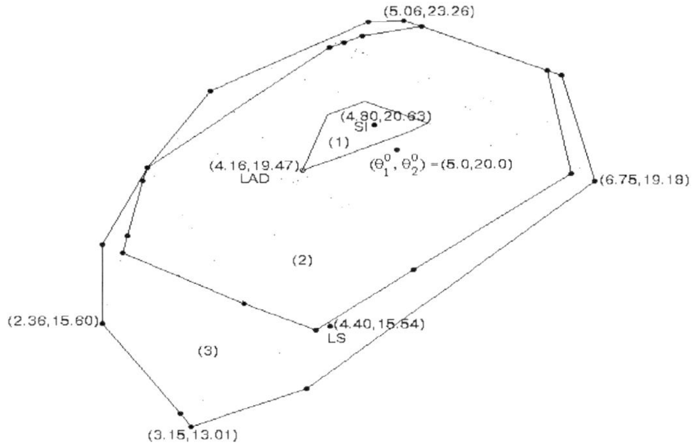

FIGURE 3.1.4. The minimal set (1) and the convex hulls of the confidence sets of level at least 0.90 (2) and 0.95 (3) in Example 2.

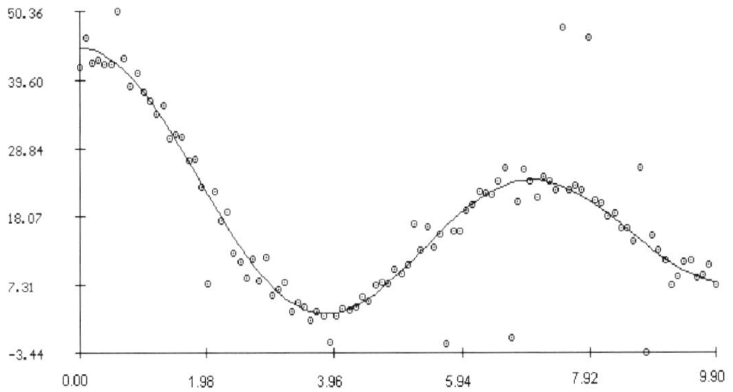

FIGURE 3.1.5. Graph of the function $x(t) = \widehat{\theta}_0 + \widehat{\theta}_1 J_0(t)$ for the sign estimate.

with contaminated distribution as in Example 1 for $\delta = 0.15$, $\sigma_1 = 3$, $\sigma_2 = 30$. This relationship is depicted in Figure 3.1.3 along with the curves obtained by the least squares (LS) and by the iterative sign procedure (SIGN).

In this example the minimum set of the function in (3.1.5) is a polygon with vertices (4.380, 20.895), (4.709, 21.227), (5.272, 20.661), (5.115, 20.464), (4.158, 19.469). The minimum set and the convex hulls of the confidence sets of level at least 0.90 and 0.95 are shown in Figure 3.1.4. The least squares estimate (LS) equals to $(4.40, 15.54)$. The sign estimate (SI) obtained by the iterative procedure (7 iterations, the LSE as the initial point) is $(4.795, 20.63)$.

EXAMPLE 3. In this example the minimization problem (3.1.5) has a unique solution. The data here correspond to the relationship $x_i = \theta_0 + \theta_1 J_0(t_i) + \xi_i$, $t_i \in [0.0, 9.9]$, $i = 1, \dots, n = 100$, for $\theta_0 = 15$, $\theta_1 = 30$, where $J_0(t)$ is the Bessel function of zero order. The i.i.d. errors again have the contaminated distribution

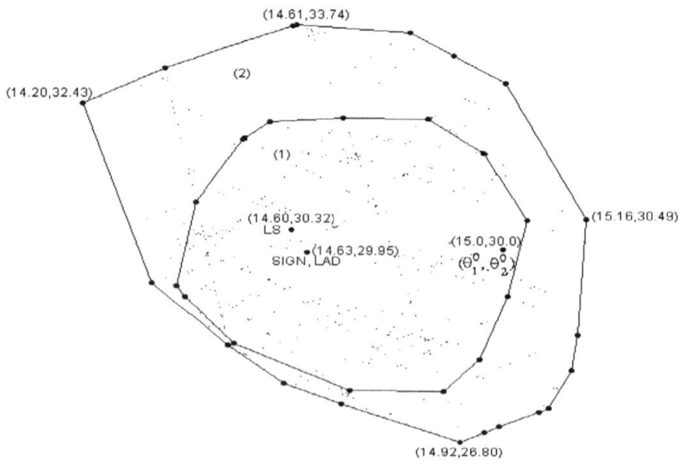

FIGURE 3.1.6. Convex hulls of the confidence regions of level at least 0.90 (1) and 0.95 (2) in Example 3.

with parameters $\delta = 0.15$, $\sigma_1 = 1.5$, $\sigma_2 = 15$. This relationship is shown in Figure 3.1.5 (on the previous page) along with the curve corresponding to the estimate $(\widehat{\theta}_0, \widehat{\theta}_1) = (14.63, 29.95)$ obtained by the iterative sign procedure (10 iterations, the LSE $(14.6, 30.32)$ as the initial point). This estimate coincides with the minimum point of (3.1.5) and the least absolute deviations estimate.

Figure 3.1.6 shows the convex hulls of the confidence regions with confidence level at least 0.90 and 0.95. These confidence regions contain all the estimates obtained, as well as the true parameter point $(\theta_0, \theta_1) = (15.0, 30.0)$.

EXAMPLE 4. Consider the harmonic oscillation model, which already appeared in the introduction. Let

$$x_i = \theta_1 + \theta_2 t_i + \theta_3 \sin(\omega t_i) + \xi_i, \qquad i = 1, \dots, n, \tag{3.1.7}$$

with a known frequency ω and unknown parameters θ_1, θ_2, θ_3 to be estimated. Such a model may be appropriate in recovery of a periodic signal from noisy observations. As a rule, the frequency ω is known, and the parameter of interest is the amplitude of the signal. Of course, in practical situations the noise ξ_i may be correlated, so that one has to describe it by some random process. However, as a first approximation, one can consider the model with independent errors. In practical applications, the normality assumption is often violated. Moreover, the data may be contaminated by the impulse noise of a different nature from that in the model. We will find sign estimators in the model (3.1.7) for various error distributions.

EXAMPLE 4A. We take ξ_i to have an asymmetric distribution with zero median. Namely, this will be the normal distribution for $x < 0$ and the (heavy tailed) Cauchy distribution for $x > 0$,

$$F(x) = \begin{cases} \frac{1}{\sqrt{2\pi}\sigma} \int_{-\infty}^{x} e^{-\frac{y^2}{2\sigma^2}}\, dy, & x \le 0, \\ \frac{1}{2} + \frac{1}{\pi} \arctan \frac{\sqrt{\pi/2}\, x}{\sigma}, & x > 0. \end{cases}$$

TABLE 3.1.1

Parameter	True	LSE	SIGN
θ_1	3.50	−13.67	3.46
θ_2	4.50	7.41	4.52
θ_3	10.00	16.66	10.12

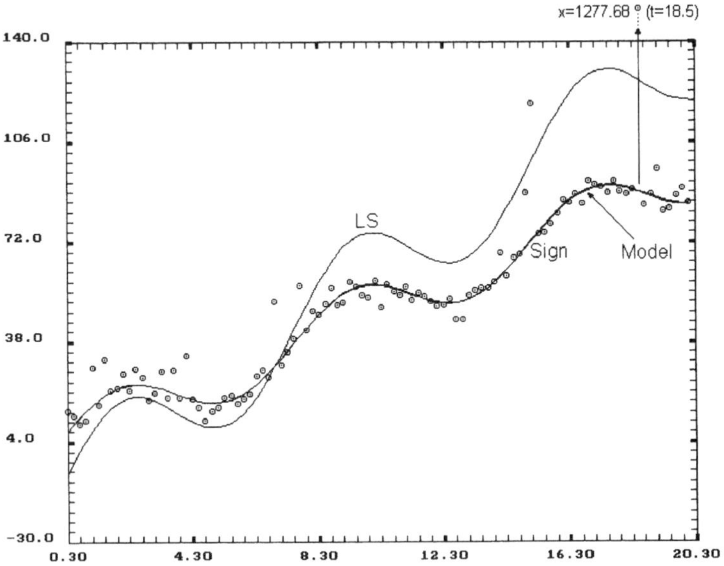

FIGURE 3.1.7. Graphs of functions $x(t) = \widehat{\theta}_1 + \widehat{\theta}_2 t + \widehat{\theta}_3 \sin \omega t$ obtained by the LS and sign methods.

Here $\sigma = 3$. Figure 3.1.7 shows $n = 101$ observations simulated according to this model with $\omega = 2\pi/7.5$, $t \in [0.3, 20.3]$ along with two curves, LS and SIGN, determined by the LSE and the sign estimate, respectively. The figure does not contain an observation 1277.68 at the point $t = 18.50$ lying far beyond the range of the picture. The true curve labelled "Model" differs very little from the SIGN curve. The true parameter values and their estimates by the two methods are given in Table 3.1.1.

The iterative sign procedure required 8 steps.

EXAMPLE 4B. The errors in this example do not satisfy the zero median assumption. Let ξ_i have a contaminated normal distribution, with contaminating distribution shifted by $\mu > 0$, so that the resulting distribution function is asymmetric with nonzero median:

$$F(x) = \frac{1-\varepsilon}{\sqrt{2\pi}\sigma_1} \int_{-\infty}^{x} e^{-\frac{y^2}{2\sigma_1^2}}\, dy + \frac{\varepsilon}{\sqrt{2\pi}\sigma_2} \int_{-\infty}^{x} e^{-\frac{(y-\mu)^2}{2\sigma_2^2}}\, dy.$$

We let $\sigma_1 = 2.0$; $\sigma_2 = 4.0$; $\mu = 15$; $\varepsilon = 0.1$.

TABLE 3.1.2

Parameter	True	LSE	SIGN
θ_1	3.50	9.18	3.42
θ_2	4.50	4.68	4.57
θ_3	10.00	8.37	10.40

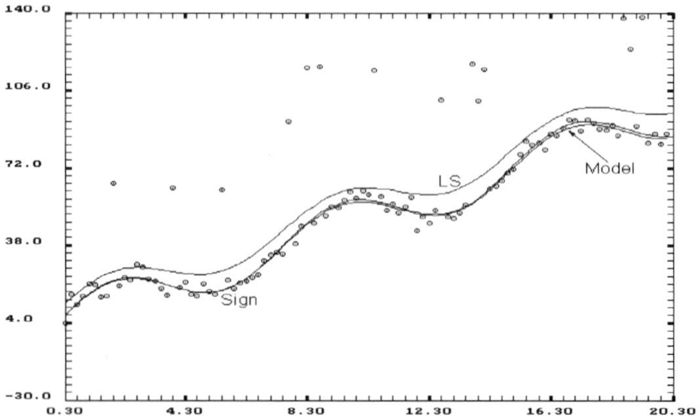

FIGURE 3.1.8. Graphs of functions $x(t) = \widehat{\theta}_1 + \widehat{\theta}_2 t + \widehat{\theta}_3 \sin(\omega t)$ obtained by the LS and sign methods.

Figure 3.1.8 gives the scatter plot of the corresponding simulated data for $n = 101$ along with the curves estimated by the LS and sign methods. The true parameter values and their estimates by the two methods are given in Table 3.1.2. It took 31 steps for the stabilization of the sign iterative process.

EXAMPLE 5. TWO-WAY LAYOUT. In §2.4 we described a locally optimal sign test in the two-way layout problem. In accordance with the general approach to deriving point estimators from test statistics, we will obtain a procedure based on this test for the estimation of the parameters μ, α_i, β_j, $i = 1, \ldots, r$, $j = 1, \ldots, t$, in the model (2.4.1)–(2.4.2). The test statistic is given by (2.4.9) and (2.4.10), where Z_{ij}, $Z_{i\cdot}$, $Z_{\cdot j}$, $Z_{\cdot\cdot}$ are defined by (2.4.6) and (2.4.8). Hence, we find the sign estimators as the solution of the extremal problem

$$\min_{\mu,\boldsymbol{\alpha},\boldsymbol{\beta}} \widetilde{T}(\mu,\boldsymbol{\alpha},\boldsymbol{\beta}) = \min_{\mu,\boldsymbol{\alpha},\boldsymbol{\beta}} \sum_{i=1}^{r} \sum_{j=1}^{t} (\widetilde{Z}_{i\cdot} + \widetilde{Z}_{\cdot j} - \widetilde{Z}_{\cdot\cdot})^2$$

under the constraints

$$\sum_{i=1}^{r} \alpha_i = 0, \qquad \sum_{j=1}^{t} \beta_j = 0,$$

where $\widetilde{Z}_{i\cdot}$ and $\widetilde{Z}_{\cdot j}$ are obtained by averaging

$$\widetilde{Z}_{ij} = \frac{1}{\sqrt{m}} \sum_{k=1}^{m} \operatorname{sign}(x_{ijk} - \mu - \alpha_i - \beta_j)$$

TABLE 3.1.3

Parameter	True	LSE	SIGN
α_1	0.50	1.22	0.65
α_2	2.00	1.78	1.93
α_3	7.80	7.70	7.82
α_4	−5.30	−5.67	−5.42
α_5	−5.00	−5.03	−4.99
β_1	−6.40	−6.40	−6.36
β_2	7.30	6.78	7.11
β_3	13.50	13.49	13.59
β_4	−7.00	−6.78	−6.93
β_5	5.40	5.54	5.50
β_6	−2.60	−2.25	−2.69
β_7	−10.20	−10.39	−10.24
μ	1.00	2.20	1.09

over j and i, respectively, and $\widetilde{Z}_{..}$ is the total average of $\widetilde{Z}_{ij}$ over i and j. The function $\widetilde{T}(\mu, \boldsymbol{\alpha}, \boldsymbol{\beta})$ can be represented in the form

$$\widetilde{T}(\mu, \boldsymbol{\alpha}, \boldsymbol{\beta}) = t\sum_{i=1}^{r} \widetilde{Z}_{i\cdot}^2 + r\sum_{j=1}^{t} \widetilde{Z}_{\cdot j}^2 - rt\widetilde{Z}_{\cdot\cdot}^2.$$

The estimates are computed by the iterative procedure above, adapted to the following model.

Step 1. For given $\beta_1, \dots, \beta_t$ the estimates for μ, $\alpha_1, \dots, \alpha_r$ are obtained as

$$\arg\min_{\mu,\boldsymbol{\alpha}} \sum_{i=1}^{r} \widetilde{Z}_{i\cdot}^2 \quad \text{given that} \quad \sum_{i=1}^{r} \alpha_i = 0.$$

Step 2. Upon substitution of $\alpha_1, \dots, \alpha_r$ from Step 1 into the functions $\widetilde{Z}_{\cdot j}^2$, the estimates for μ, $\beta_1, \dots, \beta_t$ are obtained as

$$\arg\min_{\mu,\boldsymbol{\beta}} \sum_{j=1}^{t} \widetilde{Z}_{\cdot j}^2 \quad \text{given that} \quad \sum_{j=1}^{t} \beta_j = 0.$$

Step 3. If the Euclidean distance between the estimates obtained in two consecutive iterations exceeds a preassigned (small) $\varepsilon > 0$, go to Step 1; otherwise, the procedure terminates.

EXAMPLE 6: PARAMETER ESTIMATION IN THE TWO-WAY LAYOUT. The data were simulated according to the model (2.4.1)–(2.4.2) with $r = 5$ levels of the first factor, $t = 7$ levels of the second factor, $n = 12$ observations for each combination of factors, and errors ξ_{ijk} having the contaminated Tukey distribution function

$$F(x) = (1 - \varepsilon)\Phi(x/\sigma_1) + \varepsilon\Phi((x - \mu)/\sigma_2),$$

where $\varepsilon = 0.15$, $\sigma_1 = 0.5$, $\sigma_2 = 5$, $\mu = 7.5$.

The sign estimates and the LSE are given in Table 3.1.3.

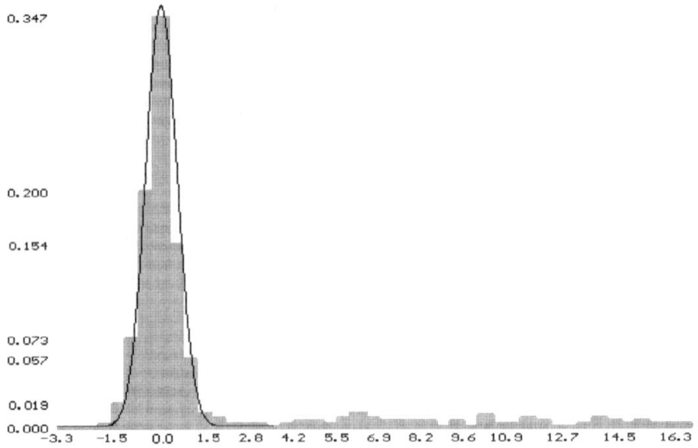

FIGURE 3.1.9. The histogram of residuals for the sign estimate and the normal density with $\sigma_1 = 0.5$ of the main part of the distribution in Example 5.

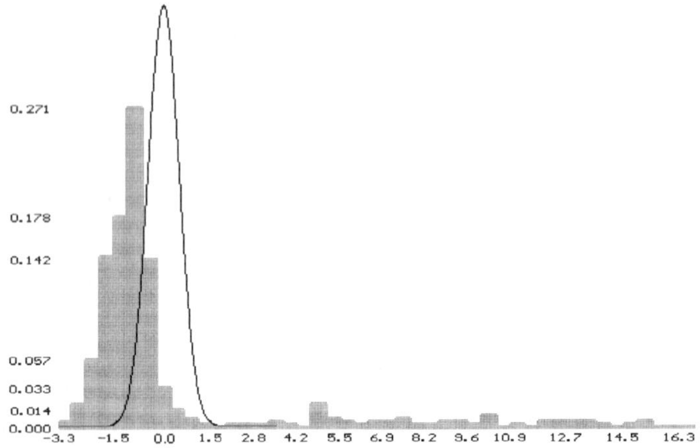

FIGURE 3.1.10. The histogram of residuals for the LSE and the normal density with $\sigma_1 = 0.5$ of the main part of the distribution in Example 5.

The histograms of the residuals for the two methods are given in Figures 3.1.9 and 3.1.10.

The above examples demonstrate that the sign method is applicable to linear models with various types of random errors: asymmetrically distributed, heavy-tailed, containing a considerable fraction of outliers, having nonzero medians, etc. The examples confirm the fact that the sign estimates are very robust.

3.2. Sign estimation: Asymptotic theory

3.2.1. The role of asymptotic theory. Asymptotic theory treats the properties of sign procedures for a large number of observations. In this case the formulae for the distributions of statistics, their critical values, confidence limits, etc., take a simpler form, which allows for better intuitive understanding of their properties. We already discussed asymptotic distributions of test statistics when dealing with hypothesis testing in Chapter 2. As regards point estimation, the two main properties of sign estimators provided by the asymptotic theory are consistency and asymptotic normality. Recall their definitions (in the form suitable for our present purpose). It is assumed thoughout that $n \to \infty$.

An estimator $\widehat{\boldsymbol{\theta}}_n$ of the unknown parameter $\boldsymbol{\theta}$ is called *consistent* if $\widehat{\boldsymbol{\theta}}_n$ converges to $\boldsymbol{\theta}$ in probability,

$$\widehat{\boldsymbol{\theta}}_n \xrightarrow{\mathbf{P}} \boldsymbol{\theta}.$$

An estimator $\widehat{\boldsymbol{\theta}}_n$ of the unknown parameter $\boldsymbol{\theta}$ is called *asymptotically normal* if the distribution of $\sqrt{n}(\widehat{\boldsymbol{\theta}}_n - \boldsymbol{\theta})$ converges to the normal distribution with zero mean and a nonzero covariance matrix.

The consistency of the sign estimators in the regression model will be proved in 3.2.2, and their asymptotic normality in 3.2.3.

In practice, consistency is interpreted as the approximate equality $\widehat{\boldsymbol{\theta}}_n \approx \boldsymbol{\theta}$ for sufficiently large number of observations n. In this connection, two questions arise: what n is sufficiently large and what is the accuracy of the approximation? We have already pointed out that the first question has no satisfactory theoretical solution. When applying limit theorems in practice onc relies on theoretical, experimental, heuristic and dogmatic recommendations, with their combination depending on the development of the subject. For this reason when dealing with computations in this book we prefer either exact methods or methods with controllable accuracy (for example, Monte Carlo).

The results on asymptotic normality provide for a more precise description of the closeness of $\widehat{\boldsymbol{\theta}}_n$ to $\boldsymbol{\theta}$ by giving an approximation for the probability distribution of the difference $\widehat{\boldsymbol{\theta}}_n - \boldsymbol{\theta}$. If we know the asymptotic covariance matrix of $\sqrt{n}(\widehat{\boldsymbol{\theta}}_n - \boldsymbol{\theta})$ we can construct confidence regions for $\boldsymbol{\theta}$ (and for linear functions of $\boldsymbol{\theta}$, if necessary). The asymptotic covariance matrix for sign estimators will be given in 3.2.4. Of course in this way we obtain only approximate coverage probabilities, but a satisfactory accuracy can be achived for sufficiently large n (so that the above questions still remain). In spite of these limitations, asymptotic results constitute a traditional and important part of statistical theory.

We will present the asymptotic properties of sign estimators for the linear regression model. Unlike the i.i.d. case, where the asymptotic process is governed by the only parameter n, in the regression scheme some other characteristics of the design matrix are also involved. In order to avoid complications, we will consider only the models with bounded design matrices (satisfying Condition 2.1(vii)) and i.i.d. errors.

3.2.2. Consistency of sign estimators. Here we show that the sign estimators of the parameter $\boldsymbol{\theta}$ in the model (2.1.5) are consistent under certain conditions on the design matrix $\mathbf{C}$ and the distribution of random errors.

THEOREM 3.2.1. *Assume that the random errors* (2.1.4) *in the model* (2.1.5) *are i.i.d. with their common distribution function satisfying Conditions* 2.1(i, iv, v). *Moreover, assume that the design matrix* $\mathbf{C}$ *satisfies Conditions* 2.1(vii, viii). *Then the sign estimators* (3.1.1) *and* (3.1.2) *are consistent.*

PROOF. For various estimators the proof of consistency proceeds in a similar manner, so that we will consider one of them, for instance, the estimator (3.1.1).

Denote by $\boldsymbol{\theta}^0$ the true value of the parameter (i.e., the parameter value upon which the distribution of the observations actually depends), whereas $\boldsymbol{\theta}$ will mean a variable running over the parameter space. It will be convenient to multiply the objective function in (3.1.1) by $1/n$. Moreover, we make the change of variables

$$\boldsymbol{\theta} = \boldsymbol{\theta}^0 + \mathbf{t}.$$

Then $\widehat{\boldsymbol{\theta}}_n = \boldsymbol{\theta}^0 + \widehat{\mathbf{t}}_n$, where

$$\widehat{\mathbf{t}}_n = \arg\min_{\mathbf{t}\in\mathbb{R}^r} \frac{1}{n} \sum_{\alpha=1}^{r} \Big(\sum_{i=1}^{n} c_{i\alpha} \operatorname{sign}(\xi_i - \mathbf{c}_i \mathbf{t}) \Big)^2, \tag{3.2.1}$$

and we are to show that $\widehat{\mathbf{t}}_n \xrightarrow{\mathbf{P}} \mathbf{0}$.

The proof consists of two parts.

1. We start with the fact that

$$\sup_{|\mathbf{t}|\le T} \left| \frac{1}{n} \sum_{i=1}^{n} c_{i\alpha} \operatorname{sign}(\xi_i - \mathbf{c}_i \mathbf{t}) - \mathbf{E}\Big(\frac{1}{n} \sum_{i=1}^{n} c_{i\alpha} \operatorname{sign}(\xi_i - \mathbf{c}_i \mathbf{t}) \Big) \right| \xrightarrow{\mathbf{P}} 0 \tag{3.2.2}$$

for $\alpha = 1, \dots, r$ and any $T > 0$. The proof of this assertion stated as Theorem 3.2.3 on the uniform law of large numbers will be given in 3.2.5. This theorem implies that

$$\arg\min_{|\mathbf{t}|\le T} \left| \frac{1}{n} \sum_{i=1}^{n} c_{i\alpha} \operatorname{sign}(\xi_i - \mathbf{c}_i \mathbf{t}) \right|^2 \tag{3.2.3}$$

approaches (in probability)

$$\arg\min_{|\mathbf{t}|\le T} \left| \frac{1}{n} \mathbf{E} \sum_{i=1}^{n} c_{i\alpha} \operatorname{sign}(\xi_i - \mathbf{c}_i \mathbf{t}) \right|^2, \tag{3.2.4}$$

as $n \to \infty$ for any $T > 0$. On the other hand,

$$\mathbf{E} \frac{1}{n} \sum_{i=1}^{n} c_{i\alpha} \operatorname{sign}(\xi_i - \mathbf{c}_i \mathbf{t}) = \frac{1}{n} \sum_{i=1}^{n} c_{i\alpha} (1 - 2F(\mathbf{c}_i \mathbf{t})), \tag{3.2.5}$$

so that (3.2.4) takes its minimal value, equal to zero, for $\mathbf{t} = \mathbf{0}$ (because $F(0) = 1/2$). Moreover, we will show now that for sufficiently large n the point $\mathbf{t} = \mathbf{0}$ is the only minimum point of (3.2.4).

Assume the contrary, i.e., assume that all the functions (3.2.5) for $\alpha = 1, \dots, r$ vanish for some $\mathbf{t} \ne \mathbf{0}$. Multiplying the right-hand side of (3.2.5) by t_α and summing up over α we obtain

$$\frac{1}{n} \sum_{i=1}^{n} (\mathbf{c}_i \mathbf{t})(1 - 2F(\mathbf{c}_i \mathbf{t})) = 0. \tag{3.2.6}$$

It is easily seen that for any $u \in \mathbb{R}^1$

$$u(1 - 2F(u)) \le 0, \tag{3.2.7}$$

and under Condition 2.1(v) the equality in (3.2.7) can hold only for $u = 0$. Therefore the equality (3.2.6) is fulfilled only if $\mathbf{c}_i \mathbf{t} = 0$, $i = 1, \dots, n$. In this case $\big(\sum_{i=1}^n \mathbf{c}_i\big)\,\mathbf{t} = 0$ and therefore $\mathbf{t}^T \big(\frac{1}{2} \mathbf{C}^T \mathbf{C}\big) \mathbf{t} = 0$ for $\mathbf{t} \ne \mathbf{0}$, which contradicts Condition 2.1(viii) for sufficiently large n.

Hence (3.2.3) converges to $\mathbf{0}$ in probability as $n \to \infty$ for any $T > 0$. Consequently, for any $T > 0$

$$\arg\min_{|\mathbf{t}|\le T} \frac{1}{n} \sum_{\alpha=1}^{r} \Big(\sum_{i=1}^{n} c_{i\alpha} \operatorname{sign}(\xi_i - \mathbf{c}_i\mathbf{t}) \Big) \xrightarrow{\mathbf{P}} 0.$$

2. In the second part of the proof we show that there exist $a > 0$ and $T > 0$ such that for sufficiently large n and for $|\mathbf{t}| > T$

$$\sum_{\alpha=1}^{r} \Big(\frac{1}{n} \sum_{i=1}^{n} c_{i\alpha} \operatorname{sign}(\xi_i - \mathbf{c}_i\mathbf{t}) \Big)^2 > a \tag{3.2.8}$$

with probability arbitrarily close to 1. This implies that for sufficiently large n the minimum point of (3.2.1) lies inside the compact set $|\mathbf{t}| \le T$ with probability arbitrarily close to 1. Together with the result of the first part this proves the consistency of the sign estimator.

For the proof of (3.2.8) consider the vector $\boldsymbol{\varphi}(\mathbf{t}) = (\varphi_1(\mathbf{t}), \dots, \varphi_r(\mathbf{t}))$, where

$$\varphi_\alpha(\mathbf{t}) = \frac{1}{n} \sum_{i=1}^{n} c_{i\alpha} \operatorname{sign}(-\xi_i + \mathbf{c}_i\mathbf{t}).$$

Let us show that for sufficiently large n there exist $a > 0$ and $T > 0$ such that $|\boldsymbol{\varphi}(\mathbf{t})\boldsymbol{\tau}|^2 > a$ for $|t| > T$ with probability arbitrarily close to 1, where $\boldsymbol{\tau} = \mathbf{t}/|\mathbf{t}|$, i.e., that

$$|\boldsymbol{\varphi}(t)\boldsymbol{\tau}|^2 = \frac{1}{|\mathbf{t}|^2} \Big(\frac{1}{n} \sum_{i=1}^{n} (\mathbf{c}_i\mathbf{t}) \operatorname{sign}(\mathbf{c}_i\mathbf{t} - \xi_i) \Big)^2 > a, \tag{3.2.9}$$

which implies (3.2.8) by the Cauchy–Schwarz inequality. Note that

$$(\mathbf{c}_i\mathbf{t}) \operatorname{sign}(\mathbf{c}_i\mathbf{t} - \xi_i) = |\mathbf{c}_i\mathbf{t}|, \quad \text{whenever } |\mathbf{c}_i\mathbf{t}| > |\xi_i|.$$

Using the inequality $|A + B| > |A| - |B|$ we obtain

$$\begin{aligned} \operatorname{sign}(\mathbf{c}_i\mathbf{t} - \xi_i)\Big| &> \frac{1}{n} \sum_{\{i:\ |\mathbf{c}_i\mathbf{t}| > |\xi_i|\}} |\mathbf{c}_i\mathbf{t}| - \frac{1}{n} \sum_{\{i:\ |\mathbf{c}_i\mathbf{t}| \le |\xi_i|\}} |\mathbf{c}_i\mathbf{t}| \\ &= \frac{1}{n} \sum_{i=1}^{n} |\mathbf{c}_i\mathbf{t}| - \frac{2}{n} \sum_{\{i:\ |\mathbf{c}_i\mathbf{t}| \le |\xi_i|\}} |\mathbf{c}_i\mathbf{t}|. \end{aligned} \tag{3.2.10}$$

Let us estimate the second term in the right-hand side of (3.2.10). Observe that for any $R > 0$

$$|\mathbf{c}_i\mathbf{t}|\ I(|\mathbf{c}_i\mathbf{t}| \le |\xi_i|) \le R I(|\xi_i| \le R) + |\mathbf{c}_i\mathbf{t}|\ I(|\xi_i| > R).$$

Choose R such that

$$\gamma = 1 - 2\mathbf{P}\{|\xi_i| > R\} > 0.$$

Hence we obtain

$$\frac{2}{n} \sum_{\{i:\ |\mathbf{c}_i\mathbf{t}| \le |\xi_i|\}} |\mathbf{c}_i\mathbf{t}| \le \frac{2}{n} \sum_{i=1}^{n} \{R\ I(|\xi_i| \le R) + |\mathbf{c}_i\mathbf{t}|\ I(|\xi_i| > R)\}. \tag{3.2.11}$$

In view of Condition 2.1(vii) the right-hand side of (3.2.11) by the law of large numbers is equal to

$$2R\mathbf{P}\{|\xi_i| \le R\} + 2\mathbf{P}\{|\xi_i| > R\}\frac{1}{n}\sum_{i=1}^{n}|\mathbf{c}_i\mathbf{t}| + o_p(1)$$

as $n \to \infty$. Putting this into (3.2.10) we obtain

$$\begin{aligned}(3.2.12)\qquad &\left|\frac{1}{n}\sum_{i=1}^{n}(\mathbf{c}_i\mathbf{t})\operatorname{sign}(\mathbf{c}_i\mathbf{t} - \xi_i)\right| \\ &\qquad \ge -2R\mathbf{P}\{|\xi_i| \le R\} + (1 - 2\mathbf{P}\{|\xi_i| > R\})\frac{1}{n}\sum_{i=1}^{n}|\mathbf{c}_i\mathbf{t}|.\end{aligned}$$

Under Conditions 2.1(vii, viii) we have

$$(3.2.13)\qquad \frac{1}{n}\sum_{i=1}^{n}|\mathbf{c}_i\mathbf{t}|^2 \le K\left(\frac{1}{n}\sum_{i=1}^{n}|\mathbf{c}_i\mathbf{t}|\right)|\mathbf{t}|,$$

and

$$\frac{1}{n}\sum_{i=1}^{n}|\mathbf{c}_i\mathbf{t}|^2 = \frac{1}{n}\sum_{i=1}^{n}(\mathbf{c}_i\mathbf{t})^T(\mathbf{c}_i\ \mathbf{t}) = \mathbf{t}^T\left(\frac{1}{n}\mathbf{C}^T\mathbf{C}\right)\mathbf{t} \to \mathbf{t}^T\mathbf{\Sigma}\ \mathbf{t} \ge k|\mathbf{t}|^2$$

for some $k > 0$. Therefore, for sufficiently large n

$$(3.2.14)\qquad \frac{1}{n}\sum_{i=1}^{n}|\mathbf{c}_i\mathbf{t}|^2 \ge \frac{k}{2}|\mathbf{t}|^2.$$

Now (3.2.13) and (3.2.14) imply that for sufficiently large n

$$(3.2.15)\qquad \frac{1}{n}\sum_{i=1}^{n}|\mathbf{c}_i\mathbf{t}| \ge \frac{k}{2K}|\mathbf{t}|.$$

Putting (3.2.15) into (3.2.12) we obtain that for sufficiently large n

$$(3.2.16)\qquad \left|\frac{1}{n}\sum_{i=1}^{n}(\mathbf{c}_i\mathbf{t})\operatorname{sign}(\mathbf{c}_i\mathbf{t} - \xi_i)\right| > \widetilde{k}|\mathbf{t}| - 2R\mathbf{P}\{|\xi_i| \le R\},$$

where $\widetilde{k} = \frac{k}{2K}\big(1 - 2\mathbf{P}\{|\xi_i| > R\}\big)$. Hence, turning to (3.2.9) we obtain that for sufficiently large $|\mathbf{t}|$ and sufficiently large n

$$|\varphi(t)\boldsymbol{\tau}|^2 > \frac{(\widetilde{k}|\mathbf{t}| - 2R\mathbf{P}\{|\xi_i| \le R\})^2}{|\mathbf{t}|^2} = a > 0$$

with probability arbitrarily close to 1, which was to be proved.

This completes the proof of consistency of the sign estimator (3.1.1). □

3.2.3. Asymptotic normality of sign estimators. We will show that under certain assumptions on the distribution of random errors (somewhat stronger than in Theorem 3.2.1) the sign estimators (3.1.1) and (3.1.2) are asymptotically normal. They are handled in a similar manner and have the same asymptotic distribution. Hence we will consider one of them, for example, estimator (3.1.1).

THEOREM 3.2.2. *Assume that the random errors* (2.1.3) *in the model* (2.1.5) *are i.i.d. with a common distribution function satisfying Conditions* 2.1(i, ii, vi). *Assume that the design matrix* $\mathbf{C}$ *satisfies Conditions* 2.1(vii, viii). *Then the sign estimator* (3.1.1) *is representable in the form*

$$\widehat{\boldsymbol{\theta}}_n = \boldsymbol{\theta}^0 + \frac{1}{2f(0)} \left(\mathbf{C}^T\mathbf{C}\right)^{-1}\mathbf{C}^T\boldsymbol{\zeta} + \mathbf{o}_p\Big(\frac{1}{\sqrt{n}}\Big), \tag{3.2.17}$$

where $\boldsymbol{\zeta} = (\zeta_1, \dots, \zeta_n)^T$ *with* $\zeta_i = \operatorname{sign}\xi_i$ *being i.i.d. random variables such that* $\mathbf{P}\{\zeta_i = 1\} = \mathbf{P}\{\zeta_i = -1\} = 1/2$. *Consequently,* $\widehat{\boldsymbol{\theta}}_n$ *is asymptotically normal with mean* $\boldsymbol{\theta}^0$ *and covariance matrix* $\frac{1}{(2f(0))^2}\boldsymbol{\Sigma}^{-1}$. *One can take also* $\frac{1}{(2f(0))^2}\left(\frac{1}{n}\mathbf{C}^T\mathbf{C}\right)^{-1}$ *as the asymptotic covariance matrix.*

PROOF. It has been proved in Theorem 3.2.1 that under the present conditions $\widehat{\boldsymbol{\theta}}_n \xrightarrow{\mathbf{P}} \boldsymbol{\theta}^0$. Let us study this convergence in more detail. Make a change of variables by putting $\boldsymbol{\tau} = \sqrt{n}\,(\boldsymbol{\theta} - \boldsymbol{\theta}^0)$. Then $\widehat{\boldsymbol{\theta}}_n = \boldsymbol{\theta}^0 + \frac{1}{\sqrt{n}}\widehat{\boldsymbol{\tau}}_n$, where

$$\widehat{\boldsymbol{\tau}}_n = \arg\min_{\boldsymbol{\tau}\in\mathbb{R}^r} \sum_{\alpha=1}^r \bigg(\frac{1}{\sqrt{n}}\sum_{i=1}^n c_{i\alpha}\operatorname{sign}\Big(\xi_i - \frac{\mathbf{c}_i\boldsymbol{\tau}}{\sqrt{n}}\Big)\bigg)^2. \tag{3.2.18}$$

The proof of the asymptotic normality of $\widehat{\boldsymbol{\tau}}_n$ consists of two parts. In the first part we show that for any $0 < \gamma < 1/4$

$$\inf_{\boldsymbol{\tau}:\ |\boldsymbol{\tau}|\ge n^\gamma} \sum_{\alpha=1}^r \bigg(\frac{1}{\sqrt{n}}\sum_{i=1}^n c_{i\alpha}\operatorname{sign}\Big(\xi_i - \frac{\mathbf{c}_i\boldsymbol{\tau}}{\sqrt{n}}\Big)\bigg)^2 \xrightarrow{\mathbf{P}} \infty. \tag{3.2.19}$$

Since by the central limit theorem, the objective function in (3.2.19) for $\boldsymbol{\tau} = \mathbf{0}$ is bounded in probability, one has

$$\mathbf{P}\{|\widehat{\boldsymbol{\tau}}_n| < n^\gamma\} \to 1.$$

Hence, $\widehat{\boldsymbol{\tau}}_n$ is equivalent to

$$\widetilde{\boldsymbol{\tau}}_n = \arg\min_{\boldsymbol{\tau}:\ |\boldsymbol{\tau}|< n^\gamma} \sum_{\alpha=1}^r \bigg(\frac{1}{\sqrt{n}}\sum_{i=1}^n c_{i\alpha}\operatorname{sign}\Big(\xi_i - \frac{\mathbf{c}_i\boldsymbol{\tau}}{\sqrt{n}}\Big)\bigg)^2 \tag{3.2.20}$$

in the sense that their difference tends to zero in probability.

In the second part we study $\widetilde{\boldsymbol{\tau}}_n$. This study relies on Theorem 3.2.4 about uniform linearity (see 3.2.5), which implies that

$$\frac{1}{\sqrt{n}}\sum_{i=1}^n \mathbf{c}_i\operatorname{sign}\Big(\xi_i - \frac{\mathbf{c}_i\boldsymbol{\tau}}{\sqrt{n}}\Big) = \frac{1}{\sqrt{n}}\sum_{i=1}^n \mathbf{c}_i\Big(\operatorname{sign}\xi_i - 2f(0)\frac{\mathbf{c}_i\boldsymbol{\tau}}{\sqrt{n}}\Big) + \omega_n(\boldsymbol{\tau}), \tag{3.2.21}$$

where, for any $0 < \gamma < 1/4$,

$$\sup_{|\boldsymbol{\tau}|<n^\gamma} |\omega_n(\boldsymbol{\tau})| \xrightarrow{\mathbf{P}} 0.$$

This linearization of the objective function in (3.2.20) allows us to explore the asymptotic properties of $\widetilde{\boldsymbol{\tau}}_n$ and to complete the proof of the theorem.

1°. Let $\mathbf{e} \in \mathbb{R}^r$ be an arbitrary vector of the unit sphere $\mathbb{S}^{r-1}$. Put $\boldsymbol{\tau} = |\boldsymbol{\tau}|\mathbf{e}$ and take the scalar product of $\mathbf{e}$ with the vector

$$\frac{1}{\sqrt{n}} \sum_{i=1}^{n} \mathbf{c}_i \operatorname{sign}\Big(\xi_i - \frac{\mathbf{c}_i \boldsymbol{\tau}}{\sqrt{n}} \Big).$$

Letting $a_i = \mathbf{c}_i \mathbf{e}$ we obtain

$$\frac{1}{\sqrt{n}} \sum_{i=1}^{n} a_i \operatorname{sign}\Big(\xi_i - \frac{a_i |\boldsymbol{\tau}|}{\sqrt{n}} \Big). \tag{3.2.22}$$

By the Cauchy–Schwarz inequality

$$\Big(\frac{1}{\sqrt{n}} \sum_{i=1}^{n} a_i \operatorname{sign}\Big(\xi_i - \frac{a_i |\boldsymbol{\tau}|}{\sqrt{n}} \Big) \Big)^2 \le \Big| \frac{1}{\sqrt{n}} \sum_{i=1}^{n} \mathbf{c}_i \operatorname{sign}\Big(\xi_i - \frac{\mathbf{c}_i \boldsymbol{\tau}}{\sqrt{n}} \Big) \Big|^2. \tag{3.2.23}$$

Therefore, to obtain (3.2.19) it suffices to prove that

$$\sup_{\boldsymbol{\tau}:\ |\boldsymbol{\tau}| \ge n^\gamma} \Big(\frac{1}{\sqrt{n}} \sum_{i=1}^{n} a_i \operatorname{sign}\Big(\xi_i - \frac{a_i |\boldsymbol{\tau}|}{\sqrt{n}} \Big) \Big)^2 \xrightarrow{\mathbf{P}} \infty. \tag{3.2.24}$$

Notice that (3.2.22) is a nonincreasing function of $|\boldsymbol{\tau}|$. Hence

$$\frac{1}{\sqrt{n}} \sum_{i=1}^{n} a_i \operatorname{sign}\Big(\xi_i - \frac{a_i |\boldsymbol{\tau}|}{\sqrt{n}} \Big) \le \frac{1}{\sqrt{n}} \sum_{i=1}^{n} a_i \operatorname{sign}\Big(\xi_i - \frac{a_i\, n^\gamma}{\sqrt{n}} \Big) \tag{3.2.25}$$

for $|\boldsymbol{\tau}| \ge n^\gamma$.

It is easily seen that

$$v_n := \frac{1}{\sqrt{n}} \sum_{i=1}^{n} a_i \operatorname{sign}\Big(\xi_i - \frac{a_i\, n^\gamma}{\sqrt{n}} \Big) \xrightarrow{\mathbf{P}} -\infty. \tag{3.2.26}$$

Indeed, $\operatorname{Var} v_n$ is bounded and

$$\mathbf{E} v_n = -\frac{2n^\gamma}{n} \sum_{i=1}^{n} a_i^2 f(\kappa_i) \to -\infty,$$

where κ_i is a point between 0 and $a_i n^{-(1/2-\gamma)}$, $i = 1, \dots, n$. The latter convergence holds because $f(\kappa_i) \to f(0) > 0$ uniformly in $i = 1, \dots, n$ and

$$\frac{1}{n} \sum_{i=1}^{n} a_i^2 \to \mathbf{e}^T \boldsymbol{\Sigma} \mathbf{e} \ge \text{const} > 0$$

for any $\mathbf{e} \in \mathbb{S}^{r-1}$ by Condition 2.1(viii). It is seen from (3.2.25) and (3.2.26) that

$$\inf_{\boldsymbol{\tau}:\ |\boldsymbol{\tau}| \ge n^\gamma} \Big(\frac{1}{\sqrt{n}} \sum_{i=1}^{n} a_i \operatorname{sign}\Big(\xi_i - \frac{a_i |\boldsymbol{\tau}|}{\sqrt{n}} \Big) \Big)^2 \ge \inf_{\mathbf{e} \in \mathbb{S}^{r-1}} v_n^2.$$

Thus (3.2.26) implies (3.2.24) and hence (3.2.19).

2°. Making use of (3.2.21) we can replace (3.2.20) by

$$\min_{\boldsymbol{\tau}:\ |\boldsymbol{\tau}| < n^\gamma} \sum_{\alpha=1}^{r} \Big(\frac{1}{\sqrt{n}} \sum_{i=1}^{n} c_{i\alpha} \Big(\operatorname{sign} \xi_i - 2f(0) \frac{\mathbf{c}_i \boldsymbol{\tau}}{\sqrt{n}} \Big) + \omega_n(\boldsymbol{\tau}) \Big)^2, \tag{3.2.27}$$

where the remainder $\omega_n(\boldsymbol{\tau})$ is uniformly small in probability over $\{\boldsymbol{\tau}\colon |\boldsymbol{\tau}| < n^\gamma\}$. Therefore, the solution of the problem (3.2.27) as $n \to \infty$ is asymptotically equivalent to that of

$$\min_{\boldsymbol{\tau}\colon |\boldsymbol{\tau}|<n^\gamma} \sum_{\alpha=1}^{r} \Big(\frac{1}{\sqrt{n}} \sum_{i=1}^{n} c_{i\alpha} \Big(\operatorname{sign} \xi_i - 2f(0) \frac{\mathbf{c}_i \boldsymbol{\tau}}{\sqrt{n}} \Big) \Big)^2 . \tag{3.2.28}$$

If we neglect the restriction $|\boldsymbol{\tau}| < n^\gamma$, the solution of (3.2.28) is obtained from the system of linear equations

$$\sum_{i=1}^{n} c_{i\alpha} \big(\operatorname{sign} \xi_i - 2f(0) \frac{\mathbf{c}_i \boldsymbol{\tau}}{\sqrt{n}} \big) = 0, \qquad \alpha = 1, \ldots, r. \tag{3.2.29}$$

By defining the vector $\boldsymbol{\zeta}$ with coordinates $\zeta_i = \operatorname{sign} \xi_i$, $i = 1, \ldots, n$, we can rewrite (3.2.29) in a matrix form

$$\frac{1}{\sqrt{n}} \mathbf{C}^T \boldsymbol{\zeta} = 2f(0) \Big(\frac{1}{n} \mathbf{C}^T \mathbf{C} \Big) \boldsymbol{\tau}, \tag{3.2.30}$$

so that its solution, $\boldsymbol{\tau}_n^*$, say, is

$$\boldsymbol{\tau}_n^* = \frac{1}{2f(0)} \Big(\frac{1}{n} \mathbf{C}^T \mathbf{C} \Big)^{-1} \Big(\frac{1}{\sqrt{n}} \mathbf{C}^T \boldsymbol{\zeta} \Big). \tag{3.2.31}$$

We have already pointed out in §2.3 that under Conditions 2.1(i, vii, viii) the random vector $\frac{1}{\sqrt{n}} \mathbf{C} \boldsymbol{\zeta}^T$ has asymptotically normal distribution $N(\mathbf{0}, \boldsymbol{\Sigma})$. Hence under the same assumptions $\boldsymbol{\tau}_n^*$ is asymptotically distributed as

$$N\Big(\mathbf{0}, \frac{1}{(2f(0))^2} \boldsymbol{\Sigma}^{-1}\Big), \qquad \text{or} \qquad N\Big(\mathbf{0}, \frac{1}{(2f(0))^2} \Big(\frac{1}{n} \mathbf{C}^T \mathbf{C}\Big)^{-1}\Big).$$

However, we are interested in the solution of the problem (3.2.28), which differs from (3.2.29) by the restriction $|\boldsymbol{\tau}| < n^\gamma$. We will show that this difference is negligible. Using the fact that $\boldsymbol{\tau}_n^*$ has a nondegenerate asymptotic distribution, we can choose a constant A such that $|\boldsymbol{\tau}_n^*| \le A$ with probability arbitrarily close to 1 for sufficiently large n. In this case $\boldsymbol{\tau}_n^*$ also solves (3.2.28). Since (3.2.28) is minimized at an interior point, $\boldsymbol{\tau}_n^*$ coincides then with $\widetilde{\boldsymbol{\tau}}_n$. This completes the proof of Theorem 3.2.2. □

REMARK. The linear model (2.1.1) admits various parametrizations of the form $\mathbf{l} = \mathbf{C}\boldsymbol{\theta}$. For example, along with the above representation the vector $\mathbf{l} \in \mathbb{L}$ can be specified as

$$\mathbf{l} = \mathbf{D}\boldsymbol{\tau},$$

where $\mathbf{D}$ is a $n \times r$ matrix and $\boldsymbol{\tau} \in \mathbb{R}^r$ is a column vector. Then there exists a nondegenerate $r \times r$ matrix $\mathbf{G}$ such that

$$\boldsymbol{\theta} = \mathbf{G}\,\boldsymbol{\tau} \qquad \text{and} \qquad \mathbf{D} = \mathbf{C}\mathbf{G}.$$

Dealing with the two parametrizations of the underlying linear model, we can obtain sign estimators both for $\boldsymbol{\theta}$ and $\boldsymbol{\tau}$. We denote them by $\widehat{\boldsymbol{\theta}}_n$ and $\widehat{\boldsymbol{\tau}}_n$, respectively. Then Theorem 3.2.2 implies that

$$\widehat{\boldsymbol{\theta}}_n = \mathbf{G}\,\widehat{\boldsymbol{\tau}}_n + \mathbf{o}_p\Big(\frac{1}{\sqrt{n}}\Big).$$

3.2.4. Asymptotic covariance of sign estimators. As we discussed in 3.2.1, in order to construct confidence sets for $\boldsymbol{\theta}^0$ we need to know the asymptotic covariance matrix of the sign estimator $\widehat{\boldsymbol{\theta}}_n$. For this purpose the formulas given in Theorem 3.2.2 can be used provided we consistently estimate $f(0)$. We propose such an estimate based on (3.2.21). Let $\widehat{\boldsymbol{\theta}}_n$ be a sign estimator for $\boldsymbol{\theta}^0$ in the model (2.1.5) or some other $\sqrt{n}$-consistent estimator for $\boldsymbol{\theta}^0$,

$$\widehat{\boldsymbol{\theta}}_n = \boldsymbol{\theta}^0 + \frac{1}{\sqrt{n}} \widehat{\boldsymbol{\tau}}_n.$$

Take an arbitrary vector $\mathbf{e} \in \mathbb{S}^{r-1}$. Put $a_i = \mathbf{c}_i\, \mathbf{e}$ and for $h \in \mathbb{R}^1$ define the statistic

$$G_n(h) := \frac{1}{\sqrt{n}} \sum_{i=1}^{n} a_i \operatorname{sign}\Big(x_i - \mathbf{e}_i \widehat{\boldsymbol{\theta}}_n + \frac{a_i h}{\sqrt{n}}\Big).$$

On multiplying (3.2.1) by $\mathbf{e}$ we obtain

$$G_n(h) = \frac{1}{\sqrt{n}} \sum_{i=1}^{n} a_i \left(\operatorname{sign} \xi_i - 2f(0) \frac{\mathbf{c}_i \widehat{\boldsymbol{\tau}}_n}{\sqrt{n}} + 2f(0) \frac{a_i h}{\sqrt{n}}\right) + o_p(1),$$

where $o_p(1) \xrightarrow{\mathbf{P}} 0$ uniformly in h for $|h| < n^\gamma$, $0 < \gamma < 1/4$. Obviously,

$$\frac{1}{2}\big(G_n(h) - G_n(-h)\big) = 2f(0) h \left(\frac{1}{n} \sum_{i=1}^{n} a_i^2\right) + o_p(1).$$

Hence letting $h = h(n) = n^\gamma$ for an arbitrary γ, $0 < \gamma < 1/4$, we obtain a consistent estimator for $\big(2f(0)\big)^{-1}$,

$$\frac{\mathbf{e}^T (\mathbf{C}^T \mathbf{C}) \mathbf{e}}{\frac{n}{2}\big(G_n(h) - G_n(-h)\big)} h \xrightarrow{\mathbf{P}} \big(2f(0)\big)^{-1}.$$

We recommend taking $\mathbf{e}$ as the eigenvector of $\mathbf{C}^T \mathbf{C}$ corresponding to the largest eigenvalue.

3.2.5. Uniform law of large numbers. Note that in the following theorem (as well as in similar situations henceforth), the design matrix $\mathbf{C} = \|c_{i\alpha}\|$, $\alpha = 1, \dots, r$, $i = 1, \dots, n$, and the coefficients d_i, $i = 1, \dots, n$, depend also on n. However, we suppress this dependence for the sake of notational simplicity.

THEOREM 3.2.3. *The random function*

$$U_n(\mathbf{t}) = \frac{1}{n} \sum_{i=1}^{n} d_i \big(\operatorname{sign}(\xi_i + \mathbf{c}_i \mathbf{t}) - \mathbf{E} \operatorname{sign}(\xi_i + \mathbf{c}_i \mathbf{t})\big)$$

of $\mathbf{t} = (t_1, \dots, t_r)^T$ *converges to zero in probability as* $n \to \infty$ *uniformly in* $\mathbf{t}$, $|\mathbf{t}| \le C$, *for any* $C > 0$, *that is for any* $\varepsilon > 0$

$$\mathbf{P}\{\sup_{|\mathbf{t}| \le C} |U_n(\mathbf{t})| > \varepsilon\} \to 0 \quad \text{as} \quad n \to \infty, \tag{3.2.32}$$

provided the following conditions are fulfilled.

(i) *The sequences* $\{d_i,\ i=1,2,\dots,n\}$ *and* $\{c_{i\alpha},\ i=1,\dots,n\}$, $\alpha=1,\dots,r$, *are bounded, i.e., there exists a constant* $K>0$ *such that*

$$\max_{1\le i\le n}\big(|d_i|,|c_{i1}|,\dots,|c_{ir}|\big)\le K.$$

(ii) *The random variables* $\xi_1,\xi_2,\dots$ *are i.i.d. with their common distribution function* $F(u)=\mathbf{P}\{\xi_1<u\}$ *satisfying Condition* 2.1(iv), *i.e.,* $|F(u_1)-F(u_2)|<L|u_1-u_2|$, *for some* $L>0$.

PROOF. Let $y_i(\mathbf{t})=\operatorname{sign}(\xi_i+\mathbf{c}_i\mathbf{t})-\mathbf{E}\operatorname{sign}(\xi_i+\mathbf{c}_i\mathbf{t})$. Note that $\mathbf{E}\operatorname{sign}(\xi_i+\mathbf{c}_i\mathbf{t})=1-2F(-\mathbf{c}_i\mathbf{t})$. It is easily seen that $U_n(\mathbf{t})=n^{-1}\sum_{i=1}^n d_iy_i(\mathbf{t})\to 0$ in probability for any fixed $\mathbf{t}$. Indeed,

$$\mathbf{E}\big(U_n(\mathbf{t})\big)^2=\operatorname{Var}U_n(\mathbf{t})=\frac{1}{n^2}\sum_{i=1}^n d_i^2\operatorname{Var}\operatorname{sign}(\xi_i+\mathbf{c}_i\mathbf{t})\le\frac{1}{n^2}\sum_{i=1}^n d_i^2,$$

because

$$\operatorname{Var}\operatorname{sign}(\xi_i+\mathbf{c}_i\mathbf{t})=\mathbf{E}\big(\operatorname{sign}(\xi_i+\mathbf{c}_i\mathbf{t})\big)^2-\big(\mathbf{E}\operatorname{sign}(\xi_i+\mathbf{c}_i\mathbf{t})\big)^2\le 1.$$

By condition (i), $\frac{1}{n}\sum_{i=1}^n d_i^2\le K^2$, hence $\mathbf{E}\big(U_n(\mathbf{t})\big)^2\to 0$. This implies the convergence $U_n(\mathbf{t})\to 0$ in probability.

Now we proceed to the proof of (3.2.32). Break up the space $\mathbb{R}^r$ into congruent cubes with diagonal h. Let T be a finite collection of such cubes which cover the compact set $\{\mathbf{t}\colon |\mathbf{t}|\le C\}$.

Consider an arbitrary cube V, $V\in T$. Observe that the linear function $\mathbf{c}_i\mathbf{t}$ of variable $\mathbf{t}$ takes the maximal and minimal values on V at the endpoints of some diagonal of V. Denote these vertices by $\boldsymbol{\mu}_i$ and $\boldsymbol{\lambda}_i$. Since $\operatorname{sign}x$ and $1-2F(-x)$ are monotone nondecreasing functions of x, $x\in\mathbb{R}^1$, we have for $\mathbf{t}\in V$

$$\begin{aligned}&\operatorname{sign}(\xi_i+\mathbf{c}_i\boldsymbol{\lambda}_i)\le\operatorname{sign}(\xi_i+\mathbf{c}_i\mathbf{t})\le\operatorname{sign}(\xi_i+\mathbf{c}_i\boldsymbol{\mu}_i),\\ &1-2F(-\mathbf{c}_i\boldsymbol{\lambda}_i)\le 1-2F(-\mathbf{c}_i\mathbf{t})\le 1-2F(-\mathbf{c}_i\boldsymbol{\mu}_i).\end{aligned}$$

Let

$$d_i^+=\begin{cases}d_i, & \text{if } d_i>0\\ 0, & \text{if } d_i\le 0,\end{cases}\qquad d_i^-=\begin{cases}0, & \text{if } d_i>0\\ -d_i, & \text{if } d_i\le 0.\end{cases}$$

Now we obtain upper and lower bounds for $U_n(\mathbf{t})$, $\mathbf{t}\in V$:

$$\begin{aligned}\frac{1}{n}\sum_{i=1}^n d_iy_i(\mathbf{t})\le{}&\frac{1}{n}\sum_{i=1}^n d_i^+\big(\operatorname{sign}(\xi_i+\mathbf{c}_i\boldsymbol{\mu}_i)-\big(1-2F(-\mathbf{c}_i\boldsymbol{\mu}_i)\big)\big)\\ &-\frac{1}{n}\sum_{i=1}^n d_i^-\big(\operatorname{sign}(\xi_i+\mathbf{c}_i\boldsymbol{\lambda}_i)-\big(1-2F(-\mathbf{c}_i\boldsymbol{\lambda}_i)\big)\big)\\ &+\frac{1}{n}\sum_{i=1}^n d_i^+\big(\big(1-2F(-\mathbf{c}_i\boldsymbol{\mu}_i)\big)-\big(1-2F(-\mathbf{c}_i\boldsymbol{\lambda}_i)\big)\big)\\ &-\frac{1}{n}\sum_{i=1}^n d_i^-\big(\big(1-2F(-\mathbf{c}_i\boldsymbol{\lambda}_i)\big)-\big(1-2F(-\mathbf{c}_i\boldsymbol{\mu}_i)\big)\big)\\ ={}&A_1-A_2+A_3-A_4.\end{aligned}$$

The lower bound has a similar form:

$$\frac{1}{n}\sum_{i=1}^{n} d_i y_i(\mathbf{t}) \geq A_1' - A_2' + A_3' - A_4',$$

where

$$A_1' = \frac{1}{n}\sum_{i=1}^{n} d_i^+ \left(\operatorname{sign}(\xi_i + \mathbf{c}_i\boldsymbol{\lambda}_i) - (1 - 2F(-\mathbf{c}_i\boldsymbol{\lambda}_i))\right),$$
$$A_2' = \frac{1}{n}\sum_{i=1}^{n} d_i^- \left(\operatorname{sign}(\xi_i + \mathbf{c}_i\boldsymbol{\mu}_i) - (1 - 2F(-\mathbf{c}_i\boldsymbol{\mu}_i))\right),$$
$$A_3' = \frac{1}{n}\sum_{i=1}^{n} d_i^+ \left((1 - 2F(-\mathbf{c}_i\boldsymbol{\lambda}_i)) - (1 - 2F(-\mathbf{c}_i\boldsymbol{\mu}_i))\right),$$
$$A_4' = \frac{1}{n}\sum_{i=1}^{n} d_i^- \left((1 - 2F(-\mathbf{c}_i\boldsymbol{\mu}_i)) - (1 - 2F(-\mathbf{c}_i\boldsymbol{\lambda}_i))\right).$$

Consider the sum A_1. Denote the vertices of the cube V by $\nu_1, \dots, \nu_k$, $k = 1, \dots, 2^{r+1}$. Put

$$a_{ik} = \begin{cases} 1, & \text{if } \mu_i = \nu_k, \\ 0, & \text{otherwise.} \end{cases}$$

Then

$$A_1 = \sum_{i=1}^{2^{r+1}} \frac{1}{n}\sum_{i=1}^{n} a_{ik} d_i^+ y_i(\nu_k).$$

On account of the pointwise convergence established above, one has for each k

$$\sum_{i=1}^{n} a_{ik} d_i^+ y_i(\nu_k) \xrightarrow{\mathbf{P}} 0.$$

Therefore $A_1 \xrightarrow{\mathbf{P}} 0$. In a similar way one shows that A_2, A_1', $A_2' \xrightarrow{\mathbf{P}} 0$. Let Ω_n be the event that for each $V \in T$ the variables A_1, A_2, A_1', A_2' do not exceed δ in absolute value. By the above, $\mathbf{P}\{\Omega_n\} \to 1$ for any fixed $h > 0$, $\delta > 0$.

Consider now A_3, A_4, A_3', A_4'. Note that by condition (ii)

$$\left|F(-\mathbf{c}_i\boldsymbol{\mu}_i) - F(-\mathbf{c}_i\boldsymbol{\lambda}_i)\right| \leq L\,h$$

for any i and any cube $V, V \in T$. Hence $|A_3|$, $|A_4|$, $|A_3'|$, $|A_4'| \leq 2LKh$ for any cube $V, V \in T$. Choose h such that $2LKh < \delta$.

Now we turn to (3.2.32). Note that

$$\begin{aligned}\mathbf{P}\big\{\sup_{|\mathbf{t}|\leq C} |v_n(\mathbf{t})| < \varepsilon\big\} &= \mathbf{P}\big\{\max_{V\in T}\sup_{\mathbf{t}\in V} |v_n(\mathbf{t})| < \varepsilon\big\} \\ &\geq \mathbf{P}\big\{\max_{V\in T}\sup_{\mathbf{t}\in V} |v_n(\mathbf{t})| < \varepsilon,\, \Omega_n\big\}.\end{aligned}$$

Take some $\delta < \frac{\varepsilon}{4}$. Then Ω_n entails that for any $V \in T$

$$\sup_{\mathbf{t}\in V} |v_n(\mathbf{t})| < 4\delta,$$

i.e., the event $\{\sup_{|\mathbf{t}|\le C}|v_n(\mathbf{t})|<\varepsilon\}$ is a consequence of the event Ω_n, and therefore

$$\{\sup_{|\mathbf{t}|\le C}|v_n(\mathbf{t})|<\varepsilon\}=\mathbf{P}\{\Omega_n\}\to 1,$$

which was to be proved. □

3.2.6. Theorem on uniform linearity.

THEOREM 3.2.4. *The random function*

$$X_n(\mathbf{t})=\frac{1}{\sqrt{n}}\sum_{i=1}^n d_i\Big(\operatorname{sign}\Big(\xi_i+\frac{\mathbf{c}_i\mathbf{t}}{\sqrt{n}}\Big)-\operatorname{sign}\xi_i-2f(0)\frac{\mathbf{c}_i\mathbf{t}}{\sqrt{n}}\Big) \tag{3.2.33}$$

of $\mathbf{t}=(t_1,\dots,t_r)^T$ *converges to zero in probability as* $n\to\infty$ *uniformly in* $\mathbf{t}$ *on the set* $\{|\mathbf{t}|\le \operatorname{const} n^\gamma\}$, *for an arbitrary* $\gamma<\frac14$, *i.e., for any* $\varepsilon>0$

$$\mathbf{P}\{\sup_{\{|\mathbf{t}|\le\operatorname{const} n^\gamma\}}|X_n(\mathbf{t})|>\varepsilon\}\to 0, \tag{3.2.34}$$

provided the following conditions are fulfilled:

(i) *The sequences* $\{d_i,\ i=1,2,\dots\}$, $\{c_{i\alpha},\ i=1,2,\dots\}$, $\alpha=1,\dots,r$, *are bounded, i.e., there exists a constant* $K>0$ *such that for all* n

$$\max_{1\le i\le n}\big(|d_i|,|c_{i1}|,\dots,|c_{ir}|\big)<K.$$

(ii) *The random variables* ξ_i, $i=1,2,\dots$, *are i.i.d. with their common distribution function satisfying Conditions* 2.1(i, ii, vi).

We will often apply this theorem in the following form:

Under the above conditions (i), (ii)

$$\frac{1}{\sqrt{n}}\sum_{i=1}^n d_i\operatorname{sign}\Big(\xi_i+\frac{\mathbf{c}_i\mathbf{t}}{\sqrt{n}}\Big)=\frac{1}{\sqrt{n}}\sum_{i=1}^n d_i\Big(\operatorname{sign}\xi_i+2f(0)\frac{\mathbf{c}_i\mathbf{t}}{\sqrt{n}}\Big)+o_p(1),$$

where $o_p(1)$ *converges to zero in probability uniformly in* $\mathbf{t}$, $|\mathbf{t}|<\operatorname{const} n^\gamma$, *for any* $0<\gamma<\frac14$.

PROOF.

1°. Split the cube $|\mathbf{t}|<Cn^\gamma$, $0<\gamma<\frac14$, in $\mathbb{R}^r$ into smaller cubes of equal size with diagonal $h=h(n)$ to be chosen later on. Let $T=T(n)$ denote the set of the centers of these cubes. Take an arbitrary cube of the partition, $V=V(\boldsymbol{\tau})$ with a center $\boldsymbol{\tau}\in T$. For the proof of the uniform convergence it suffices to show that

$$\mathbf{P}\Big\{\sup_{\boldsymbol{\tau}\in T(n)}\sup_{\mathbf{t}\in V(\boldsymbol{\tau})}|X_n(\mathbf{t})|>\varepsilon\Big\}\to 0 \qquad \text{as} \quad n\to\infty.$$

Clearly,

$$\begin{aligned}\mathbf{P}\Big\{\sup_{\boldsymbol{\tau}\in T(n)}\sup_{\mathbf{t}\in V(\boldsymbol{\tau})}|X_n(\mathbf{t})|>\varepsilon\Big\}&\le\sum_{\boldsymbol{\tau}\in T(n)}\mathbf{P}\Big\{\sup_{\mathbf{t}\in V(\boldsymbol{\tau})}|X_n(\mathbf{t})|>\varepsilon\Big\}\\&\le\sum_{\boldsymbol{\tau}\in T(n)}\mathbf{P}\Big\{\sup_{\mathbf{t}\in V(\boldsymbol{\tau})}X_n(\mathbf{t})>\varepsilon\Big\}+\sum_{\boldsymbol{\tau}\in T(n)}\mathbf{P}\Big\{\inf_{\mathbf{t}\in V(\boldsymbol{\tau})}X_n(\mathbf{t})<-\varepsilon\Big\}.\end{aligned} \tag{3.2.35}$$

The number of terms in each of the two sums is $O(n^{\gamma r}h^{-r})$.

We will estimate $\mathbf{P}\{\sup_{\mathbf{t}\in V(\boldsymbol{\tau})} X_n(\mathbf{t}) > \varepsilon\}$ and $\mathbf{P}\{\inf_{\mathbf{t}\in V(\boldsymbol{\tau})} X_n(\mathbf{t}) < -\varepsilon\}$ from above. In order to show that the right-hand side of (3.2.35) tends to zero, it will be sufficient to obtain for these probabilities a uniform upper bound of order $o\big((n^{\gamma r}h^{-r})^{-1}\big)$. Both these probabilities are handled similarly, so that we consider the first one.

Observe that the linear form (of $t_1, \dots, t_r$) $\mathbf{c}_i\mathbf{t} = \sum_{\alpha=1}^r c_{i\alpha}t_\alpha$ attains its maximal and minimal values on the cube $V(\boldsymbol{\tau})$ at the endpoints of some diagonal. We denote them by $\boldsymbol{\mu}_i$ and $\boldsymbol{\lambda}_i$. Since $\operatorname{sign} x$ is a monotone nondecreasing function, for any $\mathbf{t} \in V(\boldsymbol{\tau})$ the following inequalities hold:

$$\operatorname{sign}\Big(\xi_i + \frac{\mathbf{c}_i\boldsymbol{\lambda}_i}{\sqrt{n}}\Big) \le \operatorname{sign}\Big(\xi_i + \frac{\mathbf{c}_i\mathbf{t}}{\sqrt{n}}\Big) \le \operatorname{sign}\Big(\xi_i + \frac{\mathbf{c}_i\boldsymbol{\mu}_i}{\sqrt{n}}\Big).$$

Write d_i as $d_i = d_i^+ - d_i^-$, where

$$d_i^+ = \begin{cases} d_i, & \text{if } d_i > 0, \\ 0, & \text{if } d_i \le 0, \end{cases} \qquad d_i^- = \begin{cases} 0, & \text{if } d_i \ge 0, \\ -d_i, & \text{if } d_i < 0. \end{cases}$$

Let us estimate $X(\mathbf{t})$, $\mathbf{t} \in V(\boldsymbol{\tau})$ from above and below. First we obtain an upper bound:

$$\begin{aligned}
&\frac{1}{\sqrt{n}}\sum_{i=1}^n d_i\Big(\operatorname{sign}\Big(\xi_i + \frac{\mathbf{c}_i\mathbf{t}}{\sqrt{n}}\Big) - \operatorname{sign}\xi_i - 2f(0)\frac{\mathbf{c}_i\mathbf{t}}{\sqrt{n}}\Big) \\
&\quad \le \frac{1}{\sqrt{n}}\sum_{i=1}^n d_i^+\Big(\operatorname{sign}\Big(\xi_i + \frac{\mathbf{c}_i\boldsymbol{\mu}_i}{\sqrt{n}}\Big) - \operatorname{sign}\xi_i - 2f(0)\frac{\mathbf{c}_i\boldsymbol{\lambda}_i}{\sqrt{n}}\Big) \\
&\qquad - \frac{1}{\sqrt{n}}\sum_{i=1}^n d_i^-\Big(\operatorname{sign}\Big(\xi_i + \frac{\mathbf{c}_i\boldsymbol{\lambda}_i}{\sqrt{n}}\Big) - \operatorname{sign}\xi_i - 2f(0)\frac{\mathbf{c}_i\boldsymbol{\mu}_i}{\sqrt{n}}\Big) \\
&\quad \le \frac{1}{\sqrt{n}}\sum_{i=1}^n d_i^+\Big(\operatorname{sign}\Big(\xi_i + \frac{\mathbf{c}_i\boldsymbol{\mu}_i}{\sqrt{n}}\Big) - \operatorname{sign}\Big(\xi_i + \frac{\mathbf{c}_i\boldsymbol{\tau}}{\sqrt{n}}\Big)\Big) \qquad (3.2.36)\\
&\qquad + \frac{1}{\sqrt{n}}\sum_{i=1}^n d_i^-\Big(\operatorname{sign}\Big(\xi_i + \frac{\mathbf{c}_i\boldsymbol{\tau}}{\sqrt{n}}\Big) - \operatorname{sign}\Big(\xi_i + \frac{\mathbf{c}_i\boldsymbol{\lambda}_i}{\sqrt{n}}\Big)\Big) \\
&\qquad + \frac{1}{\sqrt{n}}\Big|\sum_{i=1}^n d_i\Big(\operatorname{sign}\Big(\xi_i + \frac{\mathbf{c}_i\boldsymbol{\tau}}{\sqrt{n}}\Big) - \operatorname{sign}\xi_i - 2f(0)\frac{\mathbf{c}_i\boldsymbol{\tau}}{\sqrt{n}}\Big)\Big| \\
&\qquad + \frac{2f(0)}{n}\Big|\sum_{i=1}^n \mathbf{c}_i d_i(\boldsymbol{\tau} - \mathbf{t})\Big| \\
&\quad = A_1 + A_2 + A_3 + A_4.
\end{aligned}$$

Clearly,

$$\begin{aligned}
\mathbf{P}\Big\{\sup_{\mathbf{t}\in V(\boldsymbol{\tau})} X_n(\mathbf{t}) > \varepsilon\Big\} \le \mathbf{P}\Big\{A_1 > \frac{\varepsilon}{4}\Big\} + \mathbf{P}\Big\{A_2 > \frac{\varepsilon}{4}\Big\} \qquad (3.2.37)\\
+ \mathbf{P}\Big\{A_3 > \frac{\varepsilon}{4}\Big\} + \mathbf{P}\Big\{A_4 > \frac{\varepsilon}{4}\Big\}.
\end{aligned}$$

Notice that by assumption (i), $A_4 \le \frac{2}{n}f(0)K^2h$. Hence, this term goes to zero as $h = h(n) \to 0$, so that the last probability in (3.2.37) equals 0 for sufficiently small h.

Let us estimate the other probabilities. Since A_1 and A_2 have a similar stucture, we present the proof for one of them.

2°. For an integer $m > 0$ one has by Chebyshev's inequality

$$\mathbf{P}\Big\{A_1 > \frac{\varepsilon}{4}\Big\} < \Big(\frac{\varepsilon}{4}\Big)^{-m} \mathbf{E}A_1^m. \tag{3.2.38}$$

Let us estimate $\mathbf{E}A_1^m$. Put

$$\omega_i = \operatorname{sign}\Big(\xi_i + \frac{\mathbf{c}_i\boldsymbol{\mu}_i}{\sqrt{n}}\Big) - \operatorname{sign}\Big(\xi_i + \frac{\mathbf{c}_i\boldsymbol{\tau}}{\sqrt{n}}\Big)$$

and note that the random variables $\omega_1, \dots, \omega_n$ are mutually independent. Further,

$$\mathbf{E}A_1^m = n^{-m/2} \sum_{i_1} \cdots \sum_{i_m} d_{i_1}^+ \dots d_{i_m}^+ \mathbf{E}\omega_{i_1} \dots \omega_{i_m}. \tag{3.2.39}$$

For $\mathbf{E}\omega_i$ one has

$$\begin{aligned}\mathbf{E}\omega_i &= \mathbf{P}\Big\{\xi_i + \frac{\mathbf{c}_i\boldsymbol{\mu}_i}{\sqrt{n}} > 0 > \xi_i + \frac{\mathbf{c}_i\boldsymbol{\tau}}{\sqrt{n}}\Big\}\\ &= \mathbf{P}\Big\{-\frac{\mathbf{c}_i\boldsymbol{\mu}_i}{\sqrt{n}} < \xi_i < -\frac{\mathbf{c}_i\boldsymbol{\tau}}{\sqrt{n}}\Big\} = f(\kappa_i)\frac{\mathbf{c}_i(\boldsymbol{\mu}_i - \boldsymbol{\tau})}{\sqrt{n}},\end{aligned}$$

where κ_i is a point between $-\frac{\mathbf{c}_i\boldsymbol{\mu}_i}{\sqrt{n}}$ and $-\frac{\mathbf{c}_i\boldsymbol{\tau}}{\sqrt{n}}$. Notice that $\kappa_i \to 0$ uniformly in i and the cubes of the partition. Hence, all $f(\kappa_i)$ are bounded from above by the same constant. Therefore,

$$|\mathbf{E}\omega_i| \le \operatorname{const} \frac{h}{\sqrt{n}}. \tag{3.2.40}$$

For a product $\omega_{i_1} \dots \omega_{i_m}$, let us call the number of different indices among $i_1, \dots, i_m$ *the level* of this product. Represent $\mathbf{E}A_1^m$ as given by (3.2.39) in the form $\mathbf{E}A_1^m = \Sigma_{(m)} + \dots + \Sigma_{(1)}$, where $\Sigma_{(p)}$, $1 \le p \le m$, contains the terms of level p. Observe that $(\omega_i/2)^p = \omega_i/2$ for any integer $p \ge 1$, so that $\mathbf{E}\omega_i^p = 2^{p-1}\mathbf{E}\omega_i$. Hence for $\mathbf{E}\omega_{i_1} \dots \omega_{i_m}$ from $\Sigma_{(p)}$ one has by (3.2.40)

$$|\mathbf{E}\omega_{i_1} \dots \omega_{i_p}| \le \operatorname{const}\Big(\frac{h}{\sqrt{n}}\Big)^p.$$

Taking into account that the number of terms in $\Sigma_{(p)}$ is $O(n^p)$ we obtain

$$|\Sigma_{(p)}| \le \operatorname{const}\Big(\frac{1}{\sqrt{n}}\Big)^m n^p \Big(\frac{h}{\sqrt{n}}\Big)^p. \tag{3.2.41}$$

Now we take $h = h(n)$ such that $h(n) \to 0$ and $\sqrt{n}h(n) \to \infty$ as $n \to \infty$. For instance, let $h(n) = n^{-1/4}$. Then $\Sigma_{(p)} \le \operatorname{const} \Sigma_{(m)}$ and hence it is sufficient to consider the sum $\Sigma_{(m)}$, which is the principal term in EA_1^m. In view of (3.2.41) we have $\mathbf{E}A_1^m \le \operatorname{const} h^m$.

Recall that an upper bound for $\mathbf{P}\{\sup_{\mathbf{t}\in V(\boldsymbol{\tau})} X_n(\mathbf{t}) > \varepsilon\}$ given by (3.2.37) and (3.2.38) involves $\mathbf{E}A_1^m$ and that the number of these terms in (3.2.35) is $O(n^{\gamma r}h^{-r})$. For $h = n^{-1/4}$ this is $O\big((\frac{n^\gamma}{\sqrt{n}})^r n^{3r/4}\big) = O\big(n^{3r/4}\big)$. Taking $m \ge 3r + 1$ we obtain that the contribution of the terms (3.2.38) into (3.2.35) goes to zero as $n \to \infty$.

The probabilities $\mathbf{P}\{A_2 > \varepsilon/4\}$ are estimated in a similar manner.

3°. Consider $\mathbf{P}\{A_3 > \varepsilon/4\}$. Let

$$\omega_i = \omega_i(\boldsymbol{\tau}) = \operatorname{sign}\Big(\xi_i + \frac{\mathbf{c}_i\boldsymbol{\tau}}{\sqrt{n}}\Big) - \operatorname{sign}\xi_i - 2f(0)\frac{\mathbf{c}_i\boldsymbol{\tau}}{\sqrt{n}}.$$

Let us estimate $\mathbf{E}\omega_i$ and $\mathbf{E}|\omega_i|^p$, $p > 0$. We have

$$\begin{aligned}
\mathbf{E}\omega_i &= \mathbf{E}\operatorname{sign}\Big(\xi_i + \frac{\mathbf{c}_i\boldsymbol{\tau}}{\sqrt{n}}\Big) - 2f(0)\frac{\mathbf{c}_i\boldsymbol{\tau}}{\sqrt{n}} \\
&= \mathbf{P}\Big\{\xi_i + \frac{\mathbf{c}_i\boldsymbol{\tau}}{\sqrt{n}} > 0\Big\} - \mathbf{P}\Big\{\xi_i + \frac{\mathbf{c}_i\boldsymbol{\tau}}{\sqrt{n}} < 0\Big\} - 2f(0)\frac{\mathbf{c}_i\boldsymbol{\tau}}{\sqrt{n}} \\
&= 1 - 2F\Big(-\frac{\mathbf{c}_i\boldsymbol{\tau}}{\sqrt{n}}\Big) - 2f(0)\frac{\mathbf{c}_i\boldsymbol{\tau}}{\sqrt{n}} \\
&= 2\Big(F(0) - F\Big(-\frac{\mathbf{c}_i\boldsymbol{\tau}}{\sqrt{n}}\Big)\Big) - 2f(0)\frac{\mathbf{c}_i\boldsymbol{\tau}}{\sqrt{n}}.
\end{aligned}$$

By (i) and the assumption $|\boldsymbol{\tau}| < \operatorname{const} n^{\gamma}$, the argument $-\mathbf{c}_i\boldsymbol{\tau}/\sqrt{n}$ tends to zero and therefore belongs to the neighborhood in Condition 2.1(ii) for sufficiently large n. Hence

$$F(0) - F\Big(-\frac{\mathbf{c}_i\boldsymbol{\tau}}{\sqrt{n}}\Big) = f(\kappa_i)\frac{\mathbf{c}_i\boldsymbol{\tau}}{\sqrt{n}},$$

where κ_i is a point between 0 and $-\mathbf{c}_i\boldsymbol{\tau}/\sqrt{n}$, so that $|\kappa_i| \le \big|\frac{\mathbf{c}_i\boldsymbol{\tau}}{\sqrt{n}}\big| \le \operatorname{const} n^{-\varepsilon}$ with $\varepsilon = \frac{1}{2} - \gamma > 0$. Thus

$$\mathbf{E}\omega_i = 2\big(f(\kappa_i) - f(0)\big)\frac{\mathbf{c}_i\boldsymbol{\tau}}{\sqrt{n}}.$$

By Condition 2.1(vi)

$$|\mathbf{E}\omega_i| \le 2L|\kappa_i|\frac{|\mathbf{c}_i\boldsymbol{\tau}|}{\sqrt{n}} \le \operatorname{const} n^{-2\varepsilon}. \tag{3.2.42}$$

Next consider $\mathbf{E}\omega_i^2$. We have
(3.2.43)

$$\begin{aligned}
\mathbf{E}\omega_i^2 &= \Big(2 - 2f(0)\frac{\mathbf{c}_i\mathbf{t}}{\sqrt{n}}\Big)^2\mathbf{P}\Big\{\xi_i + \frac{\mathbf{c}_i\mathbf{t}}{\sqrt{n}} > 0, \xi_i < 0\Big\} \\
&\quad + \Big(-2 - 2f(0)\frac{\mathbf{c}_i\mathbf{t}}{\sqrt{n}}\Big)^2\mathbf{P}\Big\{\xi_i + \frac{\mathbf{c}_i\mathbf{t}}{\sqrt{n}} < 0, \xi_i > 0\Big\} \\
&\quad + \Big(-2f(0)\frac{\mathbf{c}_i\mathbf{t}}{\sqrt{n}}\Big)^2 \\
&\qquad \times \Big(1 - \mathbf{P}\Big\{\xi_i + \frac{\mathbf{c}_i\mathbf{t}}{\sqrt{n}} > 0, \xi_i < 0\Big\} - \mathbf{P}\Big\{\xi_i + \frac{\mathbf{c}_i\mathbf{t}}{\sqrt{n}} < 0, \xi_i > 0\Big\}\Big) \\
&\le \Big(2 + 2f(0)\Big|\frac{\mathbf{c}_i\mathbf{t}}{\sqrt{n}}\Big|\Big)^2 \\
&\qquad \times \Big(\mathbf{P}\Big\{\xi_i + \frac{\mathbf{c}_i\mathbf{t}}{\sqrt{n}} > 0, \xi_i < 0\Big\} + \mathbf{P}\Big\{\xi_i + \frac{\mathbf{c}_i\mathbf{t}}{\sqrt{n}} < 0, \xi_i > 0\Big\}\Big) \\
&\quad + \big(2f(0)\big)^2\Big|\frac{\mathbf{c}_i\mathbf{t}}{\sqrt{n}}\Big|^2.
\end{aligned}$$

Notice that the events

$$\Big\{\xi_i + \frac{\mathbf{c}_i\mathbf{t}}{\sqrt{n}} > 0, \xi_i < 0\Big\} \quad \text{and} \quad \Big\{\xi_i + \frac{\mathbf{c}_i\mathbf{t}}{\sqrt{n}} < 0, \xi_i > 0\Big\}$$

are mutually exclusive; hence,

$$\mathbf{P}\Big\{\xi_i + \frac{\mathbf{c}_i\mathbf{t}}{\sqrt{n}} > 0, \xi_i < 0\Big\} + \mathbf{P}\Big\{\xi_i + \frac{\mathbf{c}_i\mathbf{t}}{\sqrt{n}} < 0, \xi_i > 0\Big\} = f(\kappa_i)\Big|\frac{\mathbf{c}_i\mathbf{t}}{\sqrt{n}}\Big|, \tag{3.2.44}$$

with κ_i as above. Indeed, if $-\mathbf{c}_i\mathbf{t}/\sqrt{n} > 0$ then only the second probability is positive and by Condition 2.1(ii)

$$\mathbf{P}\Big\{0 < \xi_i < -\frac{\mathbf{c}_i\mathbf{t}}{\sqrt{n}}\Big\} = F\Big(-\frac{\mathbf{c}_i\mathbf{t}}{\sqrt{n}}\Big) - F(0) = -f(\kappa_i)\frac{\mathbf{c}_i\mathbf{t}}{\sqrt{n}}.$$

In case $-\mathbf{c}_i\mathbf{t}/\sqrt{n} < 0$ this probability equals zero and

$$\mathbf{P}\Big\{-\frac{\mathbf{c}_i\mathbf{t}}{\sqrt{n}} < \xi_i < 0\Big\} = F(0) - F\Big(-\frac{\mathbf{c}_i\mathbf{t}}{\sqrt{n}}\Big) = f(\kappa_i)\Big|\frac{\mathbf{c}_i\mathbf{t}}{\sqrt{n}}\Big|.$$

On account of (3.2.44), assumption (i), and Condition 2.1(vi), we obtain from (3.2.43) for $|\mathbf{t}| \leq \mathrm{const}\, n^\gamma$

$$\mathbf{E}w_i^2 \leq \mathrm{const}\, n^{-2\varepsilon}.$$

Since the random variables ω_i are bounded, this implies

$$\mathbf{E}|\omega_i|^p \leq \mathrm{const}\, n^{-2\varepsilon} \tag{3.2.45}$$

for any $p \geq 1$. Recall that we write

$$\mathbf{E}A_3^m = n^{-m/2}\sum_{i_1}\cdots\sum_{i_m} d_{i_1}\dots d_{i_m}\mathbf{E}\omega_{i_1}\dots\omega_{i_m} \tag{3.2.46}$$

in the form

$$\mathbf{E}A_3^m = n^{-m/2}\{\Sigma_{(m)} + \Sigma_{(m-1)} + \cdots + \Sigma_{(p)} + \cdots + \Sigma_{(1)}\}, \tag{3.2.47}$$

where $\Sigma_{(p)}$ contains the terms from (3.2.46) with exactly p different indices among $i_1, \dots, i_m$. The terms in $\Sigma_{(p)}$ are uniformly bounded by $\mathrm{const}\, n^{-2\varepsilon p}$ and the number of terms in $\Sigma_{(p)}$ is $O(n^p)$. Therefore

$$|\Sigma_{(p)}| \leq \mathrm{const}\, n^{-m/n}n^p n^{-2\varepsilon p}. \tag{3.2.48}$$

It is seen from (3.2.48) that the main term in (3.2.47) is $\Sigma_{(m)}$, so that

$$\mathbf{E}A_3^m \leq \mathrm{const}\, n^{m(1-4\varepsilon)/2}. \tag{3.2.49}$$

Here $1 - 4\varepsilon < 0$ since $1 - 4\varepsilon = 4\gamma - 1$ and $\gamma < \frac{1}{4}$ by assumption.

In order to obtain the desired result by substituting (3.2.49) into (3.2.37) and (3.2.35) (using the inequality similar to (3.2.38)), the right-hand side of (3.2.49) multiplied by $n^{\gamma r}h^{-r}$ must tend to 0. With $h = n^{-1/4}$ this is achieved by taking $m > r\frac{4\gamma+1}{2(1-4\gamma)}$. This completes the proof. □

3.2.7. Asymptotic power of sign tests. Now we turn to the sign tests proposed in Chapter 2 for testing the hypotheses (2.1.1) or (2.1.2) in the model (2.1.5). The theorem on uniform linearity allows us to assess their power when the number of observations is large, provided the assumptions on the design matrix and the distribution of random errors made in this theorem are fulfilled. Consider local alternatives to the hypothesis $H_0\colon \boldsymbol{\theta} = \mathbf{0}$; namely, let

$$\boldsymbol{\theta} = \frac{\boldsymbol{\tau}}{\sqrt{n}}, \tag{3.2.50}$$

where $\boldsymbol{\tau} = (\tau_1, \dots, \tau_r)^T$, $|\boldsymbol{\tau}| < \text{const}$. Consider the sign tests (2.1.18) and (2.1.21) for such $\boldsymbol{\theta}$. The test statistic (2.1.14) in the form (2.2.7) is the squared norm of the vector

$$\frac{1}{\sqrt{n}}\mathbf{C}^T\mathbf{S}(\mathbf{X}) = \frac{1}{\sqrt{n}}\mathbf{C}^T\mathbf{S}\Big(\boldsymbol{\xi} + \frac{1}{\sqrt{n}}\mathbf{C}\boldsymbol{\tau}\Big). \tag{3.2.51}$$

By Theorem 3.2.4 on uniform linearity

$$\frac{1}{\sqrt{n}}\mathbf{C}^T\mathbf{S}\Big(\boldsymbol{\xi} + \frac{1}{\sqrt{n}}\mathbf{C}\boldsymbol{\tau}\Big) = \frac{1}{\sqrt{n}}\mathbf{C}^T\boldsymbol{\zeta} + \frac{1}{n}(\mathbf{C}^T\mathbf{C})\boldsymbol{\tau} + \mathbf{o}_p(1), \tag{3.2.52}$$

where $\boldsymbol{\zeta} = (\zeta_1, \dots, \zeta_n)^T$ with $\zeta_i = \operatorname{sign}\xi_i$.

Therefore under local alternatives (3.2.50), the test statistic (2.2.18) is asymptotically distributed as the squared norm of a normal vector

$$N(\boldsymbol{\Sigma}\boldsymbol{\tau}, \boldsymbol{\Sigma}).$$

By (3.2.51) and (3.2.52) the test statistic (2.1.21) is asymptotically equivalent to

$$n\Big(\frac{1}{\sqrt{n}}\mathbf{C}^T\boldsymbol{\zeta} + \frac{1}{n}(\mathbf{C}^T\mathbf{C})\boldsymbol{\tau}\Big)^T(\mathbf{C}^T\mathbf{C})^{-1}\Big(\frac{1}{\sqrt{n}}\mathbf{C}^T\boldsymbol{\zeta} + \frac{1}{n}(\mathbf{C}^T\mathbf{C})\boldsymbol{\tau}\Big).$$

Similarly to (2.3.8) this random variable has asymtotically the noncentral chi-squared distribution with r degrees of freedom and noncentrality parameter

$$\frac{1}{n}\boldsymbol{\tau}^T(\mathbf{C}^T\mathbf{C})^{-1}\boldsymbol{\tau} = \Big|\frac{1}{\sqrt{n}}\mathbf{C}\boldsymbol{\tau}\Big|^2, \quad \text{or} \quad \boldsymbol{\tau}^T\Sigma\boldsymbol{\tau}.$$

3.2.8. Sensitivity curve. We have already discussed in Chapter 1 the role of the influence function and the sensitivity curve as notions that describe the robustness properties of statistical procedures with respect to contamination. Making use of the asymptotic results of this chapter we will evaluate Tukey's sensitivity curve for sign estimators.

Suppose that along with observations x_i in the model (2.1.5), where $x_i = \mathbf{c}_i\boldsymbol{\theta}^0 + \xi_i$, $i = 1, \dots, n$, we have an additional observation $(\mathbf{b}, y)$, where $\mathbf{b} = (b_1, \dots, b_n)^T$ and $y = \mathbf{b}^T\boldsymbol{\theta}^0 + \eta$. Let $\widehat{\boldsymbol{\theta}}_n$ denote the sign estimator (3.1.1) and $\widehat{\boldsymbol{\theta}}^*_{n+1}$ the sign estimator obtained by the same rule from the extended sample including the point $(\mathbf{b}, y)$. Then the sensitivity of the sign estimator (3.1.1) as a function of $(\mathbf{b}, \eta)$ is

$$\mathrm{SC}_n(\mathbf{b}, \eta) = n\big(\widehat{\boldsymbol{\theta}}_{n+1} - \widehat{\boldsymbol{\theta}}_n\big). \tag{3.2.53}$$

We will show that under the conditions of Theorem 3.2.2

$$\mathrm{SC}_n(\mathbf{b}, \eta) = \frac{\operatorname{sign}\eta}{2f(0)}\Big(\frac{1}{n}\mathbf{C}^T\mathbf{C}\Big)^{-1}\mathbf{b} + \mathbf{o}_p(1) \qquad \text{as} \quad n \to \infty. \tag{3.2.54}$$

PROOF. Make a change of variables $\boldsymbol{\tau} = \sqrt{n}\,(\boldsymbol{\theta} - \boldsymbol{\theta}^0)$ and define $\widehat{\boldsymbol{\tau}}_n$ and $\widehat{\boldsymbol{\tau}}^*_{n+1}$ by

$$\widehat{\boldsymbol{\theta}}_n = \boldsymbol{\theta}^0 + \frac{1}{\sqrt{n}}\widehat{\boldsymbol{\tau}}_n, \qquad \widehat{\boldsymbol{\theta}}_{n+1} = \boldsymbol{\theta}^0 + \frac{1}{\sqrt{n}}\widehat{\boldsymbol{\tau}}^*_{n+1}.$$

Arguing as in the proof of Theorem 3.2.2 one shows that $\widehat{\boldsymbol{\tau}}^*_{n+1}$ is asymptotically equivalent to the solution of the following system of equations

$$\sum_{i=1}^{n} c_{i\alpha}\Big(\operatorname{sign}\xi_i - \frac{2f(0)}{\sqrt{n}}\mathbf{c}_i\boldsymbol{\tau}\Big) + b_\alpha \operatorname{sign}\Big(\eta - \frac{1}{\sqrt{n}}\mathbf{b}\boldsymbol{\tau}\Big) = 0, \quad \alpha = 1, \dots, r, \tag{3.2.55}$$

Since $\widehat{\boldsymbol{\tau}}^*_{n+1}$ is bounded in probability, $\operatorname{sign}(\eta - \frac{1}{\sqrt{n}}\mathbf{b}^T\widehat{\boldsymbol{\tau}}^*_{n+1}) \xrightarrow{\mathbf{P}} \operatorname{sign}\eta$ for $\eta \neq 0$. Using this fact and taking into account (3.2.29)–(3.2.31) we obtain

$$\widehat{\boldsymbol{\tau}}^*_{n+1} = \widehat{\boldsymbol{\tau}}_n + \frac{1}{\sqrt{n}}\frac{1}{2f(0)}\Big(\frac{1}{n}\mathbf{C}^T\mathbf{C}\Big)^{-1}\mathbf{b}\operatorname{sign}\eta. \tag{3.2.56}$$

Substituting (3.2.56) into (3.2.53) we obtain (3.2.54). □

3.3. Comparison of estimators

3.3.1. How estimators are compared. One can apply various methods to estimate the parameters of linear models. We will discuss here only those most popular. In this discussion we restrict ourselves to the methods which, like the sign-based ones, remain applicable when the distribution of random errors is unknown (in contrast to, for example, the method of maximum likelihood). We will be interested in their properties mostly from the point of view of comparison with the properties of the sign estimators.

The term "parameter estimation" comprises two different problems. The first one is obtaining approximate values of unknown parameters based on available observations, which is the problem of point estimation. The second problem is to construct a set in the parameter space, depending on the observations, which contains ("covers") the true parameter value with a preassigned probability. This specified value for the coverage probability is referred to as the confidence level, the set is called the confidence set (region, interval, etc.), and the problem itself is that of confidence estimation. When the confidence level is close to one, it is practically certain that the true value belongs to the confidence set. In general, the methodologies of obtaining point estimators and confidence sets are different, so that these two types of estimators are not directly related to each other. However, when point and interval estimation is carried out in the framework of some common approach, such as, for example, the sign-based one, then the confidence sets surround the point estimate, so that their size and shape enable one to assess the accuracy of the point estimator: the narrower the confidence set, the more accurate the estimator. Of course, this accuracy primarily depends on the statistical properties of observation errors, but it also depends on the method of estimation. The possibility of supplementing the point estimate with confidence sets is very important, because the point estimate as an approximate value is of little use for practice if its accuracy is unknown. If we can construct confidence sets with various coverage probabilities, their size and shape provide a clear idea of this accuracy.

It is not very difficult to construct asymptotic confidence sets for a large number of observations. For instance, they can be derived if one has any asymptotically normal estimator. A broad class of such estimators is provided by, say, Huber's M-estimators (see, for example, Huber [**44**] and Hampel et al. [**36**]). The class of M-estimators includes, in particular, the least squares and least absolute deviations estimators. Our sign estimators proposed in §3.1 are M-estimators as well.

It is more difficult to obtain exact confidence sets for finite samples. For linear models the solution to this problem is well known when the random errors are normally distributed. In §3.1 we showed how this can be done in a nonparametric setup by means of sign tests. The most closely related methods to the sign-based ones with regard to their foundations and potentialities are the rank methods and signed-rank methods (see, for example, Puri and Sen [**74**], Hettmansperger [**39**], Hollander and Wolfe [**42**]). These methods also yield confidence sets for unknown parameters, though they are computationally more complicated than sign-based methods. Theoretically valid results for the rank methods require the assumption that the random errors are identically distributed with a continuous distribution function, which is not necessary for the sign-based methods. For the signed-rank methods the additional assumption that the errors are symmetrically distributed is needed. Clearly, the stronger the required assumptions the more likely it is that some of them are violated in a practical application. In such a case we cannot be certain about the conclusions made under these assumptions. Therefore the less we assume about the properties of the errors, the more reliable our inference is.

The various methods mentioned above are applicable in different situations, so that neither of them is superior to the rest. But even when one can apply two methods of confidence estimation, they usually yield different and noncomparable results. This situation is similar to comparison of statistical tests, where neither of the tests is usually uniformly better than any other one.

The subject of point estimation as approximate evaluation of unknown parameters and their functions from observed data has a long history. As a statistical method, it goes back to Laplace and Gauss (see, for example, Stigler [**81**]). Hence it is no wonder that there is a great variety of methods of point estimation. We will compare the sign estimators with the most widely used least squares and least absolute deviations estimators as well as with rank estimators.

In order to compare two methods of estimation one has to compare the corresponding estimators as random variables. The accuracy of an estimator is characterized by the concentration of its distribution around the true parameter value. Unfortunately, as a rule two distributions are noncomparable with regard to their concentration. Hence for finite samples one usually cannot find the best estimation method.

Remarkably, the comparison of estimation methods becomes possible in an asymptotic setup when the number of observations is large. There are quite a number of papers and monographs on asymptotic theory of optimal estimation. For the advanced theory of asymptotic optimality and asymptotic comparison of points estimators the reader is referred to Lehmann [**61**], Chapter 6, or Ibragimov and Khas′minskii [**46**], Chapter 2. Our comparative study, however, will require no sophisticated theory, since we restrict ourselves to evaluation of asymptotic efficiency of sign estimators relative to some other estimators.

It is typical for all reasonable estimation methods that the estimators are asymptotically normal for an increasing number of observations n, or, more precisely, for an increasing amount of information, which depends also on the design of the experiment. In order to simplify the problem, we will assume that the design satisfies Conditions 2.1(vii, viii). Under these assumptions n is the only variable to control the asymptotic process. Thus we will study estimators of parameters in a linear model for $n \to \infty$.

Before proceeding to comparison of various estimators we will briefly discuss their properties. We state some basic facts about rank estimators in 3.3.2 and about least squares estimators and least absolute deviation estimators in 3.3.3. The properties of signed-rank estimators are similar to those of rank estimators. Moreover, for their validity the error distribution has to be symmetric, which is a restrictive condition. For those reasons we will not consider these estimators.

3.3.2. Rank estimation. Recall that ranks are assigned to the elements of a finite set by their ordering according to a certain rule. For example, a set of real numbers can be arranged in ascending order, so that the smallest number is assigned rank 1, the second smallest rank 2, and so on. Henceforth we will deal only with this ranking rule, though in some instances other orderings (and hence other ranks) may be appropriate. In theory, all the numbers are assumed to be different, so that the ranks are uniquely defined. In practice this is often not the case. Most frequently the tied (equal) elements are assigned the average of their ranks. For a more detailed treatment of ties see Lehmann [**60**] or Hollander and Wolfe [**42**].

For a set of real numbers $u_1, u_2, \ldots, u_n$, the ranks of its elements will be denoted by $R(u_1), R(u_2), \ldots, R(u_n)$. Note that the ranks remain unchanged under translations and scale transforms of the set $u_1, u_2, \ldots, u_n$, i.e., under transformations of this set into $u_1 + c, u_2 + c, \ldots, u_n + c$, $c \in \mathbb{R}$, or into $du_1, du_2, \ldots, du_n$, $d > 0$. These properties determine the applicability of rank methods in the analysis of linear models, which will be discussed later on.

Rank estimators and other rank procedures for linear models are based on the ranks of residuals. Before giving their formal definition it will be expedient to choose the form of the linear model (2.1.1) or (2.1.5) most appropriate for application of rank methods. We will specify the linear model as

$$x_i = \theta_0 + \sum_{\alpha=1}^{r} c_{i\alpha}\theta_\alpha + \varepsilon_i, \qquad i = 1, \ldots, n, \tag{3.3.1}$$

assuming that $\sum_{i=1}^{n} c_{i\alpha} = 0$ for each $\alpha = 1, \ldots, r$. For asymptotic analysis (as $n \to \infty$) we assume the matrix $\|c_{i\alpha}\|$ in (3.3.1) to satisfy Conditions 2.1(vii, viii). The residuals

$$x_i - \theta_0 - \sum_{\alpha=1}^{r} c_{i\alpha}\theta_\alpha, \qquad i = 1, \ldots, n,$$

in the model (3.3.1) will be regarded as functions of $\theta_0, \theta_1, \ldots, \theta_r$. Clearly, their ranks do not depend on θ_0 and are equal to $R(x_i - \sum_{\alpha=1}^{r} c_{i\alpha}\theta_\alpha)$, $i = 1, \ldots, n$. Therefore the intercept θ_0 cannot be estimated by means of ranks. This is an unavoidable loss inherent to rank procedures.

The model (3.3.1) is analyzed by means of linear rank statistics

$$\sum_{i=1}^{n} c_{i\alpha}\, a_n\Big(R\Big(x_i - \theta_0 - \sum_{\alpha=1}^{r} c_{i\alpha}\theta_\alpha\Big)\Big), \qquad \alpha = 1, \ldots, r,$$

where $a_n(1), a_n(2), \dots, a_n(n)$ is a sequence of numbers referred to as scores. Usually the scores $a_n(k)$, $k = 1, \dots, n$, are taken in the form

$$a_n(k) = \mathbf{E}\varphi(v_n^{(k)}), \qquad k = 1, \dots, n,$$

or

$$a_n(k) = \varphi\Big(\frac{k}{n+1}\Big), \qquad k = 1, \dots, n,$$

where $\varphi(\cdot)$ is a nondecreasing function on $[0,1]$ and $v_n^{(1)}, v_n^{(2)}, \dots, v_n^{(n)}$ denote the order statistics of a sample of size n from the uniform distribution on $[0,1]$.

Theoretically the choice of scores $a_n(k)$ should be made depending on the distribution of the random errors, because to each distribution there correspond the most siutable (optimal) scores. However, this cannot be done in practice since the error distribution is unknown. (Were it known, the rank methods would not be needed.) Frequently, the so-called normal scores, which are optimal for normally disributed errors, are used. Yet more common is the use of the Wilcoxon scores $a_n(k) = \frac{k}{n+1}$. In the latter case computations related to the corresponding rank statistics are particularly simple. The distributions of many rank statistics involving the Wilcoxon scores have been tabulated (see, for example, Lehmann [**60**] or Hollander and Wolfe [**42**]).

Rank estimators $\theta_1^*, \dots, \theta_r^*$ for the parameters $\theta_1, \dots, \theta_r$ can be defined, for example, as the solution of the extremal problem

$$\sum_{\alpha=1}^{r}\Big(\sum_{i=1}^{n} c_{i\alpha} a_n\big(R(x_i - c_i\theta)\big)\Big)^2 \Longrightarrow \min_{\theta_1,\dots,\theta_n}. \tag{3.3.2}$$

Note that the statistic in (3.3.2) can also be used for the construction of confidence sets for the true parameter values $\boldsymbol{\theta}^0 = (\theta_1^0, \dots, \theta_r^0)^T$ as it was done before by means of sign statistics.

We have already pointed out that the rank methods can be applied only for identically distributed errors. We will assume their common distribution function $F(\cdot)$ to satisfy the following conditions:

1. There exists an absolutely continuous density $f(x) = f(x)$ with derivative $f'(x)$.
2. The Fisher information
$$\int_{-\infty}^{\infty}\Big(\frac{f'(x)}{f(x)}\Big)^2 f(x)\,dx$$
is finite.
3. The scores are derived from a score generating the function φ by either
$$a_N(i) = \mathbf{E}\varphi(U_N^{(i)})$$
or
$$a_N(i) = \varphi\Big(\frac{i}{N+1}\Big),$$
where $U_N^{(1)}, U_N^{(2)}, \dots, U_N^{(N)}$ are order statistics of a sample of size N from the rectangular distribution on the unit interval $[0,1]$.
4. φ is nonconstant, nondecreasing and square integrable on $[0,1]$.

Now we can formulate the results on the asymptotic normality of the rank estimators in the model (3.3.1) as $n \to \infty$ as given in Heiler and Willers [**38**].

Under the above assumptions on the design, scores, error distribution, and Conditions 2.1(vii, viii), *the following results hold.*

(1) *The vector* $\boldsymbol{\theta}^*$ *of rank estimators is asymptotically distributed as*

$$N\Big(\boldsymbol{\theta}^0, \frac{A^2}{\gamma^2}(\mathbf{C}^T\mathbf{C})^{-1}\Big),$$

where $\boldsymbol{\theta}^0$ *is the true parameter value,*

$$A^2 = \int_0^1 \big(\varphi(u) - \overline{\varphi}\big)^2\,du, \qquad \overline{\varphi} = \int_0^1 \varphi(u)\,du, \qquad \gamma = \int_{-\infty}^{\infty} \varphi\big(F(x)\big)f'(x)\,dx.$$

(2) *All solutions of the problem* (3.3.2) *are asymptotically equivalent in the sense that for any two solutions* $\boldsymbol{\theta}'$ *and* $\boldsymbol{\theta}''$,

$$\sqrt{n}(\boldsymbol{\theta}' - \boldsymbol{\theta}'') \xrightarrow{\mathbf{P}} \mathbf{0} \qquad \text{as} \quad n \to \infty.$$

3.3.3. Least squares and least absolute deviations estimators. These are apparently the most widely known methods of estimation having a long history, especially the method of least squares. This method is often attributed to Gauss who, indeed, successfully applied it, though referred to it as a known one.

The least squares estimator (LSE) in the linear model is defined as the solution of the extremal problem

$$\sum_{i=1}^n \big(x_i - \mathbf{c}_i\boldsymbol{\theta}\big)^2 \Longrightarrow \min_{\boldsymbol{\theta}}. \tag{3.3.3}$$

The solution to be denoted by $\widehat{\boldsymbol{\theta}}$, is given by the explicit formula

$$\widehat{\boldsymbol{\theta}} = (\mathbf{C}^T\mathbf{C})^{-1}\mathbf{C}^T\mathbf{X}. \tag{3.3.4}$$

About the i.i.d. random errors it is usually assumed that

$$\mathbf{E}\xi_i = 0, \qquad \sigma^2 := \mathbf{E}\xi_i^2 < \infty.$$

Then under Condition 2.1(viii) on the design matrix, the estimators (3.3.4) are consistent and asymptotically normal $N\big(\boldsymbol{\theta}, \frac{\sigma^2}{n}\boldsymbol{\Sigma}^{-1}\big)$. This has been a well-known fact for so long that it is difficult to establish its authorship. For the properties of the LSE under more general conditions on the design matrix (see, for example, Eiker [**26**] and Anderson [**2**], Chapter 2).

The method of least squares is closely connected with Gaussian linear models, i.e., the linear models with normally distributed errors. In this case the LSE are at the same time the maximum likelihood estimates. For Gaussian models the method of least squares is in a certain sense optimal. It yields the best point estimators, allows for obtaining confidence sets and for testing linear hypotheses, and so on. We need not discuss this method in more detail, since it well exposed in statistical textbooks and monographs.

The method of least absolute deviations (LAD) has been known for as long as the method of least squares. In particular, it was advocated by Laplace. The LAD estimator is defined as the solution of the extremal problem

$$\sum_{i=1}^n \big|x_i - \mathbf{c}_i\boldsymbol{\theta}\big| \Longrightarrow \min_{\boldsymbol{\theta}}. \tag{3.3.5}$$

As we will see, the LAD estimators in the linear model are closely related to the sign estimators. The LAD estimators are the maximum likelihood estimators when the random errors have the bilateral exponential distribution, which is also known as the Laplace distribution. Its standardized density function is $p(x) = \frac{1}{2}e^{-|x|}$.

There is no explicit formula for the LAD estimator similar to (3.3.4). In general, the problem (3.3.5) is computationally more difficult than (3.3.3). The minimization in (3.3.5) is done either by linear programming methods or by iterative procedures using the convexity of the objective function (see, for example, Bloomfield and Steiger [**7**]).

From an asymptotic standpoint, the sign and LAD estimators are equivalent. Although the LAD estimators have been known for long time, their asymptotic properties were studied relatively recently. We quote here the theorem by Basset and Koenker [**4**], see also Pollard [**72**].

THEOREM (on asymptotic normality of the LAD estimators). *Assume that the common distribution function $F(x)$ of the i.i.d. errors in the model* (2.2.5) *has zero median and its density $f(\cdot)$ is continuous and positive at zero point. Suppose the matrix $\mathbf{C}$ satisfies Condition* 2.1(viii)*. Then, as $n \to \infty$, the least absolute deviation estimator $\widehat{\boldsymbol{\theta}}_n$ defined by* (3.3.5) *is asymptotically normal*

$$N\left(\mathbf{0}, \frac{1}{\big(2f(0)\big)^2}(\mathbf{C}^T\mathbf{C})^{-1}\right),$$

or

$$\sqrt{n}(\widehat{\boldsymbol{\theta}}_n - \boldsymbol{\theta}) \sim N\left(\mathbf{0}, \frac{1}{\big(2f(0)\big)^2}\boldsymbol{\Sigma}^{-1}\right).$$

It is seen that, except for the technical Condition 2.1(vii), the conditions and the conclusion of this theorem are the same as in Theorem 3.2.2 (though the latter tells us somewhat more about the asymptotic behavior of the sign estimators).

3.3.4. Asymptotic efficiency of sign estimators. The asymptotic results show that (under suitable regularity conditions) all the estimators considered above have asymptotically normal distributions with mean vectors equal to the true parameter vector and covariance matrices proportional to each other. The asymptotic covariance matrix of the sign estimator equals $\mathbf{J}^{-1}$, where $\mathbf{J}$ is the matrix of Fisher information about $\boldsymbol{\theta}$ contained in $\{\operatorname{sign}(x_i - \mathbf{c}_i\boldsymbol{\theta}),\ i = 1, \dots, n\}$. This assertion can be easily proved using the formula (2.2.11). The matrix $\mathbf{C}^T\mathbf{C}$ in this formula depends on the design of experiment, and the factor $\big(2f(0)\big)^2$ on the method of estimation.

The asymptotic covariance matrices of other estimators have a similar structure. They are proportional to the matrix $(\mathbf{C}^T\mathbf{C})^{-1}$, or (under Condition 2.1(viii)) to $\frac{1}{n}\boldsymbol{\Sigma}^{-1}$, with a scalar factor depending on the method of estimation. This proportionality of covariance matrices allows for the (large sample) comparison of the accuracy of various methods of estimation, provided they are applicable simultaneously. Indeed, out of two normal distributions with common center $\boldsymbol{\theta}^0$ the one with a smaller covariance matrix is more concentrated around $\boldsymbol{\theta}^0$. If the covariance matrices are proportional, this comparison is well defined. In this case the ratio of the scalar coefficients in the asymptotic covariance matrices of two estimators is called the asymptotic relative efficiency (ARE) of one estimator with respect to the other (see, for example, Lehmann [**61**], §6.2, or Bickel and Doksum [**5**], §4.4.B).

The comparison of estimation methods in terms of their asymptotic relative efficiency has a long tradition in the literature. The results of the comparison depend on the error distribution. Of course, two methods may be compared only when the error distribution allows for asymptotic results valid for both of them. Since the estimation methods discussed above differ by the areas where they are correctly applicable, their asymptotic efficiency comparison has a limited significance.

Since the sign estimators and the least absolute deviation estimators are asymptotically equivalent, all facts known for the LAD estimators turn out to hold for the sign estimators. Let us compare the sign estimators with the least squares estimators. If the errors are normally distributed $N(0,\sigma^2)$, the asymptotic efficiency of the sign (and LAD) estimators relative to the LSE equals $2/\pi$. (This means that the two methods result in estimators of equal accuracy if the LSE is applied to data with error variance larger by factor $\pi/2$ than that for the sign estimator.)

Now let the errors have the bilateral exponential distribution with variance σ^2. Its density is

$$\frac{1}{\sigma\sqrt{2}}\exp\Big\{\frac{-\sqrt{2}|x|}{\sigma}\Big\}, \qquad x \in \mathbb{R}^1.$$

Then $(\mathbf{C}^T\mathbf{C})^{-1}$ in the asymptotic covariance matrix of the sign estimator has the factor $\big(\frac{1}{2f(0)}\big)^2 = \frac{\sigma^2}{2}$. By comparison with the covariance matrix for the LSE stated in 3.3.3 we see that for this error distribution the sign estimator is twice as efficient as the LSE.

It is easy to see that the sign estimators can have arbitrarily large ARE with respect to the LSE. Let us compare them, for example, for Tukey's distribution, which is a mixture of the standard normal distribution and the normal distribution with variance τ^2 taken with weights $1-\varepsilon$ and ε, $\varepsilon > 0$. Its density is

$$(1-\varepsilon)\frac{1}{\sqrt{2\pi}}e^{-\frac{x^2}{2}} + \varepsilon\frac{1}{\tau\sqrt{2\pi}}e^{-\frac{x^2}{2\tau^2}}.$$

This is a convenient model for a "contaminated" set of normal errors containing some fraction of outliers (gross errors). This distribution has variance $\sigma^2 = (1-\varepsilon)+\varepsilon\tau^2$, and $(2f(0))^2 = \frac{2}{\pi}\big(1-\varepsilon+\frac{\varepsilon}{\tau}\big)$. In this case the ARE of the sign estimator with respect to the LSE is

$$\left(\frac{2f(0)}{\sigma}\right)^2 = \frac{2}{\pi}\frac{1-\varepsilon+\frac{\varepsilon}{\tau}}{1-\varepsilon+\varepsilon\tau^2}.$$

This ARE can become arbitrarily large for large τ.

CHAPTER 4

Testing Linear Hypotheses

4.1. Sign procedures for testing linear hypotheses

Hypotheses of the form

$$H\colon \mathbf{l} \in \mathbb{L}_1, \tag{4.1.1}$$

related to the model (2.1.1), where $\mathbb{L}_1 \subset \mathbb{L}$ is a given linear subspace, are referred to as linear hypotheses. In §4.3 we will consider some of the most common particular forms of linear hypotheses. Now we describe a general method of their testing. This method actually follows the scheme of testing linear hypotheses in the normal theory.

Let $\mathbb{L}_2$ be the orthogonal complement of $\mathbb{L}_1$ to the entire subspace $\mathbb{L}$, so that $\mathbb{L} = \mathbb{L}_1 \oplus \mathbb{L}_2$, $\mathbb{L}_1 \perp \mathbb{L}_2$. Let $r_1 = \dim \mathbb{L}_1$, $r_2 = \dim \mathbb{L}_2$, $r = r_1 + r_2 = \dim \mathbb{L}$. Let the vectors $(a_{1\alpha}, a_{2\alpha}, \dots, a_{n\alpha})^T$, $\alpha = 1, \dots, r_1$, form a basis of $\mathbb{L}_1$, and $(b_{1\beta}, b_{2\beta}, \dots, b_{n\beta})^T$, $\beta = 1, \dots, r_2$, a basis of $\mathbb{L}_2$. Combine these column vectors into $n \times r_1$ matrix $\mathbf{A}$ and $n \times r_2$ matrix $\mathbf{B}$, respectively,

$$\mathbf{A} = \|a_{i\alpha}, \quad i = 1, \dots, n; \quad \alpha = 1, \dots, r_1\|,$$
$$\mathbf{B} = \|b_{i\beta}, \quad i = 1, \dots, n; \quad \beta = 1, \dots, r_2\|.$$

The orthogonality of $\mathbb{L}_1$ and $\mathbb{L}_2$ implies

$$\mathbf{A}^T\mathbf{B} = \mathbf{0}. \tag{4.1.2}$$

In terms of matrices $\mathbf{A}$ and $\mathbf{B}$ the subspaces $\mathbb{L}_1$ and $\mathbb{L}_2$ can be represented as

$$\mathbb{L}_1 = \{\mathbf{l}\colon \mathbf{l} = \mathbf{A}\boldsymbol{\sigma},\ \boldsymbol{\sigma} \in \mathbb{R}^{r_1}\}, \qquad \mathbb{L}_2 = \{\mathbf{l}\colon \mathbf{l} = \mathbf{B}\boldsymbol{\tau},\ \boldsymbol{\tau} \in \mathbb{R}^{r_2}\}.$$

In this notation the model (2.1.1) takes the form

$$\mathbf{X} = \mathbf{A}\boldsymbol{\sigma} + \mathbf{B}\boldsymbol{\tau} + \boldsymbol{\xi}, \tag{4.1.3}$$

and the linear hypothesis (4.1.1) becomes

$$H\colon \boldsymbol{\tau} = \mathbf{0}. \tag{4.1.4}$$

The rule for testing linear hypotheses. We apply the following natural procedure for testing (4.1.4).

Step 1. Assuming $H\colon \boldsymbol{\tau} = \mathbf{0}$ is fulfilled, find a sign estimate for $\boldsymbol{\sigma}$:

$$\widehat{\boldsymbol{\sigma}}_n = \arg \min_{\boldsymbol{\sigma} \in \mathbb{R}^{r_1}} \sum_{\alpha=1}^{r_1} \Big(\sum_{i=1}^{n} a_{i\alpha} \operatorname{sign}(x_i - \mathbf{a}_i \boldsymbol{\sigma}) \Big)^2, \tag{4.1.5}$$

where $\mathbf{a}_i = (a_{i1}, \dots, a_{ir_1})$.

Step 2. Define

$$\widehat{x}_i = x_i - \sum_{\alpha=1}^{r_1} a_{i\alpha}\widehat{\sigma}_{n\alpha}, \qquad i = 1, \ldots, n. \tag{4.1.6}$$

Instead of $\widehat{\boldsymbol{\sigma}}_n$ one can use in (4.1.6) any other sign estimate for $\boldsymbol{\sigma}$, which would not affect the asymptotic properties of the rule (see §4.2).

Step 3. Apply the sign test (2.2.14) for testing (4.1.4) in the model

$$\mathbf{X} = \mathbf{B}\boldsymbol{\tau} + \boldsymbol{\xi} \tag{4.1.7}$$

to the vector $\widehat{\mathbf{X}} = (\widehat{x}_1, \ldots, \widehat{x}_n)^T$. In other words, the hypothesis (4.1.1) is rejected with (approximate) level ε if

$$\sum_{\beta=1}^{r_2}\Big(\sum_{i=1}^{n} b_{i\beta}\operatorname{sign}\widehat{x}_i\Big)^2 > q_{1-\varepsilon}, \tag{4.1.8}$$

where $q_{1-\varepsilon}$ is the $(1-\varepsilon)$-quantile of the random variable

$$\sum_{\beta=1}^{r_2}\Big(\sum_{i=1}^{n} b_{i\beta}\zeta_i\Big)^2. \tag{4.1.9}$$

The distribution of ζ_i, $i = 1, \ldots, n$, was defined in (2.3.1).

Instead of statistic (2.2.19), which was used in (4.1.8), one can use any other quadratic sign statisic, for example, (2.2.22). Then Step 3 will be as follows:

Step 3. Reject (4.1.1) with approximate level ε if

$$\mathbf{S}^T(\widehat{\mathbf{X}})\mathbf{B}\big(\mathbf{B}^T\mathbf{B}\big)^{-1}\mathbf{B}^T\mathbf{S}(\widehat{\mathbf{X}}) > \widetilde{q}_{1-\varepsilon}, \tag{4.1.10}$$

where $\widetilde{q}_{1-\varepsilon}$ is the $(1-\varepsilon)$-quantile of the random variable

$$\boldsymbol{\zeta}^T\mathbf{B}\big(\mathbf{B}^T\mathbf{B}\big)^{-1}\mathbf{B}^T\boldsymbol{\zeta}. \tag{4.1.11}$$

From the geometric viewpoint, the statistic in (4.1.10) is

$$\big|\operatorname{proj}_{\mathbb{L}_2}\mathbf{S}(\widehat{\mathbf{X}})\big|^2. \tag{4.1.12}$$

This shows that the structure of the sign procedure (4.1.5), (4.1.6), (4.1.10) is similar to that of the traditional normal theory rules for testing linear hypotheses, which are based on F-ratio. The statistic (4.1.12), up to a constant factor, is the nominator of the corresponding F-statistic with residuals replaced by their signs. We have already pointed out this similarity in §2.3.

The use of (4.1.10) instead of (4.1.8) may simplify the testing procedure, because (4.1.12) can be evaluated without explicit specification of the subspace $\mathbb{L}_2$ and its basis (i.e., matrix $\mathbf{B}$), since

$$\big|\operatorname{proj}_{\mathbb{L}_2}\mathbf{S}(\widehat{\mathbf{X}})\big|^2 = \big|\operatorname{proj}_{\mathbb{L}}\mathbf{S}(\widehat{\mathbf{X}})\big|^2 - \big|\operatorname{proj}_{\mathbb{L}_1}\mathbf{S}(\widehat{\mathbf{X}})\big|^2. \tag{4.1.13}$$

The subspaces $\mathbb{L}$ and $\mathbb{L}_1$ involved in the linear hypothesis are usually given in an explicit form, and the projections onto them can be easily evaluated.

Unfortunately, the above rules are not distribution free, because their actual level may differ from the nominal value ε. We will show, however, that the level tends to ε as $n \to \infty$ under some natural conditions on the matrices $\mathbf{A}$ and $\mathbf{B}$ and the error distribution. This approach was first proposed for rank tests by Hodges and Lehmann [**41**].

4.2. Asymptotic properties of sign tests for linear hypotheses

Here we study the asymptotic properties of the rules from §4.1 under local alternatives to the hypothesis (4.1.4). Namely, we will consider the alternatives with $\boldsymbol{\tau}$ in (4.1.3) restricted to the set

$$\{\boldsymbol{\tau}\colon |\boldsymbol{\tau}| < Cn^{-1/2+\gamma}\}, \tag{4.2.1}$$

as $n \to \infty$, where $C > 0$ is an arbitrary constant and $0 < \gamma < 1/4$.

We will assume the matrices $\mathbf{A}$ and $\mathbf{B}$ to satisfy the following analogues of Conditions 2.1(vii, viii):

CONDITION 4.2(i). $\max\limits_{1\le i\le n}\max\limits_{1\le \alpha\le r_1} |a_{i\alpha}| < K,\ \max\limits_{1\le i\le n}\max\limits_{1\le \beta\le r_2} |b_{i\beta}| < K.$

CONDITION 4.2(ii). $\dfrac{1}{n}\mathbf{A}^T\mathbf{A} \to \boldsymbol{\Sigma}_1 > \mathbf{0},\ \dfrac{1}{n}\mathbf{B}^T\mathbf{B} \to \boldsymbol{\Sigma}_2 > \mathbf{0}$ *as* $n \to \infty$.

The random errors will be assumed to satisfy the conditions of Theorem 3.2.4 (on uniform linearity), i.e., Conditions 2.1(i, ii, vi). Under these assumptions Theorem 3.2.4 in view of (4.1.2) implies for the model (4.1.3) that

$$\frac{1}{\sqrt{n}}\mathbf{A}^T\mathbf{S}\Big(\boldsymbol{\xi} + \frac{1}{\sqrt{n}}\mathbf{A}\boldsymbol{\sigma} + \frac{1}{\sqrt{n}}\mathbf{B}\boldsymbol{\tau}\Big) = \frac{1}{\sqrt{n}}\mathbf{A}^T\mathbf{S}(\boldsymbol{\xi}) + 2f(0)\frac{1}{n}(\mathbf{A}^T\mathbf{A})\boldsymbol{\sigma} + \omega_n, \tag{4.2.2}$$

where

$$\sup\{\,|\omega_n|\colon |\boldsymbol{\sigma}| < Cn^{\gamma},\ |\boldsymbol{\tau}| < Cn^{\gamma}\,\} = o_p(1).$$

Making use of (4.2.2) we prove the following theorem on the properties of the sign estimators under local alternatives.

THEOREM 4.2.1. *Assume that the random errors in the model* (4.1.3) *are i.i.d. with a common distribution function satisfying Conditions* 2.1(i, ii, vi). *Assume that the matrices* $\mathbf{A}$ *and* $\mathbf{B}$ *satisfy* (4.1.2) *and Conditions* 4.2(i, ii). *Then the sign estimator* (4.1.5) *under local alternatives* (4.2.1) *is representable as*

$$\widehat{\boldsymbol{\sigma}}_n = \boldsymbol{\sigma}^0 + \frac{1}{\sqrt{n}}\frac{1}{f(0)}\Big(\frac{1}{n}\mathbf{A}^T\mathbf{A}\Big)^{-1}\Big(\frac{1}{\sqrt{n}}\mathbf{A}^T\boldsymbol{\zeta}\Big) + \mathbf{o}_p(1),$$

where $\boldsymbol{\zeta} = (\zeta_1, \dots, \zeta_n)^T$ *with* $\zeta_i = \operatorname{sign}\xi_i$, $i = 1, \dots, n$. *This estimator is asymptotically normal with parameters*

$$\Big(\mathbf{0},\ \frac{1}{n}\frac{1}{\big(2f(0)\big)^2}\Big(\frac{1}{n}\mathbf{A}^T\mathbf{A}\Big)^{-1}\Big) \quad \text{or} \quad \Big(\mathbf{0},\ \frac{1}{n}\frac{1}{(2f(0))^2}\boldsymbol{\Sigma}_1^{-1}\Big).$$

PROOF. It has a similar structure to that of the proof of Theorem 3.2.2. In the problem (4.1.5) consider first the range $\{\boldsymbol{\sigma}\colon |\boldsymbol{\sigma}| \ge n^{-1/2+\gamma},\ \boldsymbol{\sigma} \in \mathbb{R}^{r_1}\}$. As in the proof of Theorem 3.2.2, we show that under local alternatives

$$\inf_{\boldsymbol{\sigma}\colon |\boldsymbol{\sigma}|\ge n^{-1/2+\gamma}} \frac{1}{n}\sum_{\alpha=1}^{r_1}\Big(\sum_{i=1}^{n} a_{i\alpha}\operatorname{sign}(x_i - \mathbf{a}_i\boldsymbol{\sigma})\Big)^2 \xrightarrow{\ \mathbf{P}\ } \infty. \tag{4.2.3}$$

Write $\boldsymbol{\sigma} \in \mathbb{R}^{r_1}$ in the form $\boldsymbol{\sigma} = |\boldsymbol{\sigma}|\mathbf{e}$, where $\mathbf{e} \in \mathbb{S}^{r_1-1} = \{\mathbf{e} \in \mathbb{R}^{r_1} : |\mathbf{e}| = 1\}$. Multiply the vector $\sum_{i=1}^{n} \mathbf{a}_i \operatorname{sign}(x_i - \mathbf{a}_i\boldsymbol{\sigma})$ by $\mathbf{e}$. For this scalar product we have by the Cauchy–Schwarz inequality

$$\frac{1}{n}\Big(\sum_{i=1}^{n}(\mathbf{a}_i\mathbf{e}) \operatorname{sign}(\xi_i + \mathbf{b}_i\boldsymbol{\tau} - (\mathbf{a}_i\mathbf{e})|\boldsymbol{\sigma}|)\Big)^2 \le \frac{1}{n}\Big(\sum_{i=1}^{n} \mathbf{a}_i \operatorname{sign}(x_i - \mathbf{a}_i\boldsymbol{\sigma})\Big)^2.$$

Put $d_i = \mathbf{a}_i\mathbf{e}$ and consider

$$Z_n := -\frac{1}{\sqrt{n}}\sum_{i=1}^{n} d_i \operatorname{sign}(\xi_i + \mathbf{b}_i\boldsymbol{\tau} - d_i|\boldsymbol{\sigma}|). \tag{4.2.4}$$

This is a nondecreasing function of $|\boldsymbol{\sigma}|$, so that for a fixed $\mathbf{e}$, $|\mathbf{e}| = 1$, it is minimized over $\{\boldsymbol{\sigma} : |\boldsymbol{\sigma}| \ge n^{-1/2+\gamma}\}$ by $|\boldsymbol{\sigma}| = n^{-1/2+\gamma}$. We will show that for this value of $\boldsymbol{\sigma}$, $Z_n \xrightarrow{\mathbf{P}} \infty$. Since Z_n has a bounded variance, it suffices to prove that $\mathbf{E}Z_n \to \infty$ under local alternatives. One has for $|\boldsymbol{\sigma}| = n^{-1/2+\gamma}$

$$\begin{aligned}\mathbf{E}Z_n &= -\frac{1}{\sqrt{n}}\sum_{i=1}^{n} d_i\big(1 - 2F(-\mathbf{b}_i\boldsymbol{\tau} + d_i n^{-1/2+\gamma})\big)\\ &= 2\frac{1}{\sqrt{n}}\sum_{i=1}^{n} d_i f(\kappa_i)(\mathbf{b}_i\boldsymbol{\tau} - d_i n^{-1/2+\gamma}),\end{aligned}$$

where κ_i is a point between 0 and $\mathbf{b}_i\boldsymbol{\tau} - d_i n^{-1/2+\gamma}$. Put $\omega_n := \max(|\kappa_1|, \dots, |\kappa_n|)$, then $\omega_n \to 0$. Notice that $\sum_{i=1}^{n} d_i\mathbf{b}_i\boldsymbol{\tau} = \mathbf{e}^T\big(\sum_{i=1}^{n}\mathbf{a}_i^T\mathbf{b}_i\big)\boldsymbol{\tau} = 0$ by (4.1.2). Hence we obtain

$$\mathbf{E}Z_n = \frac{2}{\sqrt{n}} f(0)n^{-1/2+\gamma}\sum_{i=1}^{n} d_i + 2\sum_{i=1}^{n}\big(f(\kappa_i) - f(0)\big)d_i(\mathbf{b}_i\boldsymbol{\tau} - d_i n^{-1/2+\gamma}). \tag{4.2.5}$$

The first term in (4.2.5) is equivalent to

$$2f(0)\mathbf{e}^T\Big(\frac{1}{n}\mathbf{A}^T\mathbf{A}\Big)\mathbf{e}\,n^{\gamma} \to \infty.$$

The second term is asymptotically negligible as compared to the first one, since (by Condition 2.1(vi)) it does not exceed in absolute value

$$\frac{2}{\sqrt{n}} n\,\omega_n\Big(\sum_{i=1}^{n}|d_i| + \sum_{i=1}^{n}|d_i|^2\Big)n^{-1/2+\gamma} \le \text{const}\;\omega_n\,n^{\gamma}.$$

Thus we obtain that $Z_n \xrightarrow{\mathbf{P}} \infty$ for any $\mathbf{e} \in \mathbb{S}^{r_1-1}$. Arguing as in the proof of Theorem 3.2.2, one shows also that $\sup_{\mathbf{e}\in\mathbb{S}^{r_1-1}} Z_n \xrightarrow{\mathbf{P}} \infty$.

Now consider the range $\{\boldsymbol{\sigma} : |\boldsymbol{\sigma}| < n^{-1/2+\gamma}\}$. Let $\mathbf{s} = \sqrt{n}\,\boldsymbol{\sigma}$, $|\mathbf{s}| < n^{\gamma}$. Consider the extremal problem (4.1.5) in this range, which now becomes

$$\frac{1}{\sqrt{n}}\Big|\mathbf{A}^T\mathbf{S}\Big(\mathbf{X} - \frac{1}{\sqrt{n}}\mathbf{A}\mathbf{s}\Big)\Big| \Longrightarrow \min_{\mathbf{s}\in\mathbb{R}^{r_1}, |\mathbf{s}|<n^{\gamma}}. \tag{4.2.6}$$

Denote the solution of (4.2.6) by $\widehat{\mathbf{s}}_n$. By (4.2.2) the solution of (4.2.6) under local alternatives is equivalent to the solution of

$$\Big|\frac{1}{\sqrt{n}}\mathbf{A}^T\mathbf{S}(\boldsymbol{\xi}) - 2f(0)\Big(\frac{1}{n}\mathbf{A}^T\mathbf{A}\Big)\mathbf{s}\Big| \Longrightarrow \min_{\mathbf{s}\in\mathbb{R}^{r_1},\, |\mathbf{s}|<n^{\gamma}}. \tag{4.2.7}$$

If we neglect the restriction $|\mathbf{s}| < n^\gamma$, the solution of (4.2.7) is

$$\mathbf{s}_n^* = \frac{1}{2f(0)}\Big(\frac{1}{n}\mathbf{A}^T\mathbf{A}\Big)^{-1}\Big(\frac{1}{\sqrt{n}}\mathbf{A}^T\mathbf{S}(\boldsymbol{\xi})\Big). \tag{4.2.8}$$

This vector is asymptotically normal with parameters $\mathbf{0}$ and $\frac{1}{(2f(0))^2}\boldsymbol{\Sigma}_1^{-1}$. Therefore for sufficiently large n one has $|\mathbf{s}_n^*| < n^\gamma$ with probability arbitrarily close to 1. Hence $\mathbf{s}_n^*$ is asymptotically equivalent to the solution of (4.2.6), $\widehat{\mathbf{s}}_n$. Moreover the minimal value of the objective function in (4.2.6) is bounded in probability. Thus when $\boldsymbol{\tau}$ satisfies (4.2.1), $\widehat{\mathbf{s}}_n$ is equivalent to the point of the global minimum of (4.1.5) in the sense that their difference tends to zero in probability. This completes the proof. □

Theorem 4.2.1 allows us to obtain the asymptotic power of various sign tests. It will be convenient to rescale the parameter $\boldsymbol{\tau}$ by putting $\mathbf{t} = \sqrt{n}\boldsymbol{\tau}$, so that the local alternatives (4.2.1) are represented by

$$\{\mathbf{t}\colon |\mathbf{t}| < Cn^\gamma,\ \mathbf{t} \in \mathbb{R}^{r_2}\}. \tag{4.2.9}$$

THEOREM 4.2.2. *Let the assumptions of Theorem* 4.2.1 *be satisfied. Then, under the local alternatives* (4.2.9), *the normalized test statistic in (4.1.8) converges in distribution to the squared norm of a normal vector,*

$$\frac{1}{n}\big|\mathbf{B}^T\mathbf{S}(\widehat{\mathbf{X}})\big|^2 \xrightarrow{\mathbf{P}} |\mathbf{Y}|^2, \qquad \text{where}\ \ \mathbf{Y} \sim N\big(2f(0)\,\boldsymbol{\Sigma}_2\mathbf{t},\ \boldsymbol{\Sigma}_2\big); \tag{4.2.10}$$

the test statistic in (4.1.10) *asymptotically has the noncentral chi-squared distribution,*

$$\mathbf{S}^T(\widehat{\mathbf{X}})\mathbf{B}\big(\mathbf{B}^T\mathbf{B}\big)^{-1}\mathbf{B}^T\mathbf{S}(\widehat{\mathbf{X}}) \xrightarrow{\mathbf{P}} \chi^2(\Delta, r), \tag{4.2.11}$$

with noncentrality parameter

$$\Delta = 2f(0)\sqrt{\mathbf{t}^T\boldsymbol{\Sigma}_2\mathbf{t}}. \tag{4.2.12}$$

Consequently, the actual levels of the tests (4.1.8), (4.1.10) *converge to the nominal ones as* $n \to \infty$.

PROOF. It suffices to note that (similarly to (4.2.2)) Theorem 3.2.4 under the present conditions implies

$$\frac{1}{\sqrt{n}}\mathbf{B}^T\mathbf{S}(\widehat{\mathbf{X}}) = \frac{1}{\sqrt{n}}\mathbf{B}^T\mathbf{S}(\boldsymbol{\xi}) + 2f(0)\Big(\frac{1}{n}\mathbf{B}^T\mathbf{B}\Big)\mathbf{t} + \omega_n,$$

where $\sup_{|\mathbf{t}|<Cn^\gamma}|\omega_n| \xrightarrow{\mathbf{P}} 0$, and that

$$\frac{1}{\sqrt{n}}\mathbf{B}^T\mathbf{S}(\boldsymbol{\xi}) \xrightarrow{\mathbf{d}} N(\mathbf{0}, \boldsymbol{\Sigma}).$$

□

Theorem 4.2.2 provides useful consequences for simple hypotheses as well. When applying the tests (2.2.18) and (2.2.21) for testing the simple hypothesis $H_0\colon \boldsymbol{\theta} = \boldsymbol{\theta}^0$ in the model (2.1.5) against local alternatives $H\colon \boldsymbol{\theta} = \boldsymbol{\theta}^0 + \frac{1}{\sqrt{n}}\mathbf{t}$, $|\mathbf{t}| < \mathrm{const}\ n^\gamma$, $0 < \gamma < 1/4$, we obtain similar results with $\mathbf{B}$ and $\boldsymbol{\Sigma}_2$ in (4.2.10)–(4.2.12) replaced by $\mathbf{C}$ and $\boldsymbol{\Sigma}$, respectively.

4.3. Examples

EXAMPLE 1. TESTING FOR HOMOGENEITY OF TWO SAMPLES WHICH MAY DIFFER BY LOCATION.

Let $x_1, \dots, x_m$ and $y_1, \dots, y_n$ be two independent samples. Write their elements as

$$x_i = a + \varepsilon_i, \quad i = 1, \dots, m, \qquad y_j = b + \xi_j, \quad j = 1, \dots, n, \tag{4.3.1}$$

where a and b are the medians of the distributions and $\varepsilon_1, \dots, \varepsilon_m, \xi_1, \dots, \xi_n$ are independent random variables such that

$$\mathbf{P}\{\varepsilon_i > 0\} = \mathbf{P}\{\varepsilon_i < 0\} = \mathbf{P}\{\xi_j > 0\} = \mathbf{P}\{\xi_j < 0\} = 1/2.$$

Then the hypothesis of interest can be written as

$$H\colon a = b. \tag{4.3.2}$$

The model (4.3.1) can be stated in the form (2.1.5) by putting

$$\mathbf{X} = (x_1, \dots, x_m, y_1, \dots, y_n)^T, \qquad \mathbb{L} = \mathbb{L}\{(\mathbf{1}, \mathbf{0}), (\mathbf{0}, \mathbf{1})\}, \tag{4.3.3}$$

where

$$(\mathbf{1}, \mathbf{0})^T = (\underbrace{1, \dots, 1}_{m}, \underbrace{0, \dots, 0}_{n}), \qquad (\mathbf{0}, \mathbf{1})^T = (\underbrace{0, \dots, 0}_{m}, \underbrace{1, \dots, 1}_{n})$$

The linear subspace $\mathbb{L}_1$ involved in the formulation of the linear hypothesis (4.1.1) is spanned by the vector $(\mathbf{1}, \mathbf{1})^T$. It is easily seen that the subspace $\mathbb{L}_2$, the orthogonal complement of $\mathbb{L}_1$ to $\mathbb{L}$, is spanned, for example, by the vector $m^{-1}(\mathbf{1}, \mathbf{0}) - n^{-1}(\mathbf{0}, \mathbf{1})$. The model (4.3.1) is represented in the from (4.1.3) as

$$x_i = \sigma + \frac{\tau}{m} + \varepsilon_i, \quad i = 1, \dots, m, \qquad y_j = \sigma - \frac{\tau}{n} + \xi_j, \quad j = 1, \dots, n, \tag{4.3.4}$$

and the hypothesis (4.3.2) in the form (4.1.4) becomes

$$H\colon \tau = 0. \tag{4.3.5}$$

Let us test the hypothesis (4.3.5) according to the procedure from §4.1.

Step 1. Estimate σ assuming $\tau = 0$. The sign estimate for σ is

$$\widehat{\sigma} = \operatorname{med}(x_1, \dots, x_m, y_1, \dots, y_n). \tag{4.3.6}$$

Step 2. Align the original observations (4.3.1) as

$$\widehat{x}_i = x_i - \widehat{\sigma}, \quad i = 1, \dots, m, \qquad \widehat{y}_j = y_j - \widehat{\sigma}, \quad j = 1, \dots, n. \tag{4.3.7}$$

Step 3. Apply the sign test for the hypothesis (4.3.5) in the model

$$\widehat{x}_i = \frac{\tau}{m} + \varepsilon_i, \qquad i = 1, \dots, m,$$
$$\widehat{y}_j = -\frac{\tau}{n} + \xi_j, \quad j = 1, \dots, n,$$

to the aligned observations (4.3.7) treated as independent variables.

The resulting test statistic is

$$\frac{mn}{m+n}\left(\frac{1}{m}\sum_{i=1}^{m}\operatorname{sign}\widehat{x}_i - \frac{1}{n}\sum_{j=1}^{n}\operatorname{sign}\widehat{y}_j\right)^2. \tag{4.3.8}$$

Thus the rule for testing (4.3.2) in the model (4.3.1) is as follows: reject the hypothesis of homogeneity of the two samples against two-sided shift alternatives with approximate level 2α when

$$\sqrt{\frac{mn}{m+n}}\left|\frac{1}{m}\sum_{i=1}^{m}\operatorname{sign}\widehat{x}_i - \frac{1}{n}\sum_{j=1}^{n}\operatorname{sign}\widehat{y}_j\right| > z_\alpha, \tag{4.3.9}$$

where z_α is the $(1-\alpha)$-quantile of the distribution of

$$\sqrt{\frac{mn}{m+n}}\left(\frac{1}{m}\sum_{i=1}^{m}\zeta_i - \frac{1}{n}\sum_{j=1}^{n}\zeta_j'\right) \tag{4.3.10}$$

with i.i.d. variables $\zeta_1, \dots, \zeta_m, \zeta_1', \dots, \zeta_n'$ such that

$$\mathbf{P}\{\zeta_i = 1\} = \mathbf{P}\{\zeta_i = -1\} = 1/2.$$

For large m and n, (4.3.10) is asymptotically normal $N(0,1)$. The test (4.3.9) can be modified for testing H against one-sided alternatives.

The statistic (4.3.8) can also be derived according to (4.1.13). Let

$$\mathbf{S}(\widehat{\mathbf{X}}) = (\operatorname{sign}\widehat{x}_1, \dots, \operatorname{sign}\widehat{x}_m, \operatorname{sign}\widehat{y}_1, \dots, \operatorname{sign}\widehat{y}_n)^T.$$

It is easily seen that

$$\left|\operatorname{proj}_{\mathbb{L}}\mathbf{S}(\widehat{\mathbf{X}})\right|^2 = \frac{1}{m}\left(\sum_{i=1}^{m}\operatorname{sign}\widehat{x}_i\right)^2 + \frac{1}{n}\left(\sum_{j=1}^{n}\operatorname{sign}\widehat{y}_j\right)^2,$$
$$\left|\operatorname{proj}_{\mathbb{L}_1}\mathbf{S}(\widehat{\mathbf{X}})\right|^2 = \frac{1}{m+n}\left(\sum_{i=1}^{m}\operatorname{sign}\widehat{x}_i + \sum_{j=1}^{n}\operatorname{sign}\widehat{y}_j\right)^2.$$

Now one easily verifies that (4.1.13) takes the form (4.3.8).

The structure of the statistic (4.3.8) is similar to that of some well-known statistics for testing (4.3.2), for example, Student's ratio or Wilcoxon's rank sum statistic.

Student's test is applicable for testing (4.3.2) when the random errors in (4.3.1) are identically distributed and their common distribution is normal. The nominator of Student's ratio is

$$|\bar{x} - \bar{y}| = \left|\frac{1}{m}\sum_{i=1}^{m}\widehat{x}_i - \frac{1}{n}\sum_{j=1}^{n}\widehat{y}_j\right|.$$

It is seen that (4.3.8) has a similar form with $\widehat{x}_i$, $\widehat{y}_j$ replaced by their signs.

Wilcoxon's rank sum test can be applied for testing (4.3.2) when the errors in (4.3.1) are identically distributed with a continuous distribution function. The traditional form of the Wilcoxon statistic is

$$\left|\sum_{j=1}^{n} \operatorname{rank} y_j - \frac{n(m+n+1)}{2}\right|,$$

where $\operatorname{rank} y_j$ denotes the rank of y_j among all $m+n$ observations in (4.3.1). It is easily seen that this statistic differs only by a constant factor from the statistic

$$\left|\frac{1}{n}\sum_{j=1}^{n} \operatorname{rank} y_j - \frac{1}{m}\sum_{i=1}^{m} \operatorname{rank} x_i\right|.$$

This statistic has the same structure as (4.3.8), with variables (4.3.7) replaced by their ranks rather than signs. (Note that the ranks of $\widehat{x}_i$, $\widehat{y}_j$ among variables (4.3.7) coincide with the ranks of $x_i\, y_j$ among variables (4.3.1)).

EXAMPLE 2. TESTING THE HYPOTHESIS THAT TWO REGRESSION LINES ARE PARALLEL.

Suppose we have two sets of data each of which follows a simple linear regression model. Our aim is to test the hypothesis that the regression lines have equal slopes. Consider a simplified setting where both regression lines pass through the origin. In this case the required estimates can be obtained in an explicit form.

Let the observations x_i, y_j have the form

$$\begin{aligned} x_i &= as_i + \varepsilon_i, \quad i = 1, \dots, m, \\ y_j &= bt_j + \xi_j, \quad j = 1, \dots, n, \end{aligned} \tag{4.3.11}$$

where a and b are unknown parameters, $s_1, \dots, s_m$, $t_1, \dots t_n$ are known constants, and $\varepsilon_1, \dots, \varepsilon_m$, $\xi_1, \dots, \xi_n$ are independent random variables with zero medians. The hypothesis to be tested is

$$H\colon a = b. \tag{4.3.12}$$

The model (4.3.11) can be written as a linear model (2.1.5) by putting

$$\mathbf{X} = (x_1, \dots, x_m, y_1, \dots, y_n)^T, \qquad \mathbb{L} = \mathbb{L}\{(\mathbf{s}, \mathbf{0})^T, (\mathbf{0}, \mathbf{t})^T\},$$

where

$$(\mathbf{s}, \mathbf{0})^T = (s_1, \dots, s_m, \underbrace{0, \dots, 0}_{n}), \qquad (\mathbf{0}, \mathbf{t})^T = (\underbrace{0, \dots, 0}_{m}, t_1, \dots, t_n).$$

The linear subspace $\mathbb{L}_1$ in (4.1.1) is spanned by the vector

$$(\mathbf{s}, \mathbf{t})^T = (s_1, \dots, s_m, t_1, \dots, t_n)^T.$$

The subspace $\mathbb{L}_2$, which is the orthogonal complement of $\mathbb{L}_1$ to the entire $\mathbb{L}$, is spanned by the vector

$$|\mathbf{t}|^2(\mathbf{s}, \mathbf{0})^T - |\mathbf{s}|^2(\mathbf{0}, \mathbf{t})^T.$$

In the parametrization generated by these subspaces, the model (4.3.11) takes the form

$$\begin{aligned} x_i &= \sigma s_i + \tau|\mathbf{t}|^2 s_i + \varepsilon_i, \quad i = 1, \dots, m, \\ y_j &= \sigma t_j - \tau|\mathbf{s}|^2 t_j + \xi_j, \quad j = 1, \dots, n. \end{aligned} \tag{4.3.13}$$

Accordingly, the hypothesis (4.3.12) becomes

$$H: \tau = 0. \tag{4.3.14}$$

Let us apply the procedure from §4.1 to its testing.

Step 1. Estimate σ assuming $\tau = 0$. By (1.2.12), the sign estimate for σ is the median of the discrete probability distribution with probability masses proportional to $|s_i|$ and $|t_j|$ at the points x_i/s_i, $i = 1, \dots, m$, and y_j/t_j, $j = 1, \dots, n$, respectively.

Step 2. Align the observations:

$$\begin{aligned} \widehat{x}_i &= x_i - \widehat{\sigma} s_i, \quad i = 1, \dots, m, \\ \widehat{y}_j &= y_j - \widehat{\sigma} t_j, \quad j = 1, \dots, n. \end{aligned} \tag{4.3.15}$$

Step 3. Test the hypothesis (4.3.14) using the observations (4.3.15) as if they were independent and subject to the model

$$\begin{aligned} \widehat{x}_i &= \tau |\mathbf{t}|^2 s_i + \varepsilon_i, \qquad i = 1, \dots, m, \\ \widehat{y}_j &= -\tau |\mathbf{s}|^2 t_j + \xi_j, \quad j = 1, \dots, n. \end{aligned}$$

According to (2.1.19) or (2.1.21) the sign test statistic has the form

$$\frac{1}{|\mathbf{s}|^2} \sum_{i=1}^m s_i \operatorname{sign} \widehat{x}_i - \frac{1}{|\mathbf{t}|^2} \sum_{j=1}^n t_j \operatorname{sign} \widehat{y}_j.$$

Thus the rule for testing the hypothesis (4.3.12) can be formulated as follows: reject the hypothesis (4.3.12) in the model (4.3.11) with approximate level 2α when

$$\left| \frac{1}{|\mathbf{s}|^2} \sum_{i=1}^m s_i \operatorname{sign} \widehat{x}_i - \frac{1}{|\mathbf{t}|^2} \sum_{j=1}^n t_j \operatorname{sign} \widehat{y}_j \right| > z_\alpha,$$

where z_α is the $1-\alpha$-quantile of the distribution of

$$\frac{1}{|\mathbf{s}|^2} \sum_{i=1}^m s_i \zeta_i - \frac{1}{|\mathbf{t}|^2} \sum_{j=1}^n t_j \zeta_j'$$

with $\zeta_1, \dots, \zeta_m, \zeta_1', \dots, \zeta_n'$ being independent random variables taking the values ± 1 with probabilities $1/2$.

4.4. Testing linear hypotheses in one- and two-way layout problems

1. One-way analysis (t samples which may differ by location). Let us have t samples x_{ij}, $i = 1, \dots, n_j$; $j = 1, \dots, t$, of size $n_1, n_2, \dots, n_t$. We will consider the traditional model

$$x_{ij} = \mu + \frac{\theta_j}{n_j} + \varepsilon_{ij}, \tag{4.4.1}$$

where μ, $\theta_1, \dots, \theta_t$ are unknown parameters such that

$$\sum_{j=1}^t \theta_j = 0. \tag{4.4.2}$$

We suppose that the i.i.d. errors ε_{ij} satisfy the basic assumption

$$\mathbf{P}\{\varepsilon_{ij} < 0\} = \mathbf{P}\{\varepsilon_{ij} > 0\} = 1/2.$$

The hypothesis that the t samples in the model (4.4.1), (4.4.2) are homogeneous is formulated as

$$H\colon \theta_1 = \cdots = \theta_t = 0. \tag{4.4.3}$$

Recall the geometric form of the model (4.4.1), (4.4.2), which will be useful in the sequel.

Write down the observations x_{ij}, $i = 1, \dots, n_j$, $j = 1, \dots, t$, as a vector of dimension $N = n_1 + \cdots + n_t$,

$$\mathbf{X} = (x_{11}, \dots, x_{n_1 1}, \dots, x_{1t}, \dots, x_{n_t t})^T.$$

Define the N-dimensional vectors $\mathbf{e}_0, \mathbf{e}_1, \dots, \mathbf{e}_t$ by

$$\begin{aligned} \mathbf{e}_0 &= (1, \dots, 1)^T, \\ \mathbf{e}_1 &= (\underbrace{1, \dots, 1}_{n_1}, 0, \dots, 0)^T, \\ &\dots\dots\dots\dots\dots\dots\dots\dots, \\ \mathbf{e}_t &= (0, \dots, \underbrace{1, \dots, 1}_{n_t})^T. \end{aligned}$$

Here the subspaces $\mathbb{L}_1$ and $\mathbb{L}_2$ discussed in §4.1 are as follows: $\mathbb{L}_1$ is the one-dimensional subspace spanned by the vector $\mathbf{e}_0$, and

$$\mathbb{L}_2 = \left\{ \mathbf{l}\colon \mathbf{l} = \sum_{j=1}^{t} \frac{\alpha_j}{n_j} \mathbf{e}_j, \quad \text{with } \sum_{j=1}^{t} \alpha_j = 0 \right\}.$$

It is easily verified that under the condition (4.4.2) the orthogonality conditions (4.1.2) are fulfilled. Let us apply the procedure from §4.2 to testing the linear hypothesis (4.4.3).

Step 1. We proceed as if the hypothesis H were true. Then the model involves only one parameter μ. The sign estimator $\widehat{\mu}$ for this parameter is the median of the combined sample,

$$\widehat{\mu} = \operatorname{med}\bigl(x_{11}, x_{21}, \dots, x_{n_1 1}, x_{12}, \dots, x_{n_2 2}, \dots, x_{1t}, \dots, x_{n_t t}\bigr).$$

Step 2. Align the observations by putting

$$\widehat{x}_{ij} = x_{ij} - \widehat{\mu}.$$

Step 3. Apply a sign test for testing the hypothesis (4.4.4) in the model (4.4.5), (4.4.2) with $\mu = 0$ to the variables $\widehat{x}_{ij}$ as if they were independent observations.

In order to specify this step, we have to select a definite sign test. The two options, which were discussed in §4.1, are the tests (2.1.18) and (2.1.21). We begin with the test of type (2.1.18). Let us neglect for a while the constraint (4.4.2) and consider t samples

$$y_{ij} = \frac{\theta_j}{n_j} + \varepsilon_{ij}, \qquad i = 1, \dots, n_j,$$

with unresticted parameters $\theta_1, \dots, \theta_t$. Proceeding as in §2.2, we obtain that under Conditions 2.1(i–iii) the likelihood function of the sign statistic

$$\mathbf{S}(\mathbf{Y}) = \big(\operatorname{sign} y_{ij},\ i = 1, \dots, n_j,\ j = 1, \dots, t\big)$$

is given by the following asymptotic formula:

$$\begin{aligned}(4.4.4)\qquad &\left(\frac{1}{2}\right)^N \Bigg\{ 1 + (2f(0)) \sum_{j=1}^{t} \frac{\theta_j}{n_j} \sum_{i=1}^{n_j} \operatorname{sign} y_{ij} \\ &+ \frac{1}{2}(2f(0))^2 \left[\left(\sum_{j=1}^{t} \frac{\theta_j}{n_j} \sum_{i=1}^{n_j} \operatorname{sign} y_{ij} \right)^2 - \sum_{j=1}^{t} \frac{\theta_j^2}{n_j} \right] + o(|\boldsymbol{\theta}|^2) \Bigg\},\end{aligned}$$

where $N = n_1 + \cdots + n_t$. Then the statistic (2.2.18) of the maximum mean curvature sign test becomes

$$(4.4.5)\qquad \sum_{j=1}^{t} \left(\frac{1}{n_j} \sum_{i=1}^{n_j} \operatorname{sign} y_{ij} \right)^2.$$

However, we are interested in the mean curvature of the power function at $\theta = 0$ under the constraint (4.4.2). Notice that this mean curvature is the sum of the mean curvatures of the power function restricted to the subspace specified by (4.4.2) and to its orthogonal complement. The latter is the one-dimensional subspace spanned by the vector $(t^{-1/2}, \dots, t^{-1/2})^T$. Consider the likelihood (4.4.5) along this direction by putting $\theta_j = u/\sqrt{t}$, $j = 1, \dots, m$; $u \in \mathbb{R}^1$. Then the test statistic of the sign test of maximum mean curvature (determined by the coefficient of u^2) is

$$(4.4.6)\qquad \frac{1}{t} \left(\sum_{j=1}^{t} \frac{1}{n_j} \sum_{i=1}^{n_j} \operatorname{sign} y_{ij} \right)^2.$$

Now the test statistic, to be used in Step 3, of the sign test (2.2.18) for the hypothesis (4.4.3) in the model (4.4.1), (4.4.2) given that $\mu = 0$ is the difference of (4.4.5) and (4.4.6). As a result, the testing rule for the hypothesis (4.4.3) is as follows.

Reject the hypothesis H as in (4.4.3) with approximate level α, $0 < \alpha < 1$, when

$$\sum_{j=1}^{t} \left(\frac{1}{n_j} \sum_{i=1}^{n_j} \operatorname{sign} \widehat{x}_{ij} \right)^2 - \frac{1}{t} \left(\sum_{j=1}^{t} \frac{1}{n_j} \sum_{i=1}^{n_j} \operatorname{sign} \widehat{x}_{ij} \right)^2 > q_{1-\alpha},$$

where $q_{1-\alpha}$ is the $(1-\alpha)$-quantile of the distribution of

$$(4.4.7)\qquad \sum_{j=1}^{t} \left(\frac{1}{n_j} \sum_{i=1}^{n_j} \zeta_{ij} \right)^2 - \frac{1}{t} \left(\sum_{j=1}^{t} \frac{1}{n_j} \sum_{i=1}^{n_j} \zeta_{ij} \right)^2.$$

Here ζ_{ij}, $i = 1, \dots, n_j$, $j = 1, \dots, t$, are independent random variables taking the values ± 1 with probabilities $1/2$.

Consider now the other option, which is to use at the third step the sign test (2.2.21). According to (4.1.12), its test statistic is the squared norm of the projection of the vector

$$\big(\operatorname{sign} \widehat{x}_{ij},\ i = 1, \dots, n_j;\ j = 1, \dots, t\big)^T$$

onto the subspace $\mathbb{L}_2$. This statistic is easily found to have the form

$$\sum_{j=1}^{t} n_j \Big(n_j^{-1} \sum_{i=1}^{n_j} \operatorname{sign} \widehat{x}_{ij} - \Big(\sum_{j=1}^{t} n_j \Big)^{-1} \sum_{j=1}^{t} \sum_{i=1}^{n_j} \operatorname{sign} \widehat{x}_{ij} \Big)^2 .$$

Note that by the definition of $\widehat{\mu}$ as the median of the set of observations x_{ij},

$$\sum_{j=1}^{t} \sum_{i=1}^{n_j} \operatorname{sign} \widehat{x}_{ij} = 0.$$

Therefore the testing rule for the hypothesis (4.4.3) in the model (4.4.1), (4.4.2) by using the test (2.1.21) can be formulated as follows:

Reject the hypothesis H as in (4.4.3) with approximate level α, $0 < \alpha < 1$, when

$$\sum_{j=1}^{t} n_j^{-1} \Big(\sum_{i=1}^{n_j} \operatorname{sign} \widehat{x}_{ij} \Big)^2 > q_{1-\alpha}, \tag{4.4.8}$$

where $q_{1-\alpha}$ is the $(1-\alpha)$-quantile of the distribution of

$$\sum_{j=1}^{t} n_j \Big(\frac{1}{n_j} \sum_{i=1}^{n_j} \zeta_{ij} - \sum_{j=1}^{t} \sum_{i=1}^{n_j} \zeta_{ij} \Big/ \sum_{j=1}^{t} n_j \Big)^2 \tag{4.4.9}$$

with ζ_{ij} as described above. For large $n_1, \dots, n_t$ one can take $q_{1-\alpha}$ to be the quantile of the chi-squared distribution with $\dim \mathbb{L}_2 = (t-1)$ degrees of freedom, which is the limiting distribution of statistic (4.4.8) under H as $n_1, \dots, n_t \to \infty$.

2. Two-way analysis. Consider the additive model of factor effects and the design with an equal number of observations for each combination of factor levels (m observations per cell). We dealt with this scheme in §2.3. We will use the notation introduced therein. Usually the analysis of a two-way layout starts with testing the hypothesis of no effect of one of the factors (no treatment effect), i.e., the hypothesis

$$H\colon \beta_1 = \dots = \beta_t = 0. \tag{4.4.10}$$

According to the scheme of §4.1, the procedure for testing of (4.4.10) begins by estimating the other parameters to use these estimates for the alignment of observations. Clearly, the sign estimate for $\mu + \alpha_i$, $i = 1, \dots, r$, is the median of the observations in the ith row, i.e.,

$$(\mu + \alpha_i)^* = \operatorname{med}(x_{i11}, \dots, x_{i1m}, x_{i21}, \dots, x_{itm}).$$

Hence $\widehat{x}_{ijk} = x_{ijk} - (\mu + \alpha_i)^*$ for all i, j, k.

We will restrict ourselves to the test (2.1.17) at the third step of the testing procedure. We pointed out in §4.1 that this statistic (up to a constant factor) has the same form as the nominator of the F-statistic for testing the hypothesis (4.4.10) in the Gaussian setup with observations replaced by their signs. As applied to observations (2.4.1), the F-ratio for testing (4.4.10) equals

$$\frac{trm - r - t - 1}{t-1} \frac{mr \sum_{j=1}^{t} (x_{\cdot j \cdot} - x_{\dots})^2}{\sum_{i=1}^{r} \sum_{j=1}^{t} \sum_{k=1}^{m} (x_{ijk} - x_{i\cdot\cdot} - x_{\cdot j \cdot} + x_{\dots})^2}, \tag{4.4.11}$$

where a dot, as usual, denotes the averaging with respect to the corresponding subscript (see, for example, Bickel and Doksum [**5**], 7.3.B). Thus it remains to replace the observations x_{ijk} by their signs $\operatorname{sign} x_{ijk}$. Put

$$z_{ij} = \frac{1}{\sqrt{m}} \sum_{k=1}^{m} \operatorname{sign} \widehat{x}_{ijk}.$$

Now the testing rule for the hypothesis H as in (4.4.10) is as follows.

Reject H with approximate level α, $0 < \alpha < 1$, when

$$r \sum_{j=1}^{t} \left(z_{\cdot j} - z_{\cdot\cdot}\right)^2 > q_{1-\alpha}, \tag{4.4.12}$$

where $q_{1-\alpha}$ is the $(1-\alpha)$-quantile of the distribution of

$$r\, m \sum_{j=1}^{t} \left(\zeta_{\cdot j \cdot} - \zeta_{\cdot\cdot\cdot}\right)^2. \tag{4.4.13}$$

As usual, ζ_{ijk} are i.i.d. with $\mathbf{P}\{\zeta_{ijk} = 1\} = \mathbf{P}\{\zeta_{ijk} = -1\} = 1/2$.

The random variable (4.4.13) has asymptotically, as $m \to \infty$, the $\chi^2(t-1)$ distribution. Hence for large m one can take $q_{1-\alpha}$ to be the $(1-\alpha)$-quantile of the chi-squared distribution with $(t-1)$ degrees of freedom.

4.5. Computation of critical values in testing linear hypotheses

Here we discuss the computation of critical values for test statistics (4.4.9) and (4.4.13) derived in the previous section. Short tables of these critical values are given at the end of this section.

One-way layout problem. Consider the computation of critical values of the statistic (4.4.9).

Let B_j, $j = 1, \ldots, t$, denote independent random variables with binomial distributions $\mathrm{Bi}(n_j, 1/2)$. Note that $\frac{1}{2}(\zeta_{ij}+1)$ takes the values 1 and 0 with probabilities $\frac{1}{2}$. Hence $\sum_{i=1}^{n_j} \zeta_{ij} = 2B_j + n_j$. Consequently, (4.4.9) is equal to

$$\zeta_t = 4 \sum_{j=1}^{t} n_j \left(\frac{B_j}{n_j} - \sum_{j=1}^{t} B_j \Big/ \sum_{j=1}^{t} n_j \right)^2. \tag{4.5.1}$$

For small and moderate $n := \sum_{j=1}^{m}(n_j+1)$ the exact quantiles of (4.5.1) can be found by enumeration. For large n one can apply the Monte Carlo method to compute approximate quantiles by simulation of the random variable (4.5.1).

Table 4.5.1 presents some upper quantiles q of the random variable (4.5.1) along with corresponding exact probabilities $\mathbf{P}\{\zeta_t \le q\}$ for $t = 2$, $n_1 = 5, \ldots, 15$, $n_2 = 5, \ldots, 15$.

Table 4.5.2 contains the upper quantiles of the random variable (4.5.1) for $t > 2$ and equal sample sizes $n = n_1 = \cdots = n_t$.

These tables can be used in the problems of one-way analysis as well as in some problems which can be reduced to them.

For example, it is easily shown that the sign test statistic (4.3.8) in the two-sample problem (see §4.3, Example 1) coincides with the statistic (4.4.8) for $t = 2$, $n_1 = m$, $n_2 = n$. Therefore the critical values can be taken from Table 4.5.1.

Two-way analysis. The random variable (4.4.12) for testing the hypothesis of no treatment effect can be written in the form

$$\frac{4}{mr}\sum_{j=1}^{t}\left(B_j - \frac{1}{t}\sum_{j=1}^{t} B_j\right)^2,$$

where B_j are i.i.d. random variables with binomial distribution $B_j \sim \mathrm{Bi}(mr, 1/2)$. This expression has the same form as the random variable (4.5.1) for $n_1 = \cdots = n_t = mr$. Hence when testing this hypothesis in two-way analysis one can use Table 4.5.2 for obtaining critical values.

The package SIGN mentioned above contains programs for computation of critical values of (4.4.6) and (4.4.8) for other m, $n_1, \ldots, n_m$ as well.

TABLE 4.5.1. Quantiles of ζ_2 as in (4.5.1)

	q	$\mathbf{P}\{\zeta_2 \le q\}$		q	$\mathbf{P}\{\zeta_2 \le q\}$
$n_1 = 5$ $n_2 = 5$	0.400	0.656	$n_1 = 5$ $n_2 = 6$	2.048	0.842
	1.600	0.891		2.376	0.915
	3.600	0.979		4.376	0.974
$n_2 = 7$	2.143	0.859	$n_2 = 8$	2.223	0.893
	3.086	0.944		2.777	0.913
	3.810	0.961		3.723	0.964
$n_2 = 9$	1.834	0.861	$n_2 = 10$	1.200	0.796
	2.800	0.928		2.133	0.901
	3.968	0.955		3.333	0.959
$n_2 = 11$	2.405	0.865	$n_2 = 12$	2.075	0.877
	2.618	0.916		3.075	0.937
	3.823	0.958		3.529	0.951
$n_2 = 13$	2.137	0.858	$n_2 = 14$	2.033	0.867
	2.492	0.907		2.890	0.930
	3.501	0.953		3.898	0.966
$n_2 = 15$	1.667	0.841	$n_2 = 16$	2.002	0.866
	2.400	0.907		2.752	0.918
	4.267	0.976		3.621	0.950
$n_1 = 6$ $n_2 = 6$	0.333	0.612	$n_1 = 6$ $n_2 = 7$	2.117	0.899
	1.333	0.854		2.374	0.908
	3.000	0.961		3.875	0.969
$n_2 = 8$	2.381	0.893	$n_2 = 9$	2.178	0.886
	2.881	0.934		2.844	0.926
	4.024	0.960		3.600	0.958
$n_2 = 10$	2.017	0.871	$n_2 = 11$	2.228	0.869
	2.817	0.922		2.410	0.907
	3.750	0.951		3.651	0.960
$n_2 = 12$	1.778	0.866	$n_2 = 13$	2.116	0.856
	2.778	0.935		2.429	0.901
	4.000	0.973		3.498	0.954
$n_2 = 14$	1.867	0.851	$n_2 = 15$	2.305	0.888
	2.752	0.922		2.743	0.913
	3.810	0.961		3.733	0.954
$n_1 = 7$ $n_2 = 7$	1.143	0.820	$n_1 = 7$ $n_2 = 8$	1.905	0.878
	2.571	0.943		2.976	0.901
	4.571	0.987		3.471	0.961
$n_2 = 9$	2.286	0.892	$n_2 = 10$	2.101	0.859
	2.683	0.919		2.827	0.925
	3.813	0.957		3.661	0.953

TABLE 4.5.1 (continued)

		q	$\mathbf{P}\{\zeta_2 \le q\}$			q	$\mathbf{P}\{\zeta_2 \le q\}$
	$n_2 = 11$	2.104	0.876		$n_2 = 12$	2.256	0.871
		2.773	0.918			2.566	0.910
		3.951	0.951			3.812	0.961
	$n_2 = 13$	1.978	0.862		$n_2 = 14$	1.524	0.834
		2.848	0.907			2.381	0.911
		4.064	0.962			3.429	0.957
	$n_2 = 15$	2.002	0.871		$n_2 = 16$	2.242	0.880
		2.771	0.911			2.484	0.902
		3.990	0.967			3.728	0.952
$n_1 = 8$	$n_2 = 8$	1.000	0.790	$n_1 = 8$	$n_2 = 9$	1.729	0.855
		2.250	0.923			2.941	0.934
		4.000	0.979			3.346	0.951
	$n_2 = 10$	2.178	0.887		$n_2 = 11$	2.299	0.884
		2.500	0.903			2.766	0.921
		3.600	0.954			3.828	0.951
	$n_2 = 12$	2.133	0.879		$n_2 = 13$	1.995	0.859
		2.700	0.917			2.645	0.912
		4.033	0.966			3.875	0.961
	$n_2 = 14$	2.104	0.866		$n_2 = 15$	1.878	0.855
		2.864	0.910			2.557	0.901
		4.058	0.960			3.623	0.955
$n_1 = 9$	$n_2 = 9$	0.889	0.762	$n_1 = 9$	$n_2 = 10$	2.395	0.860
		2.000	0.904			2.704	0.927
		3.556	0.969			4.126	0.956
	$n_2 = 11$	2.069	0.876		$n_2 = 12$	2.286	0.886
		2.766	0.901			2.683	0.914
		3.735	0.958			4.063	0.966
	$n_2 = 13$	2.127	0.863		$n_2 = 13$	2.099	0.869
		2.741	0.917			2.672	0.912
		3.885	0.952			3.877	0.961
	$n_2 = 15$	2.178	0.871		$n_2 = 16$	1.960	0.858
		2.500	0.901			2.668	0.904
		3.600	0.951			3.738	0.954

TABLE 4.5.1 (continued)

	q	$\mathbf{P}\{\zeta_2 \le q\}$		q	$\mathbf{P}\{\zeta_2 \le q\}$
$n_1 = 10$ $n_2 = 10$	0.800	0.737	$n_1 = 10$ $n_2 = 10$	2.244	0.859
	1.800	0.885		2.500	0.917
	3.200	0.959		3.825	0.954
$n_2 = 12$	1.964	0.862	$n_2 = 12$	2.259	0.884
	2.933	0.926		2.590	0.903
	4.097	0.956		3.901	0.961
$n_2 = 14$	2.100	0.858	$n_2 = 15$	2.160	0.879
	2.743	0.916		2.667	0.914
	4.005	0.962		3.840	0.960
$n_1 = 11$ $n_2 = 11$	1.636	0.866	$n_1 = 11$ $n_2 = 12$	2.108	0.861
	2.909	0.948		2.324	0.905
	4.545	0.983		3.563	0.952
$n_2 = 13$	2.155	0.853	$n_2 = 14$	2.198	0.879
	2.798	0.922		2.702	0.903
	3.921	0.952		3.740	0.955
$n_2 = 15$	2.060	0.856	$n_2 = 16$	2.189	0.870
	2.719	0.913		2.735	0.915
	3.939	0.961		3.779	0.950
$n_1 = 12$ $n_2 = 12$	1.500	0.848	$n_1 = 12$ $n_2 = 13$	1.986	0.862
	2.667	0.936		3.103	0.911
	4.167	0.977		3.450	0.955
$n_2 = 14$	2.289	0.867	$n_2 = 15$	2.141	0.872
	2.670	0.917		2.674	0.902
	3.751	0.951		3.919	0.955
$n_1 = 13$ $n_2 = 13$	1.385	0.831	$n_1 = 13$ $n_2 = 14$	1.875	0.859
	2.462	0.924		2.930	0.909
	3.846	0.971		4.220	0.950
$n_2 = 15$	2.216	0.868	$n_2 = 16$	2.080	0.861
	2.550	0.909		2.632	0.901
	3.798	0.960		3.830	0.952
$n_1 = 14$ $n_2 = 14$	1.286	0.815	$n_1 = 14$ $n_2 = 15$	1.776	0.853
	2.286	0.913		2.775	0.909
	3.571	0.964		4.099	0.955
$n_1 = 15$ $n_2 = 15$	1.200	0.800	$n_1 = 15$ $n_2 = 16$	1.809	0.850
	2.133	0.901		2.634	0.908
	3.333	0.957		3.884	0.954

TABLE 4.5.2. Quantiles of ζ_t as in (4.5.1) for $n = n_1 = \cdots = n_t$

	q	$\mathbf{P}\{\zeta_t \le q\}$		q	$\mathbf{P}\{\zeta_t \le q\}$
$t=3$ $n=2$	1.333	0.719	$t=3$ $n=3$	2.667	0.777
	4.000	0.906		3.556	0.918
	5.333	1.000		6.222	0.988
$n=4$	2.667	0.832	$n=5$	3.733	0.900
	4.667	0.949		4.800	0.943
	6.000	0.978		6.400	0.961
$n=6$	3.111	0.850	$n=7$	3.429	0.860
	4.000	0.902		4.952	0.950
	5.778	0.972		6.095	0.965
$n=8$	4.000	0.858	$n=9$	3.852	0.897
	4.333	0.924		4.741	0.921
	6.333	0.968		5.630	0.951
$n=10$	3.467	0.869	$n=11$	3.879	0.872
	5.067	0.933		4.606	0.913
	5.600	0.960		6.061	0.955
$n=12$	4.222	0.893	$n=13$	3.897	0.872
	4.667	0.929		4.308	0.912
	6.000	0.950		5.744	0.956
$n=14$	3.619	0.851	$n=15$	3.733	0.877
	4.762	0.910		4.800	0.907
	5.905	0.961		5.511	0.951
$t=4$ $n=2$	4.000	0.758	$t=4$ $n=3$	5.333	0.856
	5.500	0.945		6.333	0.909
	6.000	0.977		6.667	0.961
$n=4$	5.000	0.874	$n=5$	5.400	0.894
	6.000	0.915		6.400	0.906
	6.750	0.956		7.200	0.959
$n=6$	5.833	0.891	$n=7$	5.143	0.871
	6.000	0.918		6.143	0.907
	7.333	0.957		7.429	0.952
$n=8$	5.375	0.864	$n=9$	5.667	0.875
	6.375	0.911		6.222	0.909
	7.375	0.959		7.556	0.953
$n=10$	5.200	0.853			
	5.900	0.909			
	7.600	0.954			

TABLE 4.5.2 (continued)

	q	$\mathbf{P}\{\zeta_t \le q\}$		q	$\mathbf{P}\{\zeta_t \le q\}$
$t = 5$ $n = 2$	5.600	0.834	$t = 5$ $n = 3$	6.400	0.821
	6.400	0.922		6.933	0.920
	8.000	0.980		9.067	0.973
$n = 4$	6.800	0.889	$n = 5$	7.040	0.867
	7.200	0.914		7.360	0.910
	9.200	0.968		8.960	0.963

Part 2

Linear Models of Time Series

Introduction to Part 2

In the second part of the book we set out nonparametric sign procedures in autoregression models.

The sign-based analysis of the one-parameter stationary autoregression is the subject of Chapter 6. The detailed Introduction (§6.1) to this chapter should convey to the reader the main ideas of the approach based on signs. In Chapter 7 we extend these results to multiparameter autoregression. Its content is also exposed in the Introduction (§7.1). Before proceeding to sign procedures we present in Chapter 5 results concerning the most widely used procedures in the statistical analysis of autoregression, namely, the least squares and least absolute deviations procedures. These results will be used in Chapters 6 and 7 for comparison of these procedures with sign procedures. It is to be noted that besides well-known results on this subject, Chapter 5 contains a number of original results. These are, for example, the properties of tests and estimates under local alternatives, the approach to treatment of least absolute deviations tests and estimates, some facts about the behavior of estimates under contamination, tests for stationarity of autoregression, and some others. All previously known results are supplied with references. Otherwise, when no reference is given, the corresponding result has not appeared in the literature before.

CHAPTER 5

Least Squares and Least Absolute Deviations Procedures in the Simplest Autoregressive Model

5.1. Introduction

In the first part of this chapter we consider the most commonly used model of a stationary time series, namely, the stationary autoregression model. In the simplest one-parameter case it is determined by the stochastic difference equation

$$u_i = \beta u_{i-1} + \varepsilon_i, \qquad i \in \mathbb{Z} = \{0, \pm 1, \dots\}, \quad |\beta| < 1.$$

A random sequence $\{u_i\}$ satisfying this equation is called an *autoregression process*. We will assume that the sequence $\{\varepsilon_i\}$ is formed by i.i.d. random variables with an unknown distribution function. In this nonparametric setup we deal with estimation and testing problems about the unknown coefficient β. Our inference will be based on observations $u_0, \dots, u_n$ from a strictly stationary solution of the autoregression equation. The existence and a.s. uniqueness of such a solution are ensured by the assumption $|\beta| < 1$ and moment conditions on the variables ε_i. These matters are covered in §5.2. In §5.3 we study the simplest and widely used estimator for β, the least squares estimator (LSE), as well as related tests. Asymptotically, for large n, we consider fixed and local alternatives.

Many asymptotic results valid for stationary autoregression can be extended to the nonstationary case. There are many ways to specify a nonstationary autoregression process, and we will not treat the nonstationary case systematically in this book. Rather, we present several selected results. In particular, in §5.4 we consider the simplest nonstationary autoregression process

$$u_i = \beta u_{i-1} + \varepsilon_i, \qquad i = 1, 2, ..., \quad \beta \in \mathbb{R}^1, \quad u_0 = 0$$

(which differs from the stationary case by the fixed initial value u_0). For i.i.d. normally distributed $\{\varepsilon_i\}$ we present results on the asymptotic behavior of the LSE (which coincides in this case with the maximum likelihood estimator) based on the observations $u_1, \dots, u_n$. They show that the asymptotic properties of the LSE are completely different in cases when $|\beta| < 1$ (this case is asymptotically equivalent to the stationary one), $|\beta| = 1$ (the critical case), and $|\beta| > 1$ (explosive autoregression).

In §5.5 we again consider the stationary autoregression and study the least absolute deviations (LAD) estimator and related tests. The asymptotic study here is based on a uniform stochastic expansion for the "derivative" of the convex objective function specifying the LAD estimator (rather than for the objective function itself, as is usually done). A similar approach is systematically employed in Chapters 6 and 7 for the study of sign procedures, with which we cannot associate any convex optimization problem.

In §5.6 we study robustness of LSE and LAD estimators to contamination of data. We consider the model of contamination by independent outliers, which is specified as

$$y_i = u_i + z_i^\gamma \xi_i, \qquad i \in \mathbb{Z},$$

where $\{z_i^\gamma\}$ are Bernoulli random variables taking values 0 and 1 with probabilities $1-\gamma$ and γ, where γ is the contamination level and $\{\xi_i\}$ is a sequence of i.i.d. random variables playing the role of outliers; the sequences $\{u_i\}$, $\{z_i^\gamma\}$, $\{\xi_i\}$ are assumed to be mutually independent. The robustness of estimators based on $y_1, \ldots, y_n$ is characterized by their influence functionals. We show that the LSE and LAD estimators have unbounded gross error sensitivity over some natural classes of distributions of ξ_i. Qualitatively this means that these estimators are sensitive to even a small fraction of gross errors in the data.

Apart from the nonstationary autoregression considered in §5.4, a natural alternative to the simplest model of stationary autoregression is a nonstationary model

$$u_i = \beta_i u_{i-1} + \varepsilon_i, \qquad i \in \mathbb{Z},$$

with coefficient β_i depending on the time parameter i. We consider this scheme in §5.7 and construct tests for the hypothesis that $\{\beta_i\}$ are constant, i.e., that the autoregression process is stationary. The test statistics are Kolmogorov and ω^2 type functionals of random processes similar to the empirical process involved in the Donsker–Prohorov invariance principle. We study the power of these tests under local alternatives.

The results contained in this chapter will be needed in Chapters 6 and 7 for the study of the sign procedures. We do not touch upon many relevant aspects, in particular, the case of unknown and, in general, variable mean value of $\{\varepsilon_i\}$, the case where $\{\varepsilon_i\}$ form a moving average sequence, etc. Along the way we give references, which will help the reader to find some results of this kind in the literature.

Recall that all asymptotic studies are done for $n \to \infty$, unless otherwise stated.

5.2. The simplest stationary autoregressive equation and its solutions

Sections 5.2 and 5.3 of this chapter deal with the simplest stationary one-parameter autoregression model

$$u_i = \beta u_{i-1} + \varepsilon_i, \qquad i \in \mathbb{Z}. \tag{5.2.1}$$

Here $\{\varepsilon_i\}$ are i.i.d. r.v.'s with distribution function $G(x) = \mathbf{P}\{\varepsilon_1 < x\}$ and β is a nonrandom unknown parameter, $|\beta| < 1$. To avoid unnecessary trivialities, the distribution of ε_i's will be assumed to be nondegenerate without special mention.

In Chapters 5 and 6 we will construct distribution free procedures (valid under unknown $G(x)$) for testing hypotheses about β and for estimation of β, based on observations $u_0, u_1, \ldots, u_n$ which form a strictly stationary solution of (5.2.1). The existence and uniqueness of such a solution are guaranteed by certain moment conditions on $G(x)$. Usually these properties are proved under the following

CONDITION 5.2(i). $\mathbf{E}\varepsilon_1 = 0$, $\mathbf{E}\varepsilon_1^2 < \infty$.

This assumption ensures the existence of a strictly stationary solution of the equation (5.2.1) representable in the form

$$u_i = \sum_{j=0}^{\infty} \beta^j \varepsilon_{i-j}. \tag{5.2.2}$$

The series here converges in L^2, and this is the only strictly stationary solution a.s. We will prove a similar assertion under a weaker

CONDITION 5.2(ii). $\mathbf{E}|\varepsilon_1|^{1+\Delta} < \infty$ *for some* $\Delta > 0$.

LEMMA 5.2.1. *Under Condition* 5.2(ii) *the equation* (5.2.1) *has an a.s. unique strictly stationary solution representable in the form* (5.2.2) *with the series converging in* $L^{1+\Delta}$.

PROOF. The variables $\{u_i\}$ in (5.2.2) formally satisfy (5.2.1). Since $L^{1+\Delta}$ is complete, the series (5.2.2) converges in $L^{1+\Delta}$ provided that the partial sums

$$S_n := \sum_{j=0}^{n} \beta^j \varepsilon_{i-j}$$

form a Cauchy sequence. By Minkowski's inequality (Shiryaev [**78**], Chapter 2, §6) one has for $m > n$

$$\begin{aligned} \{\mathbf{E}_\beta |S_m - S_n|^{1+\Delta}\}^{1/(1+\Delta)} &= \left\{\mathbf{E}_\beta \left| \sum_{j=n+1}^{m} \beta^j \varepsilon_{i-j} \right|^{1+\Delta} \right\}^{1/(1+\Delta)} \\ &\le \sum_{j=n+1}^{m} \{\mathbf{E}_\beta |\beta^j \varepsilon_{i-j}|^{1+\Delta}\}^{1/(1+\Delta)} \\ &= \{\mathbf{E}|\varepsilon_1|^{1+\Delta}\}^{1/(1+\Delta)} \sum_{j=n+1}^{m} |\beta|^j = O(|\beta|^n) = o(1). \end{aligned}$$

Thus, indeed, (5.2.2) determines a solution which belongs to $L^{1+\Delta}$. It is obvious that this solution is strictly stationary, and it remains to establish its uniqueness. For any $k \in \mathbb{N}$ a sequence $\{\widetilde{u}_i\}$ satisfying (5.2.1) can be written as

$$\widetilde{u}_i = \beta(\beta \widetilde{u}_{i-2} + \varepsilon_{i-1}) = \cdots = \beta^k \widetilde{u}_{i-k} + \sum_{j=0}^{k-1} \beta^j \varepsilon_{i-j}. \tag{5.2.3}$$

Let $\{\widetilde{u}_i\}$ be strictly stationary. Then $\beta^k \widetilde{u}_{i-k} \xrightarrow{\mathbf{P}_\beta} 0$, $k \to \infty$, by the assumption $|\beta| < 1$. By (5.2.3) $\{\widetilde{u}_i\}$ is given by the right-hand side of (5.2.2), where the series converges in probability. Since the limit of a sequence convergent in probability is unique a.s., the lemma is proved. □

Under Condition 5.2(ii) the condition $\mathbf{E}\varepsilon_1 = 0$ implies by Lemma 5.2.1 that $\mathbf{E}_\beta u_1 = 0$. Moreover, when $\Delta = 1$,

$$\operatorname{Var}_\beta u_1 = \mathbf{E}_\beta \Big(\sum_{j=0}^{\infty} \beta^j \varepsilon_{i-j} \Big)^2 = \sum_{j=0}^{\infty} \mathbf{E}_\beta (\beta^j \varepsilon_{i-j})^2 = \mathbf{E}\varepsilon_1^2 (1-\beta^2)^{-1}. \tag{5.2.4}$$

It is to be noted that the treatment of statistical problems for $\Delta \geq 1$ and $\Delta < 1$ will be considerably different. These differences will be discussed in §5.3 and subsequent sections.

A natural question to ask here is what moment conditions on $G(x)$ suffice to ensure the existence and a.s. uniqueness of a strictly stationary solution of (5.2.1).

A nearly minimal condition is

CONDITION 5.2(iii). $\mathbf{E}\log^+|\varepsilon_1| < \infty$, *where* $\log^+ x = \max(0, \log x)$.

We quote the following result from Yohai and Maronna [**93**].

LEMMA 5.2.2. *Under Condition* 5.2(iii) *there exists an a.s. unique strictly stationary solution of* (5.2.1) *representable by a moving average* (5.2.2), *where the series absolutely converges a.s.*

PROOF. For the series (5.2.2) to absolutely converge a.s., it is sufficient that

$$\limsup_{j\to\infty} |\varepsilon_{i-j}|^{1/j} \leq 1 \quad \text{a.s.}$$

By the Borel–Cantelli lemma this will follow from the convergence of the series

$$\sum_{j=0}^{\infty} \mathbf{P}\{|\varepsilon_{i-j}|^{1/j} \geq 1+\delta\}$$

for every $\delta > 0$. This series is equal to

$$\sum_{j=0}^{\infty} \mathbf{P}\{\log^+|\varepsilon_1| \geq j\log(1+\delta)\},$$

which converges by Condition 5.2(iii). The uniqueness of a strictly stationary solution is proved as in Lemma 5.2.1. □

The autoregressive models and, more generally, the models of autoregression–moving average, are widely and successfully used in various fields of applications, such as economics, control, hydrology, geophysics, medicine, processing of images and audio signals etc. Some references can be found in a survey paper Al′tshuler [**1**]; we point out also the papers by Walden [**92**], Denoël and Solvay [**21**], and Maldelbrot [**62, 63**] which deal with applied problems leading to autoregressive models with infinite variance. The statistical theory for autoregression–moving average processes is systematically treated in many monographs and papers. We indicate here only widely known monographs (Anderson [**2**], Hannan [**37**], Jenkins and Watts [**48**], Box and Jenkins [**18**], Kashyap and Rao [**49**]). In subsequent exposition further references will be given.

5.3. Least squares procedures

Consider the autoregression model (5.2.1) under Condition 5.2(i). Suppose we have observations $u_0, u_1, \ldots, u_n$ from the strictly stationary solution (5.2.2). Its existence is implied by Lemma 5.2.1.

In this section we study the least squares estimator for β based on $u_0, u_1, \ldots, u_n$ as well as related tests for hypotheses about β.

5.3.1. Least squares estimator. The least squares estimator (LSE) of β, to be denoted by $\widehat{\beta}_{n,LS}$, is defined as the solution of the problem

$$\sum_{i=1}^{n}(u_i - \theta u_{i-1})^2 \Longrightarrow \inf_{\theta\in\mathbb{R}^1},$$

or, equivalently, the solution of the equation

$$\sum_{i=1}^{n}(u_i - \theta u_{i-1})u_{i-1} = 0. \tag{5.3.1}$$

Since

$$n^{-1}\sum_{i=1}^{n} u_{i-1}^2 \xrightarrow{\mathbf{P}_\beta} \mathbf{E}_\beta u_1^2 = \mathbf{E}\varepsilon_1^2(1-\beta^2)^{-1} \neq 0$$

(see the relation (5.3.6) in the proof of Theorem 5.3.1 below), with probability tending to 1 there exists a unique solution $\widehat{\beta}_{n,LS}$ of (5.3.1), which is given by the formula

$$\widehat{\beta}_{n,LS} = \frac{\sum_{i=1}^n u_i u_{i-1}}{\sum_{i=1}^n u_{i-1}^2}. \tag{5.3.2}$$

The following theorem establishes the asymptotic normality of $\widehat{\beta}_{n,LS}$ (see Anderson [**2**], Section 5.5.4).

THEOREM 5.3.1. *Let Condition* 5.2(i) *be fulfilled. Then*

$$\sqrt{n}(\widehat{\beta}_{n,LS} - \beta) \xrightarrow{d_\beta} N\big(0, \sigma_{LS}^2(\beta)\big), \tag{5.3.3}$$

where $\sigma_{LS}^2(\beta) = 1 - \beta^2$.

PROOF. By substituting $\beta u_{i-1} + \varepsilon_i$ for u_i in (5.3.2) we obtain

$$\sqrt{n}(\widehat{\beta}_{n,LS} - \beta) = \frac{n^{-1/2}\sum_{i=1}^n \varepsilon_i u_{i-1}}{n^{-1}\sum_{i=1}^n u_{i-1}^2}. \tag{5.3.4}$$

It is seen from (5.2.1) that

$$n^{-1}\sum_{i=1}^{n} u_i^2 = \beta^2 n^{-1}\sum_{i=1}^{n} u_{i-1}^2 + 2\beta n^{-1}\sum_{i=1}^{n}\varepsilon_i u_{i-1} + n^{-1}\sum_{i=1}^{n}\varepsilon_i^2. \tag{5.3.5}$$

It follows from (5.2.2) that u_i is independent of the set of r.v.'s $\{\varepsilon_j,\ j > i\}$. (This fact will be often used in the sequel.) Hence

$$\mathbf{E}_\beta n^{-1}\sum_{i=1}^{n}\varepsilon_i u_{i-1} = 0, \qquad \mathbf{E}_\beta\left(n^{-1}\sum_{i=1}^{n}\varepsilon_i u_{i-1}\right)^2 = n^{-1}\mathbf{E}\varepsilon_1^2\mathbf{E}_\beta u_0^2$$

and, therefore,

$$n^{-1}\sum_{i=1}^{n}\varepsilon_i u_{i-1} \xrightarrow{\mathbf{P}_\beta} 0.$$

Since by the law of large numbers

$$n^{-1}\sum_{i=1}^{n}\varepsilon_i^2 \xrightarrow{\mathbf{P}} \mathbf{E}\varepsilon_1^2,$$

it follows from (5.3.5) that

$$n^{-1}\sum_{i=1}^{n} u_i^2 \xrightarrow{\mathbf{P}_\beta} \mathbf{E}\varepsilon_1^2(1-\beta^2)^{-1} = \mathbf{E}_\beta u_1^2 \neq 0. \tag{5.3.6}$$

By virtue of (5.3.4) and (5.3.6) the theorem will follow if we prove that

$$S_n := n^{-1/2}\sum_{i=1}^{n} \varepsilon_i u_{i-1} \xrightarrow{d_\beta} N(0, \mathbf{E}\varepsilon_1^2 \mathbf{E}_\beta u_0^2). \tag{5.3.7}$$

Define the variables

$$u_{ik} = \sum_{j=0}^{k} \beta^j \varepsilon_{i-j}, \qquad i \in \mathbb{Z}.$$

By the central limit theorem for k-dependent random sequences (see Theorem 7.7.5 in Anderson [**2**])

$$Z_{kn} := n^{-1/2}\sum_{i=1}^{n} \varepsilon_i u_{i-1,k} \xrightarrow{d_\beta} Z_k, \tag{5.3.8}$$

where

$$Z_k \sim N\left(0, (\mathbf{E}\varepsilon_1^2)^2 \frac{1-\beta^{2(k+1)}}{1-\beta^2}\right).$$

Obviously,

$$Z_k \xrightarrow{d_\beta} Z \sim N\big(0, (\mathbf{E}\varepsilon_1^2)^2(1-\beta^2)^{-1}\big) \qquad \text{as} \quad k \to \infty.$$

Further, one has for $X_{kn} := S_n - Z_{kn}$

$$\sup_n \mathbf{E}_\beta X_{kn}^2 = \mathbf{E}\varepsilon_1^2 \mathbf{E}_\beta (u_0 - u_{0k})^2 = \mathbf{E}\varepsilon_1^2 \mathbf{E}_\beta \Big(\sum_{j=k+1}^{\infty} \beta^j \varepsilon_{-j} \Big)^2$$
$$= (\mathbf{E}\varepsilon_1^2)^2 \frac{\beta^{2(k+1)}}{1-\beta^2} \to 0 \qquad \text{as} \quad k \to \infty.$$

We will use the following assertion, which is a restatement of Theorem 7.7.1 in Anderson [**2**].

LEMMA 5.3.1. *Let* $S_n = Z_{kn} + X_{kn}$, $n, k = 1, 2, \ldots$. *Assume that*:
1. $\sup_n \mathbf{E}X_{kn}^2 \to 0$ *as* $k \to \infty$;
2. $Z_{kn} \xrightarrow{d} Z_k$ *as* $n \to \infty$;
3. $Z_k \xrightarrow{d} Z$ *as* $k \to \infty$.

Then $S_n \xrightarrow{d} Z$ *as* $n \to \infty$.

Returning to the proof of Theorem 5.3.1, note that in our case the conditions of Lemma 5.3.1 are fulfilled. This lemma implies (5.3.7), which completes the proof of Theorem 5.3.1. □

A remarkable feature of the LSE $\widehat{\beta}_{n,LS}$ is that the limiting variance $\sigma_{LS}^2(\beta) = 1 - \beta^2$ does not depend on $G(x)$.

It follows from (5.3.3) that, with $\xi_{1-p/2}$ denoting the $(1-p/2)$-quantile of the standard normal distribution, the set

$$\begin{aligned}\{\theta\colon \widehat{\beta}_{n,LS} - n^{-1/2}(1-\widehat{\beta}^2_{n,LS})^{1/2}\xi_{1-p/2} \\ < \theta < \widehat{\beta}_{n,LS} + n^{-1/2}(1-\widehat{\beta}^2_{n,LS})^{1/2}\xi_{1-p/2}\},\end{aligned} \tag{5.3.9}$$

is a confidence interval for β with asymptotic confidence level $1-p$.

The LSE $\widehat{\beta}_{n,LS}$ is a nonparametric estimator. It is of interest to compare it with the maximum likelihood estimator $\widehat{\beta}_{n,ML}$, which is an asymptotically efficient parametric estimator. Let $G(x)$ have a regular density $g(x)$ (see Cox and Hinkley [**20**], §9.1). Denote by $I(g)$ the Fisher information of $g(x)$. Using the Markov property of the sequence $\{u_i\}$ one readily obtains that the joint distribution of $u_0, u_1, \ldots, u_n$, $n \geq 1$, has a density

$$g^n(x_0, x_1, \ldots, x_n, \beta) = g_u(x_0, \beta) \prod_{i=1}^n g(x_i - \beta x_{i-1}),$$

where $g_u(x_0, \beta)$ is the density of u_0.

For regular $g(x)$ a consistent solution $\widehat{\beta}_{n,ML}$ of the likelihood equation

$$\log g^n(u_0, u_1, \ldots, u_n, \theta) = 0$$

is representable as

$$\sqrt{n}(\widehat{\beta}_{n,ML} - \beta) = \big(I(g)\mathbf{E}_\beta u_1^2\big)^{-1} n^{-1/2} \sum_{i=1}^n \frac{u_{i-1}g'(\varepsilon_i)}{g(\varepsilon_i)} + o_p(1),$$

hence

$$\sqrt{n}(\widehat{\beta}_{n,ML} - \beta) \xrightarrow{d_\beta} N\big(0, \sigma^2_{ML}(\beta)\big) \tag{5.3.10}$$

with

$$\sigma^2_{ML}(\beta) = (1-\beta^2)\big(I(g)\mathbf{E}\varepsilon_1^2\big)^{-1}$$

(see Cox and Hinkley [**20**], Section 9.2.3, Example 9.11). Now (5.3.10) and (5.3.3) imply that the asymptotic relative efficiency (ARE) of $\widehat{\beta}_{n,LS}$ with respect to $\widehat{\beta}_{n,ML}$ equals

$$e_{LS,ML} := \frac{\sigma^2_{ML}(\beta)}{\sigma^2_{LS}(\beta)} = \big(I(g)\mathbf{E}\varepsilon_1^2\big)^{-1}$$

and does not depend on β and on the scale parameter of $G(x)$. When $G(x) \sim N(0,1)$, $e_{LS,ML} = 1$ and $\widehat{\beta}_{n,LS}$ is asymptotically efficient.

5.3.2. Tests based on the LSE. One can use (5.3.3) for testing $H_0\colon \beta = \beta_0$ by taking

$$T_{n,LS}(\beta_0) = \sqrt{n}\, \sigma^{-1}_{LS}(\beta_0)(\widehat{\beta}_{n,LS} - \beta_0)$$

as the test statistic. Under Condition 5.2(i) its null distribution is asymptotically the standard normal one. For a fixed alternative $H_1\colon \beta \neq \beta_0$, $T_{n,LS}(\beta_0)$ by Theorem 5.3.1 goes to infinity, so that this test is consistent against H_1.

It is of interest to study the asymptotic power of the test based on $T_{n,LS}(\beta_0)$ under local alternatives

$$H_{1n}\colon \beta = \beta_n,\ \beta_n \to \beta_0,$$

approaching H_0.

The following theorem is useful for the study of the test statistic when the autoregression parameter varies with n.

THEOREM 5.3.2. *Let Condition* 5.2(i) *be fulfilled. Then under* H_{1n}

$$\sqrt{n}(\widehat{\beta}_{n,LS} - \beta_n) \xrightarrow{d_{\beta_n}} N(0, \sigma^2_{LS}(\beta_0)), \tag{5.3.11}$$

where $\sigma^2_{LS}(\beta) = 1 - \beta^2$.

Before giving the proof, let us discuss come consequences of this theorem.

COROLLARY 5.3.1. *Let* $\sqrt{n}(\beta_n - \beta_0) \to \infty$. *Then* $T_{n,LS}(\beta_0)$ *diverges to infinity, and hence the corresponding test is consistent against* H_{1n}.

COROLLARY 5.3.2. *Let* $\sqrt{n}(\beta_n - \beta_0) = a + o(1)$ *with some constant* a. *Then*

$$T_{n,LS}(\beta_0) \xrightarrow{d_{\beta_n}} N\big(a(1-\beta_0^2)^{-1/2}, 1\big). \tag{5.3.12}$$

We will write $H_{1n}(a)$ for such alternatives. Corollary 5.3.2 says that if $a \neq 0$ the test has a nontrivial asymptotic power against $H_{1n}(a)$ (i.e., its power is bounded away from the significance level and 1). In particular, the level α test for H_0 against

$$H^+_{1n}(a) \colon \beta = \beta_n := \beta_0 + a\,n^{-1/2} + o(n^{-1/2}),\ a > 0,$$

has power

$$\mathbf{P}_{\beta_n}\{T_{n,LS}(\beta_0) > \xi_{1-\alpha}\} \to \Phi\big(a(1-\beta_0^2)^{-1/2} - \xi_{1-\alpha}\big),$$

where $\Phi(x)$ is the standard normal d.f. If $a = 0$ then the test statistic has the same asymptotic distribution under $H_{1n}(0)$ as under H_0, so that the hypotheses H_0 and $H_{1n}(0)$ are asymptotically indistinguishable.

PROOF OF THEOREM 5.3.2. Similarly to (5.3.4) we obtain for $\beta = \beta_n$

$$\sqrt{n}(\widehat{\beta}_{n,LS} - \beta_n) = \frac{n^{-1/2}\sum_{i=1}^n \varepsilon_i u_{i-1}}{n^{-1}\sum_{i=1}^n u_{i-1}^2}. \tag{5.3.13}$$

It follows from (5.2.2) that

$$\mathbf{E}_{\beta_n} u_1^2 = \mathbf{E}\varepsilon_1^2(1-\beta_n^2)^{-1} \to \mathbf{E}\varepsilon_1^2(1-\beta_0^2)^{-1},$$

so that $n^{-1}\mathbf{E}_{\beta_n} u_1^2 = o(1)$. As in the proof of Theorem 5.3.1, this relation and (5.3.5) with β substituted by β_n imply

$$n^{-1}\sum_{i=1}^n u_i^2 \xrightarrow{\mathbf{P}_{\beta_n}} \mathbf{E}\varepsilon_1^2(1-\beta_0^2)^{-1} \neq 0. \tag{5.3.14}$$

In view of (5.3.13) and (5.3.14) for the proof of (5.3.11) it suffices to show that

$$S_n := n^{-1/2}\sum_{i=1}^n \varepsilon_i u_{i-1} \xrightarrow{d_{\beta_n}} N\big(0, (\mathbf{E}\varepsilon_1^2)^2(1-\beta_0^2)^{-1}\big). \tag{5.3.15}$$

Put $u_i^0 := \sum_{j\geq 0} \beta_0^j \varepsilon_{i-j}$. Then by (5.3.7)

$$S_{n0} := n^{-1/2}\sum_{i=1}^n \varepsilon_i u_{i-1}^0 \xrightarrow{d_{\beta_n}} N\big(0, (\mathbf{E}\varepsilon_1^2)^2(1-\beta_0^2)^{-1}\big). \tag{5.3.16}$$

Let us show that

$$S_n - S_{n0} = o_p(1). \tag{5.3.17}$$

It suffices to show that

$$\mathbf{E}_{\beta_n}(S_n - S_{n0})^2 = \mathbf{E}\varepsilon_1^2 \mathbf{E}_{\beta_n}(u_0 - u_0^0)^2 \to 0.$$

Indeed, when $\max(|\beta_0|, |\beta_n|) \le b < 1$,

$$\mathbf{E}_{\beta_n}(u_0 - u_0^0)^2 = \mathbf{E}\varepsilon_1^2 \sum_{j=1}^{\infty} (\beta_n^j - \beta_0^j)^2 = \mathbf{E}\varepsilon_1^2 (\beta_n - \beta_0)^2 \sum_{j=1}^{\infty} \Big(\sum_{s=0}^{j-1} \beta_n^{j-s-1} \beta_0^s \Big)^2$$
$$\le \mathbf{E}\varepsilon_1^2 (\beta_n - \beta_0)^2 \sum_{j=1}^{\infty} j^2 b^{2(j-1)} = o(1).$$

Now (5.3.16)–(5.3.17) imply (5.3.15) and hence the theorem. □

Note that Corollary 5.3.2 can also be established by means of general results for contiguous alternatives, though under somewhat stronger conditions than those of Theorem 5.3.2. Namely, assume additionally that $G(x)$ has an absolutely continuous density $g(x) > 0$ with finite Fisher information $I(g)$. Then (see Kreiss [**57**]) for $\beta_n = \beta_0 + an^{-1/2} + o(n^{-1/2})$

$$\log \frac{d\mathbf{P}_{n,\beta_n}}{d\mathbf{P}_{n,\beta_0}} = a\Delta_n(\beta_0) - \frac{1}{2}a^2 I(g)\mathbf{E}_{\beta_0} u_1^2 + \delta_n,$$

with

$$\Delta_n(\beta_0) \xrightarrow{d_{\beta_0}} N\big(0, I(g)\mathbf{E}_{\beta_0} u_1^2\big), \qquad \delta_n \xrightarrow{\mathbf{P}_{\beta_0}} 0.$$

Here $\mathbf{P}_{n,\beta}$ denotes the distribution of the vector of observations $(u_0, u_1, \dots, u_n)$ and

$$\Delta_n(\theta) = n^{-1/2} \sum_{i=1}^{n} \frac{\partial}{\partial \theta} \log g(u_i - \theta u_{i-1}). \tag{5.3.18}$$

Thus the distributions $\mathbf{P}_{n,\beta}$ possess the Local Asymptotic Normality (LAN) property (for the definition see, for example, Ibragimov and Khas′minskii [**46**], Chapter 2, §2). Making use of Le Cam's Third Lemma (see Hájek and Šidák [**30**], Chapter 6, §1) and the representation

$$\sqrt{n}(\widehat{\beta}_{n,LS} - \beta_0) = \frac{1 - \beta_0^2}{\mathbf{E}\varepsilon_1^2} n^{-1/2} \sum_{i=1}^{n} u_{i-1}\varepsilon_i + o_p(1),$$

which is seen to hold under H_0 from the proof of Theorem 5.3.1, we obtain the following assertion.

Subject to the above regularity conditions, the distribution of $\sqrt{n}(\widehat{\beta}_{n,LS} - \beta_0)$ under $H_{1n}(a)$ is asymptotically normal with variance $\sigma_{LS}^2(\beta_0)$ and the mean value equal to the covariance of the bivariate normal limiting distribution of the vector $\Big(\sqrt{n}(\widehat{\beta}_{n,LS} - \beta_0),\ \log \frac{d\mathbf{P}_{n,\beta_n}}{d\mathbf{P}_{n,\beta_0}}\Big)$ under H_0. One can show that this covariance equals a, so that

$$\sqrt{n}(\widehat{\beta}_{n,LS} - \beta_0) \xrightarrow{d_{\beta_n}} N\big(a, \sigma_{LS}^2(\beta_0)\big).$$

This relation is equivalent to (5.3.12).

We will use (5.3.12) to evaluate the ARE of the test based on $\widehat{\beta}_{n,LS}$ with respect to other tests. In particular, consider the parametric asymptotically optimal test with statistic

$$T_{n,ML}(\beta_0) = \big(I(g)\boldsymbol{E}_{\beta_0}u_1^2\big)^{-1/2}\Delta_n(\beta_0),$$

where $\Delta_n(\theta)$ is defined by (5.3.18). This test is asymptotically equivalent to the likelihood ratio test.

Subject to the regularity conditions needed for LAN, one obtains by Le Cam's Third Lemma that under $H_{1n}(a)$

$$T_{n,ML}(\beta_0) \xrightarrow{d_{\beta_n}} N\big(a(I(g)\boldsymbol{E}_{\beta_0}u_1^2)^{1/2},\ 1\big). \tag{5.3.19}$$

Under H_0, $T_{n,ML}(\beta_0)$ has asymptotically the standard normal distribution. The ARE of the test based on $T_{n,LS}(\beta_0)$ relative to the one based on $T_{n,ML}(\beta_0)$ can be derived from (5.3.19) and (5.3.12) as

$$\big(\delta_{LS}(\beta_0)/\delta_{ML}(\beta_0)\big)^2,$$

where $\delta_{LS}(\beta_0)$ and $\delta_{ML}(\beta_0)$ are asymptotic shifts (mean values) of $T_{n,LS}(\beta_0)$ and $T_{n,ML}(\beta_0)$ under $H_{1n}(a)$. This ARE is seen to equal $e_{LS,ML}$ and thus to coincide with the ARE of the respective estimators.

It is to be emphasized that all the properties of $\widehat{\beta}_{n,LS}$ stated above are of an asymptotic nature. Optimal procedures based on serial correlations as well as exact distributions in Gaussian models specified by difference autoregression equations and their modifications (not necessarily of first order) are treated, for example, in Anderson [**2**], Chapter 6, or Hannan [**37**], Chapter 6, §3. We will not pursue this kind of problem here. In many applications the normality assumption is not adequate. Hence, distribution free procedures valid for finite n under a broad class of distributions $G(x)$ are preferable. The next two chapters deal with procedures based on signs, which possess these properties.

Let us point out here that the assumption $\mathbf{E}\varepsilon_1^2 < \infty$ is essential for the asymptotic normality of $\sqrt{n}(\widehat{\beta}_{n,LS} - \beta)$. It is known (see Yohai and Maronna [**93**]) that if $G(x)$ is symmetric about zero and Condition 5.2(iii) is fulfilled then $\sqrt{n}(\widehat{\beta}_{n,LS} - \beta) = O_p(1)$, however the limiting distribution of $\sqrt{n}(\widehat{\beta}_{n,LS} - \beta)$ need not be normal. For instance, as it is pointed out in Yohai and Maronna [**93**], if $G(x)$ is symmetric about zero and $\lim_{x\to+\infty} x^\alpha(1 - G(x)) = k > 0$ for some $\alpha \in (0,2)$ (hence $\mathbf{E}|\varepsilon_1|^\alpha$ is infinite) then $n^\delta(\widehat{\beta}_{n,LS} - \beta) = o_p(1)$ for any $\delta < 1/\alpha$.

The anomalous rate of convergence of $\widehat{\beta}_{n,LS}$ to β is due to the fact that in this case the LAN property fails and the related asymptotic lower bound for the risks of regular estimators (see Ibragimov and Khas′minskii [**46**], Chapter 2, §11) breaks down.

It is appropriate to mention here that the properties of the least squares procedures stated above for the one-parameter model (5.2.1) remain essentially unchanged in a multiparameter model as well. The one-parameter and multiparameter models differ basically in technical details. A thorough treatment of the multiparameter case can be found, for example, in Anderson [**2**], Chapter 5; in particular, the case of $\{\varepsilon_i\}$ having nonzero mean is also considered therein. We will deal with the least squares estimator in multiparameter autoregression models in Chapter 7.

5.4. Least squares estimator in nonstationary autoregression

Consider a nonstationary autoregression relation

$$u_i = \beta u_{i-1} + \varepsilon_i, \qquad i = 1, 2, \ldots, \qquad u_0 = 0. \tag{5.4.1}$$

Here $\{\varepsilon_i\}$ form a sequence of independent $N(0,1)$ random variables, and $\beta \in \mathbb{R}^1$.

Let $u_1, \ldots, u_n$ be observations generated by the model (5.4.1). It follows from (5.4.1) that

$$u_i = \sum_{j=0}^{i-1} \beta^j \varepsilon_{i-j}, \qquad i = 1, 2, \ldots. \tag{5.4.2}$$

By (5.4.2), if $\beta \neq 0$ then $\operatorname{Var}_\beta u_i = 1 + \beta^2 + \cdots + \beta^{2(i-1)}$ depends on i; hence, the sequence $\{u_i\}$ is nonstationary.

Our object is to present for this model an interesting result about the asymptotic behavior of the LSE $\widehat{\beta}_{n,LS}$ given by (5.3.2) (which in this case is also the maximum likelihood estimator). Although this is a parametric model (the distribution of $u_1, \ldots, u_n$ is Gaussian with the only unknown parameter β), both the result and its proof will be useful for us. Later on, in §6.3, we will obtain finite-sample and asymptotic results for sign test statistics related to the model (5.4.1) in a nonparametric setup where the distribution function of ε_i's is unknown. It will be instructive to compare these results with the results of the present section.

Let $J_n(\beta)$ denote the Fisher information about β contained in $u_1, \ldots, u_n$. It is easy to show that

$$J_n(\beta) = \begin{cases} \frac{1}{1-\beta^2}\left(n - \frac{1-\beta^{2n}}{1-\beta^2}\right), & |\beta| \neq 1, \\ 2^{-1} n(n-1), & |\beta| = 1. \end{cases} \tag{5.4.3}$$

Indeed, the joint density of $u_1, \ldots, u_n$ is

$$g(x_1, \ldots, x_n, \beta) = (2\pi)^{-n/2} \exp\left\{ -\frac{1}{2} \sum_{i=1}^n (x_i - \beta x_{i-1})^2 \right\},$$

hence

$$J_n(\beta) = \mathbf{E}_\beta \left(-\frac{\partial^2 \log g(u_1, \ldots, u_n, \beta)}{\partial \beta^2} \right)^2 = \sum_{i=1}^n \mathbf{E}_\beta u_{i-1}^2.$$

Together with the equality

$$\mathbf{E}_\beta u_i^2 = \mathbf{E}_\beta \left(\sum_{j=0}^{i-1} \beta^j \varepsilon_{i-j} \right)^2 = \sum_{j=0}^{i-1} \beta^{2j} = \begin{cases} \frac{1-\beta^{2i}}{1-\beta^2}, & |\beta| \neq 1, \\ i, & |\beta| = 1, \end{cases}$$

for $i = 1, ..., n$ this implies (5.4.3).

In its turn, (5.4.3) implies that

$$J_n(\beta) \sim d_n^2(\beta) := \begin{cases} n/(1-\beta^2), & |\beta| < 1, \\ n^2/2, & |\beta| = 1, \\ \beta^{2n}/(1-\beta^2)^2, & |\beta| > 1. \end{cases} \tag{5.4.4}$$

Denote by $K(0,1)$ the Cauchy distribution with location and scale parameters 0 and 1. Let $H(\beta)$ denote the distribution of the random variable

$$\beta\,\frac{\nu^2(1)-1}{2^{2/3}\int_0^1 \nu^2(t)\,dt}, \tag{5.4.5}$$

where $\nu(t)$, $t \in [0,1]$, is a standard Wiener process. Denote the density function of (5.4.5) by $h_\beta(x)$, and the corresponding distribution function by $H_\beta(x)$. Since the case of $|\beta| = 1$ will be of special interest for us, it is worth pointing out that for $\beta = \pm 1$ the function $h_\beta(x)$ is not even. In particular (see Evans and Savin [**27**]),

$$\begin{aligned} h_1(-0.2) &\approx 0.316, & H_1(-0.2) &\approx 0.617 \\ h_1(0) &\approx 0.341, & H_1(0) &\approx 0.683 \\ h_1(0.2) &\approx 0.357, & H_1(0.2) &\approx 0.753. \end{aligned}$$

The following theorem is taken from Shiryaev and Spokoiny [**79**], Chapter 5, Section 1.

THEOREM 5.4.1. *Let $\{\varepsilon_i,\ i = 1,2,\dots\}$ in the model (5.4.1) be independent standard normal random variables. Then*

$$d_n(\beta)(\widehat{\beta}_{n,LS} - \beta) \xrightarrow{\mathbf{d}_\beta} \begin{cases} N(0,1), & |\beta| < 1, \\ H(\beta), & |\beta| = 1, \\ K(0,1), & |\beta| > 1. \end{cases}$$

PROOF. Let

$$U_n = d_n^{-1}(\beta) \sum_{i=1}^n \varepsilon_i u_{i-1}, \qquad V_n = d_n^{-2}(\beta) \sum_{i=1}^n u_{i-1}^2.$$

Then by (5.3.4)

$$d_n(\beta)(\widehat{\beta}_{n,LS} - \beta) = U_n V_n^{-1}. \tag{5.4.6}$$

Denote by $f_n(s,t)$ the joint characteristic function of U_n, V_n. It is known (see Rao [**75**]) that

$$f(s,t) := \lim_{n\to\infty} f_n(s,t) = \begin{cases} \exp\{it - s^2/2\}, & |\beta| < 1, \\ \exp\{-\frac{is\beta}{\sqrt{2}}\}\big[\cos(2\sqrt{it}) - \frac{\beta it}{\sqrt{2is}} \sin(2\sqrt{is})\big]^{-1/2}, & |\beta| = 1, \\ (1 + s^2 - 2it)^{-1/2}, & |\beta| > 1. \end{cases}$$

For $|\beta| < 1$, $f(s,t)$ is the characteristic function of the pair $(\xi, 1)$, where $\xi \sim N(0,1)$. Consequently, $(U_n, V_n) \xrightarrow{d_\beta} (\xi, 1)$, which by (5.4.6) implies

$$d_n(\beta)(\widehat{\beta}_{n,LS} - \beta) = U_n V_n^{-1} \xrightarrow{d_\beta} \xi \sim N(0,1).$$

For $|\beta| > 1$, $f(s,t)$ is the characteristic function of the pair $(\xi\eta, \eta^2)$, where ξ and η are independent standard normal random variables, which is shown by a straightforward calculation. Hence $(U_n, V_n) \xrightarrow{d_\beta} (\xi\eta, \eta^2)$ and $U_n V_n^{-1} \xrightarrow{d_\beta} \xi/\eta \sim K(0,1)$. Consequently, $d_n(\beta)(\widehat{\beta}_{n.LS} - \beta) \xrightarrow{d_\beta} K(0,1)$.

Now, let $\beta = 1$; the case $\beta = -1$ is handled similarly. Then by (5.4.2)

$$U_n = \frac{\sqrt{2}}{n}\sum_{i=1}^{n}\varepsilon_i u_{i-1} = \frac{\sqrt{2}}{n}\sum_{i=2}^{n}\sum_{j=1}^{i-1}\varepsilon_j\varepsilon_i, \tag{5.4.7}$$

$$V_n = \frac{2}{n^2}\sum_{i=1}^{n}u_{i-1}^2 = \frac{2}{n^2}\sum_{i=2}^{n}\Big(\sum_{j=1}^{i-1}\varepsilon_j\Big)^2. \tag{5.4.8}$$

Consider the random process

$$\nu_n(t) = n^{-1/2}\sum_{i\le nt}\varepsilon_i, \qquad t\in[0,1],$$

and let $\Delta\nu_n\big(\frac{i}{n}\big) := \nu_n\big(\frac{i}{n}\big) - \nu_n\big(\frac{i-1}{n}\big)$. Then (5.4.7), (5.4.8) imply

$$U_n = \sqrt{2}\sum_{i=1}^{n}\nu_n\Big(\frac{i-1}{n}\Big)\Delta\nu_n\Big(\frac{i}{n}\Big), \qquad V_n = 2\sum_{i=1}^{n}\nu_n^2\Big(\frac{i-1}{n}\Big)\frac{1}{n}.$$

By the invariance principle (see, for example, Billingsley [**6**], Chapter 5) $\nu_n(t)$ weakly converges to $\nu(t)$ in the Skorokhod space $D[0,1]$. Therefore, the general results on weak convergence (see, for example, Jacod and Shiryaev [**47**]) imply

$$(U_n, V_n) \xrightarrow{d_1} \Big(\sqrt{2}\int_0^1 \nu(t)d\nu(t),\, 2\int_0^1 \nu^2(t)dt\Big).$$

Consequently,

$$d_n(1)(\widehat{\beta}_{n,LS} - 1) = U_nV_n^{-1} \xrightarrow{d_1} \frac{\int_0^1\nu(t)d\nu(t)}{\sqrt{2}\int_0^1\nu^2(t)dt} = \frac{\nu^2(1)-1}{2\sqrt{2}\int_0^1\nu^2(t)dt},$$

since $\int_0^1\nu(t)d\nu(t) = 2^{-1}(\nu^2(1)-1)$. This completes the proof. □

It is worth noting that for $|\beta|\le 1$ the conclusion of Theorem 5.4.1 remains valid for the model (5.4.1) when $\{\varepsilon_i\}$ are i.i.d. random variables with zero mean and a finite variance and u_0 is any random variable with a finite second moment independent of $\{\varepsilon_i,\, i=1,2,\dots\}$. This can be proved for $|\beta|<1$ quite similarly to Theorem 5.3.1, and for $|\beta|=1$ similarly to Theorem 5.4.1.

For $|\beta|>1$ the limiting distribution of $d_n(\widehat{\beta}_{n,LS}-\beta)$ depends on the distribution of ε_i (see Koul and Pflug [**55**]).

In view of Theorem 5.4.1, $d_n(\beta_0)(\widehat{\beta}_{n,LS}-\beta_0)$ can be used as a test statistic for $H_0\colon \beta=\beta_0$ for any $\beta_0\in\mathbb{R}^1$. In case $|\beta_0|<1$, it has the same asymptotic distribution under local alternatives as in 5.3.2. This distribution can be found for $|\beta_0|\ge 1$ as well, but we will not take up this matter here.

5.5. Least absolute deviations procedures

We continue to consider the difference autoregression relation (5.2.1). In this section we study the least absolute deviations estimator for β and related tests for hypotheses about β.

5.5.1. Least absolute deviations estimator. We impose the following conditions.

CONDITION 5.5(i). $\mathbf{E}\varepsilon_1 = 0$, $\mathbf{E}|\varepsilon_1|^{2+\Delta} < \infty$ *for some* $\Delta > 0$.

CONDITION 5.5(ii). *$G(x)$ has a density $g(x)$ such that $\sup_x g(x) < \infty$, $g(x)$ satisfies the Lipschitz condition at $x = 0$, and $g(0) > 0$.*

By Lemma 5.2.1 under Condition 5.5(i) there exists a stationary solution of equation (5.2.1) which has the form (5.2.2), and the series in (5.2.2) converges in $L^{2+\Delta}$.

Let $u_0, u_1, \ldots, u_n$ be observations satisfying (5.2.2). We will study here the asymptotic behavior of the least absolute deviations (LAD) estimator $\widehat{\beta}_{n,LD}$ for β, which is defined as a solution to the problem

$$L_n^{LD}(\theta) := \sum_{k=1}^{n} |u_k - \theta u_{k-1}| \Longrightarrow \inf_{\theta \in \mathbb{R}^1}. \tag{5.5.1}$$

Obviously, $L_n^{LD}(\theta)$ is a convex function of θ, linear on each interval $(z_{(k-1)}, z_{(k)})$. Here $z_{(k)}$ denotes the kth order statistic among $z_1, \ldots, z_n$, where $z_k = u_k/u_{k-1}$.

The derivative of $L_n^{LD}(\theta)$ at any point θ different from z_k, $k = 1, \ldots, n$, equals

$$l_n^{LD}(\theta) = \sum_{k=1}^{n} u_{k-1} \operatorname{sign}(u_k - \theta u_{k-1}). \tag{5.5.2}$$

Since $G(x)$ is continuous, $l_n^{LD}(\theta) \neq 0$ a.s. in each interval $(z_{(k-1)}, z_{(k)})$. Therefore the solution of the problem (5.5.1) is a.s. unique and is given by one of the points z_k. It can be found, for example, by enumeration. It can be also presented in an explicit form. If we place probability masses $p_k = |u_k| / \sum_{i=1}^{n} |u_i|$ at points z_k, then the solution of (5.5.1) is the median of the distribution $(z_k, p_k, k = 1, \ldots, n)$.

Rewriting $l_n^{LD}(\theta)$ in the form

$$l_n^{LD}(\theta) = \sum_{k=1}^{n} |u_{k-1}| \operatorname{sign}(u_k/u_{k-1} - \theta)$$

we see that the trajectories of $l_n^{LD}(\theta)$ are nonincreasing step functions with jumps at the points z_k. For a continuous $G(x)$ the solution $\widehat{\beta}_{n,LD}$ of (5.5.1) coincides with the a.s. unique solution of the equation

$$l_n^{LD}(\theta) \div 0 \tag{5.5.3}$$

(with $\div$ in (5.5.3) denoting the crossing of zero level), which we will study henceforth.

Under Conditions 5.5(i, ii)

$$\mathbf{E}_\beta n^{-1} l_n^{LD}(\theta) = \Lambda_{LD}(\theta),$$

where $\Lambda_{LD}(\theta) = -2\mathbf{E}_\beta u_1 G((\theta - \beta)u_1)$ is differentiable with respect to θ with the derivative at $\theta = \beta$ equal to

$$\lambda_{LD}(\beta) := \Lambda'_{LD}(\beta) = -2g(0)\mathbf{E}_\beta u_1^2 = -2g(0)\mathbf{E}\varepsilon_1^2(1 - \beta^2)^{-1}.$$

Indeed,

$$\begin{aligned}\Lambda_{LD}(\theta) &= \mathbf{E}_\beta u_1 \operatorname{sign}(u_2 - \theta u_1) = \mathbf{E}_\beta u_1\{1 - 2I(\varepsilon_2 < (\theta - \beta)u_1)\} \\ &= -2\mathbf{E}_\beta \mathbf{E}_\beta\{u_1 I(\varepsilon_2 < (\theta - \beta)u_1)/u_1\} = -2\mathbf{E}_\beta u_1 G\big((\theta - \beta)u_1\big)\end{aligned}$$

due to the independence of ε_2 and u_1. Further, by the Taylor formula,

$$\Lambda_{LD}(\theta) = -2\mathbf{E}_\beta u_1\{G(0) + g(\theta_1)u_1(\theta - \beta)\} = -2\mathbf{E}_\beta u_1^2 g(\theta_1)(\theta - \beta),$$

where $|\theta_1| \leq |(\theta - \beta)u_1|$. The function $g(x)$ is bounded and satisfies the Lipschitz condition at $x = 0$; hence, it satisfies Hölder's condition of any order $0 < \delta \leq 1$. Take $\delta = \min(\Delta, 1)$. Then

$$\Lambda_{LD}(\theta) = -2g(0)\mathbf{E}_\beta u_1^2(\theta - \beta) + O(|\theta - \beta|^{1+\delta}),$$

which implies the expression for $\Lambda'_{LD}(\beta)$.

The following Theorem 5.5.1 provides a linear stochastic expansion for the process $n^{-1/2}l_n^{LD}(\theta)$ which holds uniformly in θ over a vicinity of $\theta = \beta$ of size $O(n^{-1/2})$.

THEOREM 5.5.1. *Let Conditions* 5.5(i, ii) *be fulfilled. Then for any* $0 < \Theta < \infty$

$$\sup_{|\theta| \leq \Theta} \left| n^{-1/2} l_n^{LD}(\beta + n^{-1/2}\theta) - n^{-1/2} l_n^{LD}(\beta) - \lambda_{LD}(\beta)\,\theta \right| = o_p(1).$$

The proof of this theorem will be given in §5.8.

The following Theorem 5.5.2 is easily obtained from Theorem 5.5.1. Its proof is also given in §5.8.

THEOREM 5.5.2. *Let Conditions* 5.5(i, ii) *be fulfilled. Then:*
1. $\sqrt{n}(\widehat{\beta}_{n,LD} - \beta) = -\lambda_{LD}^{-1}(\beta) n^{-1/2} l_n^{LD}(\beta) + o_p(1)$;
2. $\sqrt{n}(\widehat{\beta}_{n,LD} - \beta) \xrightarrow{d_\beta} N\big(0, \sigma_{LD}^2(\beta)\big)$, *where* $\sigma_{LD}^2(\beta) = (1 - \beta^2)\big((2g(0))^2\mathbf{E}\varepsilon_1^2\big)^{-1}$.

Theorem 5.5.2 remains valid with condition $\mathbf{E}\varepsilon_1 = 0$ in Condition 5.5(i) replaced by $G(0) = 1/2$. In this case $\sigma_{LD}^2(\beta) = \big((2g(0))^2\mathbf{E}_\beta u_1^2\big)^{-1}$.

Regarding Theorem 5.5.2 it is to be noted that the asymptotic normality of the LAD estimator in autoregression was rigorously justified relatively recently. In particular, the asymptotic normality of the LAD estimator was established in Pollard [**72**] by a unified method for several models, such as linear regression with independent errors, linear regression with random regressors, autoregression of an arbitrary order. Furthermore, the asymptotic normality of $\widehat{\beta}_{n,LD}$ in autoregression is established under weaker assumptions than 5.5(i, ii), in particular, under moment conditions $\mathbf{E}\varepsilon_1 = 0$, $\mathbf{E}\varepsilon_1^2 < \infty$. The argument in Pollard [**72**] hinges on the convexity of the objective function to be minimized (which is $L_n^{LD}(\theta)$ in the one-parameter autoregression). A pointwise expansion obtained for this function implies the uniform one on account of convexity, which leads to asymptotic normality. Our approach is different. We treat $\widehat{\beta}_{n,LD}$ as a root of the equation (5.5.3) and analyze the "derivative" $l_n^{LD}(\theta)$ of the objective function, which possesses no convexity. Nevertheless we are able to establish a uniform stochastic expansion for $l_n^{LD}(\theta)$ which yields Theorem 5.5.2 as a simple consequence. We will see in Chapters 6 and 7 that our approach is applicable to the study of estimators (based on signs or ranks) which cannot be associated with any convex minimization problem. Moreover, a similar approach will be used in §5.7 to analyse a nonstationary autoregression model. The definition of $\widehat{\beta}_{n,LD}$ as a root of equation (5.5.3) is also used in §5.6 to derive the influence functional of this estimator.

Our last comment on Theorem 5.5.2 concerns the behavior of $\widehat{\beta}_{n,LD}$ when $\mathbf{E}\varepsilon_1^2 = \infty$. It is quite similar to that of the least squares estimator discussed in

5.3.2. For example, for ε_i's having the Cauchy distribution,

$$n(\widehat{\beta}_{n,LD} - \beta) = O_p(1)$$

(see Pollard [**72**]). Other results and references can be found in Bloomfield and Steiger [**7**], §3.2.

Now we continue to consider the model (5.2.1) under Conditions 5.5(i, ii). The limiting variance $\sigma^2_{LD}(\beta)$ obtained in Theorem 5.5.2 can be estimated from the observations by making use of the following

COROLLARY 5.5.1. *If* $\sqrt{n}(\widehat{\beta}_n - \beta) = O_p(1)$, *then under Conditions* 5.5(i, ii)

$$n^{-1/2} l_n^{LD}(\widehat{\beta}_n) = n^{-1/2} l_n^{LD}(\beta) + \lambda_{LD}(\beta)\sqrt{n}(\widehat{\beta}_n - \beta) + o_p(1). \tag{5.5.4}$$

PROOF. For any $\varepsilon > 0$ and $0 < \Theta < \infty$

$$\begin{aligned}
&\mathbf{P}_\beta\{|n^{-1/2} l_n^{LD}(\widehat{\beta}_n) - n^{-1/2} l_n^{LD}(\beta) - \lambda_{LD}(\beta)\sqrt{n}(\widehat{\beta}_n - \beta)| > \varepsilon\} \\
&\quad \le \mathbf{P}_\beta\{|n^{-1/2} l_n^{LD}(\widehat{\beta}_n) - n^{-1/2} l_n^{LD}(\beta) - \lambda_{LD}(\beta)\sqrt{n}(\hat{\beta}_n - \beta)| > \varepsilon, \\
&\qquad\qquad\qquad\qquad\qquad\qquad\qquad\qquad |\sqrt{n}(\widehat{\beta} - \beta)| \le \Theta\} \\
&\qquad\qquad + \mathbf{P}_\beta\{|\sqrt{n}(\widehat{\beta}_n - \beta)| > \Theta\} \\
&\quad \le \mathbf{P}_\beta\{\sup_{|\theta| \le \Theta} |n^{-1/2} l_n^{LD}(\beta + n^{-1/2}\theta) - n^{-1/2} l_n^{LD}(\beta) - \lambda_{LD}(\beta)\,\theta| > \varepsilon\} \\
&\qquad\qquad + \mathbf{P}_\beta\{|\sqrt{n}(\widehat{\beta}_n - \beta)| > \Theta\}.
\end{aligned}$$

On account of Theorem 5.5.1 and assumption $\sqrt{n}(\widehat{\beta}_n - \beta) = O_p(1)$ both probabilities in the right-hand side become arbitrarily small for sufficiently large n and Θ. □

Letting $\widehat{\beta}_n = \widehat{\beta}_{n,LD}$ in (5.5.4), we obtain for any constant $h \ne 0$

$$\widehat{\lambda}_n = \big(l_n^{LD}(\widehat{\beta}_{n,LD} + hn^{-1/2}) - n^{-1/2} l_n^{LD}(\widehat{\beta}_{n,LD})\big)(\sqrt{n}h)^{-1} = \lambda_{LD}(\beta) + o_p(1).$$

Since $\sigma^2_{LD}(\beta) = \mathbf{E}_\beta u_1^2/\big(\lambda_{LD}(\beta)\big)^2$, a consistent estimate for $\sigma^2_{LD}(\beta)$ is given by

$$\widehat{\sigma}^2_{n,LD} = s_n^2/\widehat{\lambda}_n^2,$$

where $s_n^2 = n^{-1}\sum_{k=1}^n u_k^2$ is a consistent estimate for $\mathbf{E}_\beta u_1^2$ (see (5.3.6)). Hence, with $\xi_{1-p/2}$ denoting the $(1-p/2)$-quantile of the standard normal distribution, the set

$$\left\{\theta\colon \widehat{\beta}_{n,LD} - \frac{\xi_{1-p/2}\widehat{\sigma}_{n,LD}}{\sqrt{n}} < \theta < \widehat{\beta}_{n,LD} + \frac{\xi_{1-p/2}\widehat{\sigma}_{n,LD}}{\sqrt{n}}\right\}$$

is a confidence interval for β of asymptotic level $1 - p$.

5.5.2. Tests based on the LAD estimator. Consider testing the hypothesis $H_0\colon \beta = \beta_0$. Take

$$T_{n,LD}(\beta_0) = s_n^{-1} n^{-1/2} l_n^{LD}(\beta_0)$$

as the test statistic for this problem. This statistic is computationally simpler than $\widehat{\beta}_{n,LD}$. Conditions 5.5(i, ii) imply the strong mixing condition for the process $\{u_i\}$ with an exponentially decreasing mixing coefficient (see Mokkadem [**69**]). Obviously, the process $\{u_{i-1}\operatorname{sign}\varepsilon_i,\ i \in \mathbb{Z}\}$ also satisfies the strong mixing condition with an exponentially decreasing mixing coefficient; hence, by the central limit theorem for such processes (see, for example, Ibragimov and Linnik [**45**], Theorem 18.5.3) one has under H_0

$$n^{-1/2} l_n^{LD}(\beta_0) = n^{-1/2} \sum_{k=1}^{n} u_{k-1} \operatorname{sign} \varepsilon_k \xrightarrow{d_{\beta_0}} N(0, \mathbf{E}_{\beta_0} u_1^2).$$

Therefore the test statistic $T_{n,LD}(\beta_0)$ has asymptotically the standard normal distribution under H_0.

Consider now the behavior of the test statistic under alternatives. The simplest case is when the alternative $H_1\colon \beta \neq \beta_0$ is fixed. Since $\{u_{i-1} \operatorname{sign}(u_i - \beta_0 u_{i-1}),\ i \in \mathbb{Z}\}$ is a strongly mixing stationary sequence, one has by the law of large numbers for such sequences (see, for example, Hannan [**37**], Chapter 4, §2, Theorem 2)

$$n^{-1} l_n^{LD}(\beta_0) \xrightarrow{\mathbf{P}_\beta} \Lambda_{LD}(\beta_0) = -2\mathbf{E}_\beta u_1 G\big((\beta_0 - \beta)u_1\big). \tag{5.5.5}$$

Under Conditions 5.5(i, ii) the function $\Lambda_{LD}(\beta_0)$ is differentiable with respect to β_0 at any point β_0, its derivative is nonpositive with $\Lambda'_{LD}(\beta) = \lambda_{LD}(\beta) = -2g(0)\mathbf{E}_\beta u_1^2 < 0$ and $\Lambda_{LD}(\beta) = 0$.

From the above properties we see that $\Lambda_{LD}(\beta_0)$ vanishes only at $\beta_0 = \beta$. Together with (5.5.5) this implies that $n^{-1/2} l_n^{LD}(\beta_0)$ under H_1 diverges to infinity. Consequently, the test based on $T_{n,LD}(\beta_0)$ is consistent against H_1.

Consider now alternatives approaching H_0,

$$H_{1n}\colon \beta = \beta_n, \qquad \beta_n \to \beta_0.$$

Under H_{1n}

$$\sqrt{n}\big[n^{-1} l_n^{LD}(\beta_0) - \Lambda_{LD}(\beta_0)\big] = O_p(1). \tag{5.5.6}$$

Indeed, the random variable in the left-hand side of (5.5.6) is equal a.s. to

$$n^{-1/2} \sum_{i=1}^{n} u_{i-1} + n^{-1/2} \sum_{i=1}^{n} \big[-2u_{i-1} I\big(\varepsilon_i < (\beta_0 - \beta_n)u_{i-1}\big) - \Lambda_{LD}(\beta_0)\big]. \tag{5.5.7}$$

It follows from (5.2.1) with $\beta = \beta_n$ that

$$n^{-1/2} \sum_{i=1}^{n} u_i = \beta_n n^{-1/2} \sum_{i=1}^{n} u_{i-1} + n^{-1/2} \sum_{i=1}^{n} \varepsilon_i,$$

which implies

$$n^{-1/2} \sum_{i=1}^{n} u_{i-1} = (1 - \beta_n)^{-1} n^{-1/2} \sum_{i=1}^{n} \varepsilon_i + o_p(1) \xrightarrow{d_{\beta_n}} N\big(0, \mathbf{E}\varepsilon_1^2 (1 - \beta_0)^{-2}\big).$$

Therefore, the first sum in (5.5.7) is $O_p(1)$. The second sum in (5.5.7) is centered and its terms are uncorrelated (cf. a similar argument concerning the sequence $\xi_1, \dots, \xi_n$ in the proof of Lemma 5.8.1). This implies that the variance of this sum is bounded uniformly in n and hence this sum is $O_p(1)$. Thus (5.5.6) holds true.

Moreover, we have $n^{-1/2} l_n^{LD}(\beta_0) = \sqrt{n}\big[n^{-1} l_n^{LD}(\beta_0) - \Lambda_{LD}(\beta_0)\big] + \sqrt{n}\Lambda_{LD}(\beta_0)$, so that, whenever $\sqrt{n}(\beta_n - \beta_0) \to \infty$,

$$\sqrt{n}\Lambda_{LD}(\beta_0) \sim -2g(0)\sqrt{n}(\beta_0 - \beta_n)\mathbf{E}_{\beta_0} u_1^2 \to \infty.$$

Consequently, if $\sqrt{n}(\beta_n - \beta_0) \to \infty$, the statistic $n^{-1/2} l_n^{LD}(\beta_0)$ diverges to infinity and the corresponding test is consistent against such alternatives H_{1n}.

Consider now the alternatives

$$H_{1n}(a)\colon \beta = \beta_n := \beta_0 + a\,n^{-1/2} + o(n^{-1/2}) \quad \text{with some constant } a.$$

The analysis of the test statistic under the alternative $H_{1n}(a)$ is based on the following Theorem 5.5.3, which in fact is valid under more general alternative H_{1n}.

THEOREM 5.5.3. *Assume that Conditions* 5.5(i, ii) *are fulfilled and let the alternative* H_{1n} *hold. Then for any* $0 < \Theta < \infty$

$$\sup_{|\theta| \le \Theta} \left| n^{-1/2} l_n^{LD}(\beta_n + n^{-1/2}\theta) - n^{-1/2} l_n^{LD}(\beta_n) - \lambda_{LD}(\beta_0)\theta \right| = o_p(1).$$

The proof of Theorem 5.5.3 is quite similar to that of Theorem 5.5.1 and hence will be omitted.

Theorem 5.5.3 implies

COROLLARY 5.5.2. *Let the alternative* $H_{1n}(a)$ *hold. If* $\sqrt{n}(\widehat{\beta}_n - \beta_n) = O_p(1)$, *then under Conditions* 5.5(i, ii)

$$n^{-1/2} l_n^{LD}(\widehat{\beta}_n) = n^{-1/2} l_n^{LD}(\beta_n) + \lambda_{LD}(\beta_0)\sqrt{n}(\widehat{\beta}_n - \beta_n) + o_p(1).$$

The proof is similar to the proof of Corollary 5.5.1.

THEOREM 5.5.4. *Assume that Conditions* 5.5(i, ii) *are fulfilled and let the alternative* $H_{1n}(a)$ *hold. Then:*

1.

$$T_{n,LD}(\beta_0) = (\mathbf{E}_{\beta_0} u_1^2)^{-1/2} n^{-1/2} \sum_{k=1}^{n} u_{k-1} \operatorname{sign} \varepsilon_k + 2g(0)(\mathbf{E}\varepsilon_1^2)^{1/2}(1-\beta_0^2)^{-1/2} a + o_p(1)$$

and

$$T_{n,LD}(\beta_0) \xrightarrow{d_{\beta_n}} N(2g(0)(\mathbf{E}\varepsilon_1^2)^{1/2}(1-\beta_0^2)^{-1/2} a, 1);$$

2.

$$\sqrt{n}(\widehat{\beta}_{n,LD} - \beta_0) = -\lambda_{LD}^{-1}(\beta_0) n^{-1/2} l_n^{LD}(\beta_n) + a + o_p(1)$$

and

$$\sqrt{n}(\widehat{\beta}_{n,LD} - \beta_0) \xrightarrow{d_{\beta_n}} N\big(a, \sigma_{LD}^2(\beta_0)\big),$$

where $\sigma_{LD}^2(\beta) = (1-\beta^2)\big((2g(0))^2 \mathbf{E}\varepsilon_1^2\big)^{-1}$.

PROOF. Assertion 1 follows from (5.3.14), Corollary 5.5.2 (with $\widehat{\beta}_n = \beta_0$) and the fact that under H_{1n}

$$(\mathbf{E}_{\beta_0} u_1^2)^{-1/2} n^{-1/2} l_n^{LD}(\beta_n) = (\mathbf{E}_{\beta_0} u_1^2)^{-1/2} n^{-1/2} \sum_{k=1}^{n} u_{k-1} \operatorname{sign} \varepsilon_k \xrightarrow{d_{\beta_n}} N(0,1),$$

which is established similarly to (5.3.15). Assertion 2 is shown similarly to the proof of Theorem 5.5.2. □

Theorem 5.5.4 implies, in particular, that for $a = 0$, i.e., when $\sqrt{n}(\beta_n - \beta_0) = o(1)$, the hypotheses H_0 and $H_{1n}(0)$ are asymptotically indistinguishable.

It is worth noting that the asymptotic distributions in Theorem 5.5.4 could be established by means of Le Cam's Third Lemma using Theorem 5.5.2, since under Condition 5.5(i) and some additional conditions on $G(x)$ (see 5.3.2) H_0 and $H_{1n}(a)$ are contiguous.

Let us find the ARE $e_{LD,LS}$ of the test based on $T_{n,LD}(\beta_0)$ relative to the test based on $T_{n,LS}(\beta_0) = (1-\beta_0^2)^{-1/2}\sqrt{n}(\widehat{\beta}_{n,LS} - \beta_0)$. Since under $H_{1n}(a)$

$$T_{n,LS}(\beta_0) \xrightarrow{d_{\beta_n}} N\big(a(1-\beta_0^2)^{-1/2}, 1\big)$$

(see (5.3.12)), one has in view of part 1 of Theorem 5.5.4 $e_{LD,LS} = \big(2g(0)\big)^2 \mathbf{E}\varepsilon_1^2$, which does not depend on β and on the scale parameter of $G(x)$. The expression for $e_{LD,LS}$ is the same as for the ARE of the sample median relative to the sample mean in an i.i.d. sample (see Lehmann [**61**], Chapter 5, §3). Therefore, Theorem 3.3 in Lehmann [**61**], Chapter 5, says that within the class $\Im$ of distribution functions having a symmetric density function with maximum at zero, the problem

$$e_{LD,LS} = \big(2g(0)\big)^2 \mathbf{E}\varepsilon_1^2 \Longrightarrow \inf_{G\in\Im}$$

is solved by $G(x) \sim R(-1/2, 1/2)$, in which case $e_{LD,LS} = 1/3$, with $R(a,b)$ denoting the uniform distribution on $[a,b]$.

For $G(x)$ normal, $e_{LD,LS} = 2/\pi \approx 0.64$; for $G(x)$ a Laplace d.f., $e_{LD,LS} = 2$; for $G(x)$ a logistic d.f., $e_{LD,LS} = \pi^2/12 \approx 0.82$.

For heavy-tailed $G(x)$, $e_{LD,LS}$ may become arbitrarily large. For example, for Tukey's d.f. $G(x) = (1-\varepsilon)\Phi(x) + \varepsilon\Phi(x/\tau)$,

$$e_{LD,LS} = \frac{2}{\pi}(1-\varepsilon+\varepsilon/\tau)^2(1-\varepsilon+\varepsilon\tau^2) \to \infty \qquad \text{as} \quad \tau \to \infty.$$

Of course, the above statements hold for the ARE of the estimators $\widehat{\beta}_{n,LD}$ and $\widehat{\beta}_{n,LS}$, which also equals $e_{LD,LS}$. If $G(x)$ is the Laplace d.f., then the ARE of $\widehat{\beta}_{n,LD}$ with respect to the maximum likelihood estimator $\widehat{\beta}_{n,ML}$ equals one. In this case $\widehat{\beta}_{n,LD}$ is an asymptotically efficient estimator.

5.5.3. Weighted least absolute deviations estimators. We conclude §5.5 with a brief discussion of estimators which may be termed weighted LAD estimators. Namely, define an estimator for β (to be denoted by $\widehat{\beta}_{n,LDW}$) as a (measurable) solution of the problem

$$L_n^{LDW}(\theta) := \sum_{k=1}^{n} |\varphi(u_{k-1})(u_k - \theta u_{k-1})| \Longrightarrow \inf_{\Theta\in\mathbb{R}^1}. \tag{5.5.8}$$

Like $L_n^{LD}(\theta)$, the function $L_n^{LDW}(\theta)$ is convex; hence, this solution always exists. One may apply different approaches to its study. For example, we can consider the equivalent equation

$$l_n^{LDW}(\theta) := \sum_{k=1}^{n} |\varphi(u_{k-1})| u_{k-1} \operatorname{sign}(u_k - \theta u_{k-1}) \div 0. \tag{5.5.9}$$

Assume that the weight function $\varphi(u_{k-1})$ and $G(x)$ satisfy the following assumptions.

CONDITION 5.5(iii). $\mathbf{E}_\beta|u_1|^{2+\Delta}|\varphi(u_1)| < \infty$ *for some* $\Delta > 0$ *and, moreover,* $0 < \mathbf{E}_\beta u_1^2|\varphi(u_1)| < \infty$, $\mathbf{E}_\beta u_1^2\varphi^2(u_1) < \infty$.

CONDITION 5.5(iv). $\mathbf{E}_\beta u_1|\varphi(u_1)| = 0$ *or* $G(0) = 1/2$.

CONDITION 5.5(v). $n^{-1/2}\max_{1\le k\le n}|u_{k-1}\varphi(u_{k-1})| = o_p(1)$.

Similarly to the proof of Theorem 5.5.2 we obtain that under Conditions 5.2(iii) and 5.5(ii–v)

$$\sqrt{n}(\widehat{\beta}_{n,LDW} - \beta) = \frac{1}{2g(0)\mathbf{E}_\beta u_1^2|\varphi(u_1)|}n^{-1/2}l_n^{LDW}(\beta) + o_p(1), \tag{5.5.10}$$

$$\sqrt{n}(\widehat{\beta}_{n,LDW} - \beta) \xrightarrow{d_\beta} N\big(0, \sigma_{LDW}^2(\beta)\big), \tag{5.5.11}$$

where

$$\sigma_{LDW}^2(\beta) = \mathbf{E}_\beta[u_1^2\varphi^2(u_1)]\big\{[2g(0)]^2[\mathbf{E}_\beta u_1^2|\varphi(u_1)|]^2\big\}^{-1}.$$

If $\mathbf{E}\varepsilon_1^2 < \infty$ and we are only interested in point estimation of β from $u_1, \dots, u_n$, then the weighted LAD estimators are of no need. Indeed,

$$\big(\mathbf{E}_\beta u_1^2|\varphi(u_1)|\big)^2 \le \mathbf{E}_\beta u_1^2\mathbf{E}_\beta u_1^2\varphi^2(u_1)$$

which implies

$$\sigma_{LDW}^2(\beta) \ge \big\{[2g(0)]^2\mathbf{E}_\beta u_1^2\big\}^{-1} = \sigma_{LD}^2(\beta).$$

The equality holds only for a constant $\varphi(u_1)$, in which case $\widehat{\beta}_{n,LDW} = \widehat{\beta}_{n,LD}$.

However, in certain situations the weighted estimators are useful. One of these situations arises when $\mathbf{E}\varepsilon_1^2 = \infty$. In this case the sufficient conditions for LAN (see 5.3.2) implying $\sqrt{n}$-asymptotic normality of commonly used estimators fail, which makes it difficult to apply these estimators for interval estimation and hypothesis testing. At the same time it is easy to construct a $\sqrt{n}$-asymptotically normal $\widehat{\beta}_{n,LDW}$ estimator. For example, for $\varphi(u_{k-1}) = 1/u_{k-1}$ the equation (5.5.9) becomes

$$\sum_{k=1}^{n} \operatorname{sign} u_{k-1} \operatorname{sign}(u_k - \theta u_{k-1}) \div 0, \tag{5.5.12}$$

and one of the solutions of (5.5.12) which makes it an exact equality is the median $\widehat{\beta}_{n,M}$ of the array $u_k/u_{k-1}, k = 1, \dots, n$. For odd n $\widehat{\beta}_{n,M}$ is a unique solution of (5.5.12), while for even n the solutions form an interval, with its midpoint being taken as $\widehat{\beta}_{n,M}$.

Conditions $G(0) = 1/2$, $\mathbf{E}|\varepsilon_1|^{1+\Delta} < \infty$ imply Conditions 5.2(iii) and 5.5(iii–v), so that assuming additionally 5.5(ii) we obtain

$$\sqrt{n}(\widehat{\beta}_{n,M} - \beta) = \frac{1}{2g(0)\mathbf{E}_\beta|u_1|}n^{-1/2}\sum_{k=1}^{n}\operatorname{sign} u_{k-1}\operatorname{sign}\varepsilon_k + o_p(1) \xrightarrow{d_\beta} N\big(0, \sigma_M^2(\beta)\big),$$

where

$$\sigma_M^2(\beta) = \big(2g(0)\mathbf{E}_\beta|u_1|\big)^{-2}.$$

The median estimate $\widehat{\beta}_{n,M}$ has been thoroughly investigated in Boldin [**11**]. It was pointed out therein that

$$2^{-1}\left(\sum_{k=1}^{n} \operatorname{sign} u_{k-1} \operatorname{sign} \varepsilon_k + n\right) \sim \mathrm{Bi}(n, 1/2),$$

under mild assumptions

$$\mathbf{P}(\varepsilon_1 < 0) = \mathbf{P}(\varepsilon_1 > 0) = 1/2, \qquad \mathbf{E}\log^+ |\varepsilon_1| < \infty.$$

Therefore, under these assumptions

$$\sum_{k=1}^{n} \operatorname{sign} u_{k-1} \operatorname{sign}(u_k - \theta u_{k-1})$$

is a pivotal statistic, which can be used for interval estimation of β for finite n. Confidence intervals based on commonly used estimators $\widehat{\beta}_{n,LS}$ and $\widehat{\beta}_{n,LD}$ are free of $G(x)$ only in an asymptotic sense under a stronger condition $\mathbf{E}\varepsilon_1^2 < \infty$.

In the next section the weighted LAD estimators will be considered in the context of robustness against outliers in observations $u_0, \dots, u_n$. It will be shown that the estimators $\widehat{\beta}_{n,LDW}$ with suitably chosen φ have bounded influence functionals (gross error sensitivity) in a contamination model with independent outliers. This is an important and appealing property of the weighted LAD estimators. The commonly used estimators $\widehat{\beta}_{n,LS}$ and $\widehat{\beta}_{n,LD}$ have unbounded influence functionals.

5.6. Influence functionals of least squares and least absolute deviations estimators

The aim of the present section is to characterize quantitatively the robustness of the estimators $\widehat{\beta}_{n,LS}$, $\widehat{\beta}_{n,LD}$, $\widehat{\beta}_{n,LDW}$ against outliers in observations $u_0, u_1, \dots, u_n$. There are different ways to do that, and we will apply the approach based on the influence functional. For i.i.d. observations a related concept of the influence curve was introduced by Hampel [**35**]. This concept turned out to be a very useful heuristic tool in the theory of robustness (see Huber [**44**], §1.5). We will follow the approach presented in Martin and Yohai [**66**].

Suppose we have contaminated observations $\mathbf{Y}_n = (y_0, \dots, y_n)$, where

$$y_i = u_i + z_i^\gamma \xi_i, \qquad i \in \mathbb{Z}. \tag{5.6.1}$$

We assume that the variables $\{u_i\}$ in (5.6.1) satisfy (5.2.1), $\{z_i^\gamma\}$ form a Bernoulli sequence of i.i.d. random variables taking values 1 and 0 with probabilities γ (contamination level) and $1-\gamma$, $0 \le \gamma \le 1$, and $\{\xi_i\}$ are i.i.d. random variables with distribution μ_ξ from a class of distributions $\mathfrak{M}_\xi$; the sequences $\{u_i\}$, $\{z_i^\gamma\}$, $\{\xi_i\}$ are mutually independent. Thus we consider the simplest model of independent outliers. More complicated models, for example, those of patchy outliers (see Martin and Yohai [**66**]) can be treated in a similar way.

For contaminated observations, commonly used estimators fail to be not only asymptotically normal (with natural normalization), but even consistent. In order to characterize the performance of an estimator $\widehat{\beta}_n$ for β based on observations $\mathbf{Y}_n$, assume that there exists a limit $\widehat{\beta}_n \xrightarrow{\mathbf{P}_\beta} \theta_\gamma$, and $\theta_0 = \beta$. The simplest infinitesimal characteristic of the estimator robustness against contamination of data is the derivative

$$\mathrm{IF}(\boldsymbol{\theta}_\gamma, \mu_\xi) = \lim_{\gamma\to 0} \frac{\boldsymbol{\theta}_\gamma - \boldsymbol{\theta}_0}{\gamma}, \tag{5.6.2}$$

which is called the influence functional of the estimator $\widehat{\beta}_n$ (provided, of course, this derivative does exist). This quantity determines the main (linear) term of the asymptotic bias

$$\theta_\gamma - \theta_0 = \mathrm{IF}(\theta_\gamma, \mu_\xi)\gamma + o(\gamma). \tag{5.6.3}$$

.

It is preferable to use the estimators with finite *gross error sensitivity*

$$\mathrm{GES}(\mathfrak{M}_\xi, \theta_\gamma) := \sup_{\mu_\xi \in \mathfrak{M}_\xi} \big|\mathrm{IF}(\theta_\gamma, \mu_\xi)\big|.$$

In this case the main term of the asymptotic bias in (5.6.3) is uniformly small over all possible contamination distributions and small γ. Qualitatively, this means that for small γ even large outliers have little effect on the estimator.

We will consider the estimators determined by an equation of the form

$$\sum_{i=1}^{n} \psi_i(\mathbf{Y}_n, \boldsymbol{\theta}) = \mathbf{0}. \tag{5.6.4}$$

The class of estimators determined by (5.6.4) is fairly wide. Let us point out some particular cases. If

$$\psi_i(\mathbf{Y}_n, \theta) = \eta\big((1-\theta^2)^{1/2} y_{i-1}, y_i - \theta y_{i-1}\big),$$

where η is a bounded function, then (5.6.4) yields generalized M-estimators (GM-estimators) (see Martin [**64**], Krasker and Welsch [**56**]). The choice $\eta(\xi_1, \xi_2) = \xi_1 \psi(\xi_2)$ leads to ordinary M-estimators (see Martin [**64, 65**], Polyak and Tsypkin [**73**]), which include the least squares and least absolute deviations estimators. When

$$\psi_i(\mathbf{Y}_n, \theta) = \eta(y_i - \theta y_{i-1}, y_{i+1} - \theta y_i) + \theta\eta(y_i - \theta y_{i-1}, y_{i+2} - \theta y_{i+1}) + \dots$$
$$\dots + \theta^{n-1}\eta(y_i - \theta y_{i-1}, y_n - \theta y_{n-1}),$$

$i = 1, \dots, n-1$, $\psi_n(\mathbf{Y}_n, \theta) = 0$, one obtains RA-estimators (see Bustos and Yohai [**19**]).

We will assume the following conditions.

CONDITION 5.6(i). *There exists a function $\Lambda(\gamma, \theta)$ such that for some $\delta > 0$ and for all $0 \le \gamma \le \delta$, $|\theta - \beta| \le \delta$*

$$n^{-1} \sum_{i=1}^{n} \psi_i(\mathbf{Y}_n, \theta) \xrightarrow{\mathbf{P}_\beta} \Lambda(\gamma, \theta).$$

CONDITION 5.6(ii). $\Lambda(0, \beta) = 0$.

CONDITION 5.6(iii). *The partial derivatives $\frac{\partial}{\partial\gamma}\Lambda(\gamma, \theta)$ and $\frac{\partial}{\partial\theta}\Lambda(\gamma, \theta)$ exist and are continuous in (γ, θ) for $0 \le \gamma \le \delta$, $|\theta - \beta| \le \delta$.*

CONDITION 5.6(iv). $\lambda(\beta) := \frac{\partial}{\partial\theta}\Lambda(0,\beta) \neq 0$.

Under Conditions 5.6(i–iv) by the implicit function theorem the equation

$$\Lambda(\gamma,\theta) = 0 \tag{5.6.5}$$

in a neighborhood of $(0,\beta)$ determines a single-valued continuous function $\theta_\gamma = \theta(\gamma)$ such that $\theta_\gamma \to \theta_0 = \beta$ as $\gamma \to 0$. Moreover, θ_γ is differentiable and

$$\frac{d\theta_\gamma}{d\gamma}\bigg|_{\gamma=0} = -\lambda^{-1}(\beta)\frac{\partial}{\partial\gamma}\Lambda(0,\beta). \tag{5.6.6}$$

It is easy to see that under Conditions 5.6(i–iv) for small γ with probability tending to 1 there exists a root $\widehat{\beta}_n$ of equation (5.6.4) such that $\widehat{\beta}_n \xrightarrow{\mathbf{P}_\beta} \theta_\gamma$. Then (5.6.6) shows that the influence functional corresponding to $\widehat{\beta}_n$ is given by

$$\mathrm{IF}(\theta_\gamma,\mu_\xi) = -\lambda^{-1}(\beta)\frac{\partial}{\partial\gamma}\Lambda(0,\beta). \tag{5.6.7}$$

The influence functionals for M-, GM-, and RA-estimators have been obtained by Martin and Yohai [**66**] who dealt with smooth functions ψ_i in (5.6.4). We will consider nonsmooth functions ψ_i as well.

5.6.1. Influence functional of the least squares estimator. As an example, consider the least squares estimator $\widehat{\beta}_{n,LS}$. We assume here that

$$\mathbf{E}\varepsilon_1 = 0, \qquad \mathbf{E}\varepsilon_1^2 < \infty, \qquad \mathbf{E}\xi_1^2 < \infty.$$

The estimator $\widehat{\beta}_{n,LS}$ is defined by (5.6.4) with

$$\psi_i^{LS}(\mathbf{Y}_n,\theta) = y_{i-1}(y_i - \theta y_{i-1}). \tag{5.6.8}$$

The ergodicity of $\{y_i\}$ implies the convergence

$$n^{-1}\sum_{i=1}^n \psi_i^{LS}(\mathbf{Y}_n,\theta) \xrightarrow{\mathbf{P}_\beta} \Lambda_{LS}(\gamma,\theta)$$

for any γ, θ, where

$$\Lambda_{LS}(\gamma,\theta) = \mathbf{E}_\beta y_0(y_1 - \theta y_0) = \mathbf{E}_\beta u_0u_1 - \theta(\mathbf{E}_\beta u_0^2 + \gamma\mathbf{E}\xi_0^2) + \gamma^2\big(\mathbf{E}\xi_0\big)^2.$$

It is easily seen that

$$\Lambda_{LS}(0,\beta) = \mathbf{E}_\beta u_0\varepsilon_1 = 0$$

because u_0 and ε_1 are independent. Thus Conditions 5.6(i, ii) are fulfilled. Furthermore, Conditions 5.6(iii, iv) are easily seen to hold for any γ, θ with

$$\lambda_{LS}(\beta) := \frac{\partial}{\partial\theta}\Lambda_{LS}(0,\beta) = -\mathbf{E}_\beta u_0^2 = -\mathbf{E}\varepsilon_1^2(1-\beta^2)^{-1} \neq 0,$$

$$\frac{\partial}{\partial\gamma}\Lambda_{LS}(0,\beta) = -\beta\mathbf{E}\xi_0^2.$$

The equation (5.6.5) has a unique solution

$$\theta_\gamma^{LS} = \mathbf{E}_\beta y_0y_1/\mathbf{E}_\beta y_0^2.$$

The equation (5.6.4) for $\psi_i^{LS}(\mathbf{Y}_n, \theta)$ as given by (5.6.8) with probability tending to one has a unique solution

$$\widehat{\beta}_{n,LS} = \sum_{i=1}^{n} y_{i-1}y_i \Big/ \sum_{i=1}^{n} y_{i-1}^2 \quad \text{and} \quad \widehat{\beta}_{n,LS} \xrightarrow{\mathbf{P}_\beta} \theta_\gamma^{LS}.$$

By (5.6.7) we obtain

$$\text{IF}(\theta_\gamma^{LS}, \mu_\xi) = -\frac{\beta(1-\beta^2)}{\mathbf{E}\varepsilon_1^2}\mathbf{E}\xi_0^2. \tag{5.6.9}$$

Let $\text{IF}(\theta_\gamma^{LS}, \xi) := \text{IF}(\theta_\gamma^{LS}, \mu_\xi)$ when $\xi_i = \xi$ with a constant ξ. Then

$$\text{IF}(\theta_\gamma^{LS}, \xi) = -\frac{\beta(1-\beta^2)}{\mathbf{E}\varepsilon_1^2}\xi^2, \tag{5.6.10}$$

which is an unbounded continous function of ξ.

Let $\mathfrak{M}_i$, $i = 1, 2$, be the class of distributions μ_ξ with a finite ith absolute moment. Then

$$\text{GES}(\mathfrak{M}_2, \theta_\gamma^{LS}) = \infty, \qquad \beta \neq 0.$$

Hence even a small fraction of outiers in data may strongly affect $\widehat{\beta}_{n,LS}$.

5.6.2. Influence functional of the LAD estimator. Next consider the LAD estimator $\widehat{\beta}_{n,LD}$. Assume that $g(x)$ is continuous, $\sup_x g(x) < \infty$, $g(0) > 0$,

$$\mathbf{E}\varepsilon_1 = 0, \qquad \mathbf{E}\varepsilon_1^2 < \infty, \qquad \mathbf{E}\xi_1^2 < \infty.$$

In the contaminated case $\widehat{\beta}_{n,LD}$ is given by (5.6.4), where

$$\psi_i^{LD}(\mathbf{Y}_n, \theta) = y_{i-1}\,\text{sign}(y_i - \theta y_{i-1})$$

and the equality has to be replaced by level crossing. Condition 5.6(i) is fulfilled for any γ, θ by the ergodic theorem with

$$\Lambda_{LD}(\gamma, \theta) = \mathbf{E}_\beta y_0\,\text{sign}(y_1 - \theta y_0) = \mathbf{E}_\beta y_0\big(1 - 2G\big((\theta - \beta)u_0 - z_1^\gamma\xi_1 + \theta z_0^\gamma\xi_0\big)\big).$$

Obviously, $\Lambda_{LD}(0, \beta) = \mathbf{E}_\beta u_0\,\text{sign}\,\varepsilon_1 = 0$, and Condition 5.6(ii) is fulfilled as well. In order to verify Conditions 5.6(iii, iv), consider the events H_i, $i = 0, 1, 2$, that exactly i variables among z_0^γ and z_1^γ differ from zero. By the formula for total probability we can write

$$\Lambda_{LD}(\gamma, \theta) = \sum_{i=0}^{2} \mathbf{E}_\beta\big\{y_0\big(1 - 2G\big((\theta - \beta)u_0 - z_1^\gamma\xi_1 + \theta z_0^\gamma\xi_0\big)\big) \mid H_i\big\}\mathbf{P}\{H_i\}. \tag{5.6.11}$$

The conditional expectations in (5.6.11) do not depend on γ and are continuously differentiable with respect to θ, while $\mathbf{P}\{H_i\}$ are polynomials in γ. Therefore $\Lambda_{LD}(\gamma, \theta)$ has continuous partial derivatives with respect to γ and θ. Hence we find from (5.6.11)

$$\frac{\partial}{\partial\gamma}\Lambda_{LD}(0, \beta) = \mathbf{E}_\beta\xi_0\big(1 - 2G(\beta\xi_0)\big),$$

$$\frac{\partial}{\partial\theta}\Lambda_{LD}(0, \beta) = \lambda_{LD}(\beta) = -2g(0)\mathbf{E}_\beta u_0^2 = -2g(0)\mathbf{E}\varepsilon_1^2(1 - \beta^2)^{-1} \neq 0.$$

Recall that $\lambda_{LD}(\beta)$ was defined in 5.5.1.

Thus we have verified Conditions 5.6(i–iv). Due to monotonicity of $\psi_i^{LD}(\mathbf{Y}_n, \theta)$ in θ the equation (5.6.4) has an a.s. unique solution $\widehat{\beta}_{n,LD}$, and for small γ the equation (5.6.5) has a unique solution θ_γ^{LD}, $\widehat{\beta}_{n,LD} \xrightarrow{\mathbf{P}_\beta} \theta_\gamma^{LD}$. By (5.6.7) we obtain

$$\mathrm{IF}(\theta_\gamma^{LD}, \mu_\xi) = \frac{1-\beta^2}{2g(0)\mathbf{E}\varepsilon_1^2} \mathbf{E}_\beta \xi_0 \big(1 - 2G(\beta\xi_0)\big). \tag{5.6.12}$$

If $\xi_i = \xi$ with a constant ξ, then

$$\mathrm{IF}(\theta_\gamma^{LD}, \xi) = \frac{1-\beta^2}{2g(0)\mathbf{E}\varepsilon_1^2} \xi \big(1 - 2G(\beta\xi)\big), \tag{5.6.13}$$

which is an unbounded continous function of ξ.

Since

$$\mathrm{GES}(\mathfrak{M}_2, \theta_\gamma^{LD}) = \infty, \qquad \beta \neq 0,$$

the estimator $\widehat{\beta}_{n,LD}$ is not robust against outliers in observations $u_0, \dots, u_n$.

We see from (5.6.13) and (5.6.10) that the influence functional corresponding to $\widehat{\beta}_{n,LS}$ grows as a quadratic function of ξ as compared to the linear rate for $\widehat{\beta}_{n,LD}$. A qualitative interpretation of this fact is that $\widehat{\beta}_{n,LD}$ is less influenced by outliers than $\widehat{\beta}_{n,LS}$.

5.6.3. Influence functional of weighted LAD estimators.

CONDITION 5.6(v). $\mathbf{E}\log^+|\varepsilon_1| < \infty$, $\mathbf{E}_\beta y_1^2|\varphi(y_1)| < \infty$ *for all* γ *small, and* $\mathbf{E}_\beta u_1^2|\varphi(u_1)| \neq 0$.

CONDITION 5.6(vi). $\mathbf{E}_\beta u_1|\varphi(u_1)| = 0$ *or* $G(0) = 1/2$; $\sup_x g(x) < \infty$, $g(x)$ *is continuous,* $g(0) > 0$.

The weighted LAD estimator $\widehat{\beta}_{n,LDW}$ is obtained from the equation

$$\sum_{k=1}^n y_{k-1}|\varphi(y_{k-1})| \operatorname{sign}(y_k - \theta y_{k-1}) \div 0. \tag{5.6.14}$$

Arguing as for the LAD estimator, we obtain that under Conditions 5.6(v, vi) any solution of this equation for small γ converges to θ_γ^{LDW} and

$$\begin{aligned}\mathrm{IF}&(\theta_\gamma^{LDW}, \mu_\xi)\\ &= \frac{\mathbf{E}_\beta\{u_0|\varphi(u_0)|[1-2G(-\xi_1)] + (u_0+\xi_0)|\varphi(u_0+\xi_0)|[1-2G(\beta\xi_0)]\}}{2g(0)\mathbf{E}_\beta u_1^2|\varphi(u_1)|}.\end{aligned}$$

If $\varphi(x) = 1$ and $\mathbf{E}\varepsilon_1 = 0$, then this expression is equal to the influence functional of the LAD estimator (5.6.12).

Obviously, the requirement $\sup_x |x\varphi(x)| < \infty$ implies

$$\mathrm{GES}(\mathfrak{M}_\xi, \theta_\gamma^{LDW}) < \infty$$

for $\mathfrak{M}_\xi$ the class of distributions μ_ξ such that $\mathbf{E}_\beta y_1^2|\varphi(y_1)| < \infty$. In particular, $\widehat{\beta}_{n,LDW}$ for $\varphi(y_{k-1}) = 1/y_{k-1}$ becomes the median estimator $\widehat{\beta}_{n,M}$, $\theta_\gamma^{LDW} = \theta_\gamma^M$.

Then under the conditions $0 < \mathbf{E}|\varepsilon_1| < \infty$, $\mathbf{E}|\xi_1| < \infty$ (which ensure 5.6(v)) and Condition 5.6(vi)

$$\text{(5.6.15)}\quad \operatorname{IF}(\theta_\gamma^M, \mu_\xi) = \frac{\mathbf{E}_\beta\{\operatorname{sign} u_0[1-2G(-\xi_1)] + \operatorname{sign}(u_0+\xi_0)[1-2G(\beta\xi_0)]\}}{2g(0)\mathbf{E}_\beta|u_1|}.$$

It is seen from (5.6.15) that

$$\operatorname{GES}(\mathfrak{M}_1, \theta_\gamma^M) \le \big(g(0)\mathbf{E}_\beta|u_1|\big)^{-1}.$$

5.7. Testing for stationarity of the autoregression process

As a natural extension of the model (5.2.1) consider a nonstationary autoregressive relation

$$\text{(5.7.1)}\qquad u_i = \beta_i u_{i-1} + \varepsilon_i, \qquad i \in \mathbb{Z}.$$

If Condition 5.5(i) is satisfied and $\sup_i |\beta_i| < 1$, the equation (5.2.1) has a solution of the form

$$\text{(5.7.2)}\qquad u_i = \sum_{j\ge 0} \gamma_{ij}\varepsilon_{i-j},$$

where $\gamma_{i0} = 1$, $\gamma_{ij} = \beta_i\beta_{i-1}\dots\beta_{i-j+1}$ for $j \ge 1$ and the series (5.7.2) converges in $L^{2+\Delta}$.

Note that $\{u_i\}$ given by (5.7.2) satisfy the condition $\sup_i \mathbf{E}u_i^2 < \infty$, and any solution of (5.7.1) satisfying this condition is representable in the form (5.7.2).

Now let $u_0, u_1, \dots, u_n$ satisfying (5.7.2) be observed. The aim of this section is to construct tests for the stationarity hypothesis

$$\text{(5.7.3)}\qquad H_0\colon \beta_i = \beta, \qquad i \in \mathbb{Z},$$

with β unknown, and to study their power under alternatives

$$\text{(5.7.4)}\qquad H_{1n}\colon \beta_k = \beta + a_{kn}\, n^{-1/2}, \qquad k = 1, \dots, n, \qquad \sup_{k,n} |a_{kn}| \le A < \infty.$$

The tests will involve the estimators $\widehat{\beta}_{n,LS}$ and $\widehat{\beta}_{n,LD}$ for β constructed under the assumption that H_0 is true.

We begin with the technically more complicated case of the LAD estimator $\widehat{\beta}_{n,LD}$, which is obtained from equation (5.5.3)

$$\sum_{k=1}^n u_{k-1}\operatorname{sign}(u_k - \theta u_{k-1}) \div 0.$$

Consider the related random process

$$w_n^{LD}(t) = \sigma_u^{-1}(\beta) n^{-1/2} \sum_{k\le nt} u_{k-1}\operatorname{sign}(u_k - \widehat{\beta}_{n,LD}u_{k-1}), \qquad t\in[0,1],$$

where $\sigma_u^2(\beta) = \mathbf{E}\varepsilon_1^2(1-\beta^2)^{-1}$. Observe that $\mathbf{E}_\beta u_1^2 = \sigma_u^2(\beta)$ under H_0.

Our immediate task is to describe the asymptotic behavior of $w_n^{LD}(t)$ under H_0.

A heuristic derivation is fairly simple. Let $w(t)$, $t \in [0,1]$, denote a Brownian bridge, i.e., $w(t) = \nu(t) - t\nu(1)$, where $\nu(t)$ is a standard Wiener process. Let $D[0,1]$ be the metric space of functions on $[0,1]$ with discontinuities of the first kind equipped with the Skorokhod metric (for the definitions see, for example, Billingsley [**6**], Chapter 3).

Let, further,

$$\nu_n^{LD}(t) = \sigma_u^{-1}(\beta) n^{-1/2} \sum_{k \le nt} u_{k-1} \operatorname{sign} \varepsilon_k, \qquad t \in [0,1]. \tag{5.7.5}$$

Subject to Condition 5.5(i) the random sequence $\{u_{i-1} \operatorname{sign} \varepsilon_i, i \in \mathbb{Z}\}$ under H_0 satisfies the assumptions of Theorem 21.1 in Billingsley [**6**] (which follows by the argument of Example 1 in Chapter 4, §21). This theorem implies that $\nu_n^{LD}(t)$ weakly converges to $\nu(t)$ in $D[0,1]$. (This is a version of the well-known Donsker–Prohorov invariance principle, see Billingsley [**6**], §16.) Now on account of Corollary 5.5.1 we obtain that under H_0 for any fixed $t > 0$

$$\begin{aligned} w_n^{LD}(t) &= \sigma_u^{-1}(\beta) n^{-1/2} \sum_{k \le nt} u_{k-1} \operatorname{sign} \varepsilon_k \\ &\quad + \sigma_u^{-1}(\beta) \lambda_{LD}(\beta) \sqrt{t} \sqrt{nt} (\widehat{\beta}_{n,LD} - \beta) + o_p(1). \end{aligned} \tag{5.7.6}$$

The function $\lambda_{LD}(\beta) = -2g(0) \mathbf{E} \varepsilon_1^2 (1-\beta^2)^{-1}$ was introduced in 5.5.1.

By part 1 of Theorem 5.5.2, under Conditions 5.5(i, ii),

$$\sqrt{n}(\widehat{\beta}_{n,LD} - \beta) = -\lambda_{LD}^{-1}(\beta) n^{-1/2} \sum_{k=1}^{n} u_{k-1} \operatorname{sign} \varepsilon_k + o_p(1).$$

Substituting this expression into (5.7.6) we obtain in view of the definition (5.7.5)

$$w_n^{LD}(t) = \nu_n^{LD}(t) - t \nu_n^{LD}(1) + o_p(1). \tag{5.7.7}$$

By the relationship $w(t) = \nu(t) - t\nu(1)$, the formula (5.7.7) (established so far only for a fixed t) suggests that $w_n^{LD}(t)$ under H_0 must converge in $D[0,1]$ to a Brownian bridge $w(t)$ (which is still a heuristic argument).

It is to be expected that, as it usually happens, for large n the process $w_n^{LD}(t)$ under H_{1n} differs from that under H_0 by a shift. To describe the shift, put

$$a_n(t) = n^{-1} \sum_{k \le nt} a_{kn}$$

and assume that these functions satisfy the following

CONDITION 5.7(i). *For some $a(t) \in D[0,1]$, $\sup_{0 \le t \le 1} |a_n(t) - a(t)| \to 0$.*

Now we state a theorem on the asymptotic behavior of $w_n^{LD}(t)$ under H_0 and H_{1n}. For brevity, we write $\sigma_{LD}(\beta)$ for $-\sigma_u(\beta)/\lambda_{LD}(\beta)$, so that

$$\sigma_{LD}^{-1}(\beta) = 2g(0)(\mathbf{E}\varepsilon_1^2)^{1/2}(1-\beta^2)^{-1/2}. \tag{5.7.8}$$

The function $\sigma_{LD}^2(\beta)$ has already appeared in Theorem 5.5.2. It is the asymptotic variance of $\widehat{\beta}_{n,LD}$.

THEOREM 5.7.1. *Assume Conditions* 5.5(i, ii). *Then* $w_n^{LD}(t)$ *under the hypothesis* H_0 *given by* (5.7.3) *weakly converges in* $D[0,1]$ *to a Brownian bridge* $w(t)$ *and, under an alternative* H_{1n} *specified by* (5.7.4) *and Condition* 5.7(i), *to* $w(t) + \sigma_{LD}^{-1}(\beta)\delta(t)$, *where* $\delta(t) = a(t) - ta(1)$.

Before giving the proof, we state some statistical implications of this theorem. Let

$$\widehat{w}_n^{LD}(t) = s_n^{-1} n^{-1/2} \sum_{k \le nt} u_{k-1} \operatorname{sign}(u_k - \widehat{\beta}_{n,LD} u_{k-1}),$$

where $s_n^2 = n^{-1} \sum_{k=1}^n u_k^2$. It is easily seen that s_n^2 is a consistent estimator for $\sigma_u^2(\beta)$ both under H_0 and H_{1n}; hence, for $\widehat{w}_n^{LD}(t)$ the same assertions hold as for $w_n^{LD}(t)$. Therefore under the conditions of Theorem 5.7.1 the statistics

$$\widehat{D}_n^{LD} = \sup_{0 \le t \le 1} |\widehat{w}_n^{LD}(t)| = s_n^{-1} \max_{m \le n} \Big| n^{-1/2} \sum_{k \le m} u_{k-1} \operatorname{sign}(u_k - \widehat{\beta}_{n,LD} u_{k-1}) \Big|$$

and

$$(\widehat{\omega}_n^{LD})^2 = s_n^{-2} \int_0^1 \big(\widehat{w}_n^{LD}(t)\big)^2 dt = s_n^{-2} n^{-2} \sum_{m \le n} \Big(\sum_{k \le m} u_{k-1} \operatorname{sign}(u_k - \widehat{\beta}_{n,LD} u_{k-1}) \Big)^2$$

converge in distribution under H_0 given by (5.7.3) to $\sup_{0 \le t \le 1} |w(t)|$ and $\int_0^1 w^2(t)dt$, which have well-known asymptotic distributions of the Kolmogorov and ω^2 statistics (see, for example, Bol′shev and Smirnov [**17**], Tables 6.1 and 6.4a). Thus the statistics $\widehat{D}_n^{LD}$ and $(\widehat{\omega}_n^{LD})^2$ can be used for testing the hypothesis H_0.

Subject to Conditions 5.5(i, ii) and 5.7(i), $\widehat{D}_n^{LD}$ and $(\widehat{\omega}_n^{LD})^2$ under the alternative (5.7.4) converge to $\sup_{0 \le t \le 1} |w(t) + \sigma_{LD}^{-1}(\beta)\delta(t)|$ and $\int_0^1 \big(w(t) + \sigma_{LD}^{-1}(\beta)\,\delta(t)\big)^2 dt$, respectively.

PROOF OF THEOREM 5.7.1. We sketch the proof under H_{1n}. The detailed proof can be found in Boldin [**13**].

The following relation can be proved similarly to the proof of Theorem 5.5.1: under the conditions of Theorem 5.7.1 for any $0 < \Theta < \infty$

$$\begin{aligned} \sup_{|\theta| \le \Theta} \sup_{1 \le m \le n} \Big| n^{-1/2} & \sum_{k=1}^m \operatorname{sign}\big(u_k - (\beta + \theta n^{-1/2}) u_{k-1}\big) \\ & - n^{-1/2} \sum_{k=1}^m u_{k-1} \operatorname{sign} \varepsilon_k \\ & - \lambda_{LD}(\beta)\, \theta\, m/n + \lambda_{LD}(\beta) n^{-1} \sum_{k=1}^m a_{kn} \Big| = o_p(1). \end{aligned} \tag{5.7.9}$$

Similarly to Corollary 5.5.1, we infer from (5.7.9) that, for any sequence $\widehat{\beta}_n$ such that $\sqrt{n}(\widehat{\beta}_n - \beta) = O_p(1)$,

$$\begin{aligned} \sup_{1 \le m \le n} \Big| n^{-1/2} \sum_{k=1}^m & \operatorname{sign}\big(u_k - \widehat{\beta}_n u_{k-1}\big) - n^{-1/2} \sum_{k=1}^m u_{k-1} \operatorname{sign} \varepsilon_k \\ & - \lambda_{LD}(\beta) \sqrt{n}(\widehat{\beta}_n - \beta)\, m/n + \lambda_{LD}(\beta)\, n^{-1} \sum_{k=1}^m a_{kn} \Big| = o_p(1). \end{aligned} \tag{5.7.10}$$

Let $\nu_n^{LD}(t)$ be defined by (5.7.5). Under H_{1n} the following representation holds, which can be obtained similarly to part 2 of Theorem 5.5.4:

$$\sqrt{n}\,(\widehat{\beta}_{n,LD} - \beta) = \sigma_{LD}(\beta)\,\nu_n^{LD}(1) + a_n(1) + o_p(1) \tag{5.7.11}$$

with $\sigma_{LD}(\beta)$ defined by (5.7.8).

By (5.7.11), $\sqrt{n}(\widehat{\beta}_{n,LD} - \beta) = O_p(1)$. Hence making use of (5.7.10) with $\widehat{\beta}_n = \widehat{\beta}_{n,LD}$ one readily obtains that

$$\sup_{0\le t\le 1} \left|w_n^{LD}(t) - \nu_n^{LD}(t) + \sigma_{LD}^{-1}(\beta)\sqrt{n}\,(\widehat{\beta}_{n,LD} - \beta)t - \sigma_{LD}^{-1}(\beta)a_n(t)\right| = o_p(1). \tag{5.7.12}$$

Substitution of (5.7.11) into (5.7.12) yields

$$\sup_{0\le t\le 1} \left|w_n^{LD}(t) - \big(\nu_n^{LD}(t) - t\,\nu_n^{LD}(1)\big) - \sigma_{LD}^{-1}(\beta)\big(a_n(t) - ta_n(1)\big)\right| = o_p(1). \tag{5.7.13}$$

By the invariance principle for φ-mixing processes, $\nu_n^{LD}(t)$ weakly converges to a standard Wiener process $\nu(t)$. Therefore $\nu_n^{LD}(t) - t\nu_n^{LD}(1)$ converges to a Brownian bridge $w(t) = \nu(t) - t\,\nu(1)$, which together with (5.7.13) proves the theorem. □

Consider now the LSE $\widehat{\beta}_{n,LS}$, which is determined by the equation (5.3.1),

$$\sum_{k=1}^{n} u_{k-1}(u_k - \theta u_{k-1}) = 0.$$

Define the related process

$$w_n^{LS}(t) = (\mathbf{E}\varepsilon_1^2)^{-1/2}\sigma_u^{-1}(\beta)n^{-1/2}\sum_{k\le nt} u_{k-1}(u_k - \widehat{\beta}_{n,LS}u_{k-1}), \qquad t \in [0,1].$$

Similarly to Theorem 5.7.1 one obtains that, subject to conditions $\mathbf{E}\varepsilon_1 = 0$, $0 < \mathbf{E}\varepsilon_1^2 < \infty$, $w_n^{LS}(t)$ under H_0 weakly converges in $D[0,1]$ to a Brownian bridge $w(t)$, and under H_{1n} subject to Condition 5.7(i), to $w(t) + \sigma_{LS}^{-1}(\beta)\,\delta(t)$, where $\sigma_{LS}^2(\beta) = 1 - \beta^2$ is the asymptotic variance of $\widehat{\beta}_{n,LS}$.

It is to be noted that $\big(\sigma_{LD}(\beta)/\sigma_{LS}(\beta)\big)^2$ equals the ARE $e_{LD,LS}$ obtained in 5.5.2.

It is clear how to proceed further in order to construct the process $\widehat{w}_n^{LS}(t)$ and statistics $\widehat{D}_n^{LS}$, $(\widehat{\omega}_n^{LS})^2$.

Of course, instead of $\widehat{\beta}_{n,LS}$ and $\widehat{\beta}_{n,LD}$ one can use other estimators for β, which will determine the corresponding processes. For example, in Boldin [**11**] the estimator was taken to be the median $\widehat{\beta}_{n,M}$ of the array $\{u_k/u_{k-1},\ k = 1,\dots,n\}$, to yield the process

$$w_n^M(t) = n^{-1/2}\sum_{k\le nt} \operatorname{sign} u_{k-1}\operatorname{sign}(u_k - \widehat{\beta}_{n,M}u_{k-1}).$$

5.8. Proofs

PROOF OF THEOREM 5.5.1. Let

$$\Delta_k(x) = \begin{cases} 1, & \varepsilon_k < x, \\ 1/2, & \varepsilon_k = x, \\ 0, & \varepsilon_k > x, \quad k = 1, \dots, n. \end{cases}$$

It is easily verified that $l_n^{LD}(\theta)$ given by (5.5.2) is representable as

$$n^{-1/2} l_n^{LD}(\beta + n^{-1/2}\theta) - n^{-1/2} l_n^{LD}(\beta) = -2z_{1n}(\theta) - 2z_{2n}(\theta), \tag{5.8.1}$$

where

$$z_{1n}(\theta) = n^{-1/2} \sum_{k=1}^{n} u_{k-1}\big(\Delta_k(n^{-1/2}\theta u_{k-1}) - G(n^{-1/2}\theta u_{k-1}) - \Delta_k(0) + G(0)\big),$$

$$z_{2n}(\theta) = n^{-1/2} \sum_{k=1}^{n} u_{k-1}\Big(G(n^{-1/2}\theta u_{k-1}) - G(0)\Big).$$

The theorem is a direct consequence of the following two lemmas.

LEMMA 5.8.1. *Under Conditions* 5.5(i, ii)

$$\sup_{|\theta| \le \Theta} |z_{1n}(\theta)| = o_p(1).$$

PROOF. Rewrite $z_{1n}(\theta)$ as

$$z_{1n}(\theta) = p_{1n}(\theta) + p_{2n}(\theta),$$

where, with $I(\cdot)$ denoting the indicator of the event stated in parentheses,

$$p_{1n}(\theta) = n^{-1/2} \sum_{k=1}^{n} u_{k-1} I(u_{k-1} > 0)\big(\Delta_k(n^{-1/2}\theta u_{k-1}) \\ - G(n^{-1/2}\theta u_{k-1}) - \Delta_k(0) + G(0)\big)$$

and $p_{2n}(\theta)$ has a similar form with $I(u_{k-1} > 0)$ replaced by $I(u_{k-1} \le 0)$.

Break up the interval $[-\Theta, \Theta]$ into 3^{m_n} parts by the points

$$\eta_{sn} = -\Theta + 2\Theta 3^{-m_n} s, \qquad s = 0, 1, \dots, 3^{m_n},$$

with a sequence $\{m_n\}$ such that $3^{m_n} \sim n^{\delta/4}$ as $n \to \infty$, where $\delta = \min(\Delta, 1)$. Denote by j the value of s for which

$$0 \le \eta_{jn} - \theta \le 2\Theta 3^{-m_n}. \tag{5.8.2}$$

Let

$$\widehat{u}_{k-1,s} = u_{k-1}\big(1 - 2\Theta 3^{-m_n} \eta_{sn}^{-1} I(u_{k-1} \le 0)\big), \tag{5.8.3}$$

$$\widetilde{u}_{k-1,s} = u_{k-1}\big(1 - 2\Theta 3^{-m_n}\, \eta_{sn}^{-1} I(u_{k-1} \ge 0)\big), \quad k = 1, \dots, n. \tag{5.8.4}$$

By (5.8.2) these quantities satisfy the inequalities

$$\eta_{jn}\widetilde{u}_{k-1,j} \le \theta u_{k-1} \le \eta_{jn}\widehat{u}_{k-1,j}, \qquad k = 1, \dots, n. \tag{5.8.5}$$

Put

$$\widehat{U}_{sn} = (\widehat{u}_{0s}, \dots, \widehat{u}_{n-1,s}), \qquad \widetilde{U}_{sn} = (\widetilde{u}_{0s}, \dots, \widetilde{u}_{n-1,s}).$$

Let, furthermore,

$$q_n(\eta_{jn}, \widehat{U}_{jn}) = n^{-1/2} \sum_{k=1}^{n} u_{k-1} I(u_{k-1} > 0) \times \big(\Delta_k(n^{-1/2}\eta_{jn}\hat{u}_{k-1,j}) - G(n^{-1/2}\eta_{jn}\hat{u}_{k-1,j}) - \Delta_k(0) + G(0)\big).$$

Due to monotonicity in y of the functions $\Delta_k(y)$ and $G(y)$, (5.8.5) implies the inequalities

$$p_{1n}(\theta) \le q_n(\eta_{jn}, \widehat{U}_{jn}) + n^{-1/2} \sum_{k=1}^{n} u_{k-1} I(u_{k-1} > 0) \times \big(G(n^{-1/2}\eta_{jn}\widehat{u}_{k-1,j}) - G(n^{-1/2}\eta_{jn}\widetilde{u}_{k-1,j})\big),$$

$$p_{1n}(\theta) \ge q_n(\eta_{jn}, \widetilde{U}_{jn}) - n^{-1/2} \sum_{k=1}^{n} u_{k-1} I(u_{k-1} > 0) \times \big(G(n^{-1/2}\eta_{jn}\widehat{u}_{k-1,j}) - G(n^{-1/2}\eta_{jn}\widetilde{u}_{k-1,j})\big).$$

Therefore

$$\sup_{|\theta| \le \Theta} |p_{1n}(\theta)| \le \sup_{0 \le j \le 3^{m_n}} \{|q_n(\eta_{jn}, \widehat{U}_{jn})| + |q_n(\eta_{jn}, \widetilde{U}_{jn})|\} \tag{5.8.6}$$

$$+ \sup_{0 \le j \le 3^{m_n}} n^{-1/2} \sum_{k=1}^{n} u_{k-1} I(u_{k-1} > 0) \times \big(G(n^{-1/2}\eta_{jn}\widehat{u}_{k-1,j}) - G(n^{-1/2}\eta_{jn}\widetilde{u}_{k-1,j})\big). \tag{5.8.7}$$

Applying Taylor's formula and making use of Conditions 5.5(i, ii) and definitions (5.8.3), (5.8.4) we see that (5.8.7) is $O_p(3^{-m_n}) = o_p(1)$.

Now consider the term $q_n(\eta_{jn}, \widehat{U}_{jn})$ in (5.8.6). For the sake of brevity, put

$$\xi_k = \xi_k(j,n) = u_{k-1} I(u_{k-1} > 0)\big(\Delta_k(n^{-1/2}\eta_{jn}\widehat{u}_{k-1,j}) - G(n^{-1/2}\eta_{jn}\hat{u}_{k-1,j}) - \Delta_k(0) + G(0)\big).$$

Let $\Omega_{\le i}$ be the σ-algebra generated by $\{\varepsilon_s,\ s \le i\}$. Obviously,

$$\mathbf{E}_\beta \xi_k = \mathbf{E}\big\{\mathbf{E}(\xi_k \,|\, \Omega_{\le k-1})\big\} = 0,$$

because $\mathbf{E}_\beta(\xi_k \,|\, \Omega_{\le k-1}) = 0$ a.s. In a similar way, we have for $k < j$

$$\mathbf{E}_\beta \xi_k \xi_j = \mathbf{E}_\beta\big\{\mathbf{E}_\beta(\xi_k\xi_j \mid \Omega_{\le j-1})\big\} = \mathbf{E}_\beta\big\{\xi_k \mathbf{E}_\beta(\xi_j \mid \Omega_{\le j-1})\big\} = 0.$$

Moreover, one has for any $x_1, x_2 \in \mathbb{R}^1$ and $\delta = \min(\Delta, 1)$

$$\mathbf{E}\big\{\Delta_k(x_1) - G(x_1) - \Delta_k(x_2) + G(x_2)\big\}^2 \le |G(x_1) - G(x_2)| \le |G(x_1) - G(x_2)|^\delta.$$

Hence, by virtue of Conditions 5.5(i, ii),

$$\mathbf{E}_\beta \xi_k^2 \le \mathbf{E}_\beta u_{k-1}^2 \big|G(n^{-1/2}\eta_{jn}\widehat{u}_{k-1,n}) - G(0)\big|^\delta = O(n^{-\delta/2})$$

uniformly in k, j. Since $q_n(\eta_{jn}, \widehat{U}_{jn}) = n^{-1/2}\sum_{k=1}^{n}\xi_k$, this implies

$$\sup_{0\le j\le 3^{m_n}} \mathbf{E}_\beta q_n^2(\eta_{jn}, \widehat{U}_{jn}) = O(n^{-\delta/2}). \tag{5.8.8}$$

Now by the Chebyshev inequality one has for any $\varepsilon > 0$

$$\begin{aligned}
\mathbf{P}_\beta\Big\{\sup_{0\le j\le 3^{m_n}} |q_n(\eta_{jn}, \widehat{U}_{jn})| > \varepsilon\Big\}
&\le \sum_{j=0}^{3^{m_n}} \mathbf{P}_\beta\{|q_n(\eta_{jn}, \widehat{U}_{jn})| > \varepsilon\}\\
&\le \varepsilon^{-2}\sum_{j=0}^{3^{m_n}} \mathbf{E}_\beta q_n^2(\eta_{jn}, \widehat{U}_{jn})\\
&= O(3^{m_n} n^{-\delta/2}) = O(n^{-\delta/4}) = o(1)
\end{aligned}$$

due to the choice of m_n. In a similar way one shows that

$$\sup_{0\le j\le 3^{m_n}} \big|q_n(\eta_{jn}, \widetilde{U}_{jn})\big| = o_p(1).$$

Hence (5.8.6) is $o_p(1)$. Therefore

$$\sup_{|\theta|\le\Theta} |p_{1n}(\theta)| = o_p(1).$$

By a similar argument one shows that

$$\sup_{|\theta|\le\Theta} |p_{2n}(\theta)| = o_p(1)$$

which completes the proof. □

LEMMA 5.8.2. *Under Conditions* 5.5(i, ii)

$$z_{1n}(\theta) = g(0)\,\theta\mathbf{E}\varepsilon_1^2(1-\beta^2)^{-1} + \varepsilon_n(\theta),$$

where, for any $\Theta > 0$,

$$\sup_{|\theta|\le\Theta} |\varepsilon_n(\theta)| = o_p(1).$$

PROOF. This lemma immediately follows from the Taylor formula and the law of large numbers. □

PROOF OF THEOREM 5.5.2. It is easy to see that $\mathbf{E}_\beta\big(n^{-1/2}l_n^{LD}(\beta)\big)^2 = \mathbf{E}_\beta u_1^2$, hence $n^{-1/2}l_n^{LD}(\beta) = O_p(1)$. On account of Theorem 5.5.1, by taking $A > 0$ sufficiently large we can ensure the inequalities $l_n^{LD}(\beta - n^{-1/2}A) > 0$ and $l_n^{LD}(\beta + n^{-1/2}A) < 0$ to hold with probability arbitrarily close to one uniformly over all sufficiently large n. Due to monotonicity of $l_n^{LD}(\theta)$ this implies that $\widehat{\beta}_{n,LD} \in (\beta - n^{-1/2}A, \beta + n^{-1/2}A)$ also holds with arbitrarily high probability for all large enough n. Hence

$$\sqrt{n}(\widehat{\beta}_{n,LD} - \beta) = O_p(1).$$

Observe now that $n^{-1/2}l_n^{LD}(\widehat{\beta}_{n,LD}) = o_p(1)$. Indeed, the jumps of the function $l_n^{LD}(\theta)$ do not exceed

$$\max_{1\le k\le n} |u_{k-1}| \le \max_{0\le k\le n-1} |\varepsilon_k|\big(1-|\beta|\big)^{-1} + \beta^k|u_0|.$$

Since $n^{-1/2}\max_{1\le k\le n}|\varepsilon_k| = o_p(1)$ for $\mathbf{E}\varepsilon_1^2 < \infty$, we have $n^{-1/2}\max_{1\le k\le n}|u_{k-1}| = o_p(1)$.

To complete the proof, we put $\widehat{\beta}_n = \widehat{\beta}_{n,LD}$ in (5.5.4), resolve (5.5.4) for $\sqrt{n}(\widehat{\beta}_{n,LD} - \beta)$, and make use of the fact that $n^{-1/2}l_n^{LD}(\beta) \xrightarrow{d_\beta} N(0, \mathbf{E}_\beta u_1^2)$. $\square$

CHAPTER 6

Sign-based Analysis of One-parameter Autoregression

6.1. Introduction to sign-based autoregression analysis

In the largest part of this chapter we develop the nonparametric sign-based analysis of the simplest one-parameter stationary autoregression model

$$u_i = \beta u_{i-1} + \varepsilon_i, \qquad i \in \mathbb{Z}, \tag{6.1.1}$$

where β is an unknown parameter, $|\beta| < 1$, and $\{\varepsilon_i\}$ are i.i.d. random variables with an unknown (nondegenerate) distribution function $G(x)$. In various instances we impose some of the following conditions.

CONDITION 6.1(i). $\mathbf{P}\{\varepsilon_1 > 0\} = \mathbf{P}\{\varepsilon_1 < 0\} = 1/2$.

CONDITION 6.1(ii). $\mathbf{E}\varepsilon_1 = 0$, $\mathbf{E}|\varepsilon_1|^{1+\Delta} < \infty$ *for some* $\Delta > 0$.

CONDITION 6.1(iii). *There exists a density* $g(x) = G'(x)$ *such that* $g(x)$ *satisfies the Lipschitz condition at* $x = 0$, $g(0) > 0$, *and* $\sup_x g(x) < \infty$.

As we pointed out in §5.2 (Lemma 5.2.2), under condition $\mathbf{E}\log^+|\varepsilon_1| < \infty$ the equation (6.1.1) has a solution of the form

$$u_i = \sum_{j=0}^{\infty} \beta^j \varepsilon_{i-j} \tag{6.1.2}$$

which is the only stationary solution of (6.1.1) a.s. The series in (6.1.2) absolutely converges a.s. If $\mathbf{E}|\varepsilon_1|^{1+\Delta} < \infty$ this series converges in $L^{1+\Delta}$ (Lemma 5.2.1).

Let variables $u_0, u_1, \ldots, u_n$ satisfying (6.1.2) be observed. Our goal is to construct nonparametric sign procedures based on these observations for testing hypotheses about β and for estimation of β.

We begin §6.2 with the hypothesis testing of

$$H_0\colon \beta = \beta_0 \tag{6.1.3}$$

against one-sided and two-sided alternatives. In order to construct a nonparametric test, we form first a new sequence of observations

$$S_k(\theta) = \operatorname{sign}(u_k - \theta u_{k-1}), \qquad \theta \in \mathbb{R}^1, \quad k = 1, \ldots, n, \tag{6.1.4}$$

which are combined into the vector

$$\mathbf{S}^n(\theta) = \big(S_1(\theta), \ldots, S_n(\theta)\big). \tag{6.1.5}$$

Subject to Condition 6.1(i), $\{S_k(\beta_0) = \operatorname{sign} \varepsilon_k,\ k = 1, \ldots, n\}$ under H_0 are i.i.d. random variables with

$$\mathbf{P}_{\beta_0}\{S_k(\beta_0) = -1\} = \mathbf{P}_{\beta_0}\{S_k(\beta_0) = 1\} = 1/2,$$

so that the statistic $\mathbf{S}^n(\beta_0)$ is distributed free of $G(x)$ and hence H_0 becomes a simple hypothesis for the vector $\mathbf{S}^n(\beta_0)$. We will find an expansion for the likelihood function of the vector $\mathbf{S}^n(\beta_0)$ as $\beta \to \beta_0$ which will enable us to construct locally most powerful (LMP) tests based on $\mathbf{S}^n(\beta_0)$.

Their test statistic has the form

$$l_n^S(\beta_0) = \sum_{t=1}^{n-1} \beta_0^{t-1} \sum_{k=t+1}^{n} S_{k-t}(\beta_0) S_k(\beta_0).$$

Subject to Condition 6.1(i), it is distributed free of $G(x)$ under H_0 and, furthermore,

$$n^{-1/2} l_n^S(\beta_0) \overset{d_{\beta_0}}{\to} N\big(0, (1-\beta_0^2)^{-1}\big).$$

(As in Chapter 5, we omit the specification $n \to \infty$.)

In §6.3 we construct sign tests in the nonstationary autoregression model

$$u_i = \beta u_{i-1} + \varepsilon_i, \qquad i = 1, 2, \ldots, \qquad u_0 = 0,$$

with unrestricted unknown parameter β, $\beta \in \mathbb{R}^1$. For finite n we obtain in this model similar results to the stationary case treated in §6.2, whereas asymptotic results essentially depend on the magnitude of β and have an analogous form to those in §5.4.

From §6.4 on we consider again the stationary model (6.1.1). Though the construction of an LMP sign test based on $\mathbf{S}^n(\beta_0)$ may be of interest in its own right, it is more important to study the efficiency of this test relative to known tests. Consider a sequence of local alternatives

$$H_{1n}(a)\colon \beta = \beta_n := \beta_0 + a n^{-1/2} + o(n^{-1/2}), \tag{6.1.6}$$

with some constant a.

Under Condition 6.1(ii) the variables $u_0, \ldots, u_n$ can have infinite variance, in which case they do not satisfy sufficient conditions for the LAN property (see 5.3.2). Therefore we cannot find the limiting distribution of $n^{-1/2} l_n^S(\beta_0)$ under $H_{1n}(a)$ in the usual way via Le Cam's Third Lemma. Of course, one could try to establish the LAN condition at the point $\beta = \beta_0$ directly for the new observations $\mathbf{S}^n(\beta_0)$ to use then Le Cam's Third Lemma. However, we will not pursue this line of argument for two reasons. First, to establish LAN for $\mathbf{S}^n(\beta_0)$ at $\beta = \beta_0$, stronger moment conditions than 6.1(ii) are required. And more importantly, we are going to obtain estimators for β by minimizing the test statistic $l_n^S(\theta)$, and for that we need uniform results about the behavior of $l_n^S(\theta)$ in a vicinity of $\theta = \beta$.

Thus, assuming Conditions 6.1(i–iii), we obtain in §6.4 a uniform stochastic expansion under $H_{1n}(a)$, which has the form

$$\sup_{|\theta| \le \Theta_n} \big| n^{-1/2} l_n^S(\beta_n + n^{-1/2}\theta) - n^{-1/2} l_n^S(\beta_n) - \lambda_S(\beta_0)\,\theta \big| = o_p(1),$$

where $\Theta_n \to \infty$ at a polynomial rate depending on Δ, and

$$\lambda_S(\beta_0) = -2g(0)\mathbf{E}|\varepsilon_1|(1-\beta_0^2)^{-1}.$$

This is one of the main results of this chapter.

This uniform expansion implies that under $H_{1n}(a)$

$$n^{-1/2} l_n^S(\beta_0) \overset{d_{\beta_n}}{\to} N\big(-\lambda_S(\beta_0)a,\ (1-\beta_0^2)^{-1}\big),$$

where the asymptotic shift (mean value) equals

$$-\lambda_S(\beta_0)a = 2g(0)\mathbf{E}|\varepsilon_1|(1-\beta_0^2)^{-1}a.$$

Notice that when $\mathbf{E}\varepsilon_1^2 < \infty$ and other conditions which ensure LAN for the sequence $u_0, \dots, u_n$ are fulfilled, the hypotheses H_0 and $H_{1n}(a)$ are contiguous. In this (particular) case the asymptotic shift can be obtained by means of Le Cam's Third Lemma.

With the knowledge of the asymptotic shift, the ARE's of the sign test with respect to other known tests are readily obtainable (see §6.5). In particular, the ARE, $e_{S,LS}$, with respect to the test based on $\sqrt{n}(\widehat{\beta}_{n,LS} - \beta_0)$ equals

$$e_{S,LS} = \big(2g(0)\mathbf{E}|\varepsilon_1|\big)^2.$$

As was mentioned above, the stochastic expansion for $n^{-1/2} l_n^S(\beta_n + \theta n^{-1/2})$ is used not only for obtaining asymptotic distributions of test statistics under local alternatives, but it also provides a tool for the study of nonparametric sign estimators for β. In §6.6 we construct several such estimators. Here we describe one of them, $\widehat{\beta}_{n,S}$. We start by showing that $n^{-1} l_n^S(\theta)$ converges in probability to a function $\Lambda_S(\theta)$ uniformly in $|\theta| \le 1-\delta$, where $\Lambda_S(\beta) = 0$ and $\Lambda_S'(\beta) = \lambda_S(\beta)$, which is the function involved in the stochastic expansion. This suggests that β can be estimated by a root of the equation

$$l_n^S(\theta) \div 0.$$

The stochastic expansion for $n^{-1/2} l_n^S(\beta + n^{-1/2}\theta)$ enables us to establish the existence of a solution $\widehat{\beta}_{n,S}$ of this equation and to show that

$$\sqrt{n}(\widehat{\beta}_{n,S} - \beta) \overset{d_\beta}{\to} N\big(0,\ (1-\beta^2)(2g(0)\mathbf{E}|\varepsilon_1|)^{-2}\big).$$

The sign estimator turns out to have the same ARE, $e_{S,LS}$, with respect to $\widehat{\beta}_{n,LS}$ as that of the corresponding tests.

In §6.7 we consider the contamination model of independent (individual) outliers. We find the influence functional of $\widehat{\beta}_{n,S}$ and show that it is bounded over some natural class of contaminations. It means that the gross-error sensitivity (GES) of the sign estimator $\widehat{\beta}_{n,LS}$ is finite. The qualitative meaning of this property is that this estimator is robust against individual outliers in observations $u_0, \dots, u_n$.

In §6.8 we present and discuss numerical results.

Now we set out the reasons for using the sign procedures described above. Some of them have been mentioned already. For definiteness, we discuss the stationary model. The first reason is connected with Condition 6.1(ii). This condition does not ensure the LAN property for $u_0, \dots, u_n$, under which asymptotic distributions for standard procedures are obtained (see §§5.3, 5.5). At the same time, our sign procedures are asymptotically normal under condition $\mathbf{E}|\varepsilon_1|^{1+\Delta} < \infty$. Secondly, the sign test statistics are distributed free of $G(x)$ for finite n, which enables us to test hypotheses and to set confidence intervals for moderate sample sizes, whereas the classical procedures are distribution free only in an asymptotic sense. We regard this as the most important property of sign procedures. Thirdly, the ARE $e_{S,LS}$ can be greater than one and can become arbitrarily large for heavy-tailed $G(x)$. Moreover, we will show that it is bounded from below over some natural class of distributions $G(x)$. Next, the GES of $\widehat{\beta}_{n,S}$ is finite, unlike, for example, the GES of commonly used estimators $\widehat{\beta}_{n,LS}$ and $\widehat{\beta}_{n,LD}$. Finally, it is advantageous that the function $l_n^S(\theta)$ is invariant with respect to the scale parameter of $G(x)$, so that β can be estimated independently of this scale parameter. In fact, the only serious drawback of sign procedures is their relatively low ARE for distributions $G(x)$ with "ordinary" tails, like, for example, the normal one. In this case a good alternative is provided by rank procedures, which will be only briefly touched upon here, since this book does not envisage their detailed treatment. However, sign procedures have an important advantage over rank procedures: in many cases the null distributions of sign statistics remain unchanged for independent, but not identically distributed $\{\varepsilon_i\}$, provided that

$$\mathbf{P}\{\varepsilon_i < 0\} = \mathbf{P}\{\varepsilon_i > 0\} = 1/2;$$

whereas, rank procedures require $\{\varepsilon_i\}$ to be identically distributed. Some of the above properties hold for nonstationary autoregression as well. In particular, sign test statistics are distribution free for finite n.

We conclude this section with a remark concerning the assumption $\mathbf{E}\varepsilon_1 = 0$ in Condition 6.1(ii). This assumption may be too restrictive in some applications. As a simple alternative model, consider the autoregression process

$$v_i = \beta v_{i-1} + \delta_i, \qquad i \in \mathbb{Z}, \quad |\beta| < 1,$$

where $\delta_i = \varepsilon_i + \nu$ with an unknown constant ν. It will be convenient to rewrite this model in the form

$$v_i = \mu + u_i, \qquad i \in \mathbb{Z}, \tag{6.1.7}$$

where $\mu = (1-\beta)^{-1}\nu$ and the sequence $\{u_i\}$ satisfies (6.1.1). Let $v_0, \dots, v_n$ be observations from a strictly stationary solution of (6.1.7) and let $\widehat{\mu}_n$ be an estimator for μ based on these observations. Let us estimate the unobservable variables $u_0, \dots, u_n$ by $\widehat{u}_k = v_k - \widehat{\mu}_n$, $k = 0, \dots, n$. A natural approach is to construct sign statistics based on these estimates in a similar way to the construction of sign statistics based on $u_0, \dots, u_n$ in the model (6.1.1) for testing H_0 and estimation of β. What are the properties of such procedures? It turns out that if $\sqrt{n}(\widehat{\mu}_n - \mu) = O_p(1)$ and Conditions 6.1(i–iii) are fulfilled, they have the same asymptotic properties as the sign procedures in the zero mean stationary autoregression (6.1.1) described in §6.2 and §§6.4–6.6. This is a quite expectable result, similar to the well-known property of the least squares procedures (see Anderson [**2**], Section 5.5.5).

Unfortunately, the sign tests do not retain the important properties of being locally most powerful and distribution free under the null hypothesis for a finite sample size. For contaminated data the robustness of the sign estimators depends on the robustness of $\widehat{\mu}_n$.

We omit the detailed treatment of the autoregression model with nonzero mean, since methodologically it is quite similar to the study of the zero mean autoregression given in the subsequent sections.

6.2. Sign tests

We begin with the hypothesis testing of $H_0\colon \beta = \beta_0$ (see (6.1.3)) in the model (6.1.1) based on observations $u_0, u_1, \dots, u_n$ satisfying (6.1.2). We will construct a locally most powerful test for testing H_0 against one-sided alternatives

$$H_1^+ : \beta > \beta_0 \tag{6.2.1}$$

within the class of tests based on the vector of signs $\mathbf{S}^n(\beta_0)$ defined by (6.1.4), (6.1.5). With probability 1 the possible realizations of $\mathbf{S}^n(\beta_0)$ are vectors $\mathbf{s}^n = (s_1, \dots, s_n)$ with $s_k = \pm 1$, $k = 1, \dots, n$. For an arbitrary vector of this form put

$$\gamma_{tn}(\mathbf{s}^n) = \sum_{k=t+1}^{n} s_{k-t} s_k, \qquad t = 1, \dots, n-1. \tag{6.2.2}$$

THEOREM 6.2.1. *For $n \geq 2$, under Conditions* 6.1(i–iii)

$$\begin{aligned} &P_\beta\{\mathbf{S}^n(\beta_0) = \mathbf{s}^n\} \\ &=(\tfrac{1}{2})^n \left(1 + 2g(0)\mathbf{E}|\varepsilon_1| \sum_{t=1}^{n-1} \beta_0^{t-1} \gamma_{tn}(\mathbf{s}^n)(\beta - \beta_0)\right) + o(\beta - \beta_0) \\ &\qquad\qquad\qquad\qquad\qquad\qquad\qquad\qquad \textit{as} \quad \beta \to \beta_0. \end{aligned}$$

PROOF. Let $\Omega_{\leq i}$ be the σ-algebra generated by $\{\varepsilon_j, j \leq i\}$, and let $I(\cdot)$ denote the indicator of an event. Then, for $n = 1$,

$$\begin{aligned} P_\beta\{S_1(\beta_0) = s_1\} &= \mathbf{E}_\beta\Big\{I\big(S_1(\beta_0) = 1\big)\frac{1+s_1}{2} + I\big(S_1(\beta_0) = -1\big)\frac{1-s_1}{2}\Big\} \\ &= \frac{1+s_1}{2} - s_1 \mathbf{E}_\beta I\big(S_1(\beta_0) = -1\big) \\ &= \frac{1+s_1}{2} - s_1 \mathbf{E}_\beta I\big(u_1 - \beta_0 u_0 < 0\big) \\ &= \frac{1+s_1}{2} - s_1 \mathbf{E}_\beta I\big(\varepsilon_1 < (\beta_0 - \beta)u_0\big) \\ &= \frac{1+s_1}{2} - s_1 \mathbf{E}_\beta\Big\{\mathbf{E}_\beta I\big(\varepsilon_1 < (\beta_0 - \beta)u_0\big) \mid \Omega_{\leq 0}\Big\} \\ &= \frac{1+s_1}{2} - s_1 \mathbf{E}_\beta G\big((\beta_0 - \beta)u_0\big), \end{aligned}$$

since ε_1 and $\Omega_{\leq 0}$ are independent. Thus we have

$$P_\beta\big\{S_1(\beta_0) = s_1\big\} = \frac{1+s_1}{2} - s_1 \mathbf{E}_\beta G\big((\beta_0 - \beta)u_0\big). \tag{6.2.3}$$

By Condition 6.1(iii), $g(x)$ satisfies Hölder's condition at zero of an arbitrary order $0 < \delta \leq 1$ with some constant $h = h(\delta)$. Take $\delta = \min(\Delta, 1)$. Then

$$G(x) = G(0) + g(0)x + \varepsilon(x),$$

where $|\varepsilon(x)| \leq h|x|^{1+\delta}$, with h independent of x. Therefore

$$\begin{aligned} \mathbf{E}_\beta G\big((\beta_0 - \beta)u_0\big) &= \mathbf{E}\Big\{\frac{1}{2} - g(0)(\beta - \beta_0)u_0 + \varepsilon\big((\beta_0 - \beta)u_0\big)\Big\} \\ &= \frac{1}{2} + o(\beta - \beta_0), \end{aligned} \tag{6.2.4}$$

since $G(0) = \frac{1}{2}$, $\mathbf{E}_\beta u_0 = 0$, and $\mathbf{E}_\beta |u_0|^{1+\delta} < \infty$ by Conditions 6.1(i, ii).

It follows from (6.2.3)–(6.2.4) that

$$P_\beta\{S_1(\beta_0) = s_1\} = \frac{1}{2} + o(\beta - \beta_0). \tag{6.2.5}$$

For $n \geq 2$, we obtain by similar, though more cumbersome calculations

$$\begin{aligned} P_\beta\{\mathbf{S}^n(\beta_0) = \mathbf{s}^n\} =& \frac{1}{2} P_\beta\{\mathbf{S}^{n-1}(\beta_0) = \mathbf{s}^{n-1}\} \\ &+ (\tfrac{1}{2})^n 2g(0)\mathbf{E}|\varepsilon_1| \sum_{t=1}^{n-1} \beta_0^{t-1} s_n s_{n-t}(\beta - \beta_0) + o(\beta - \beta_0). \end{aligned} \tag{6.2.6}$$

The recurrence relation (6.2.6) with initial condition (6.2.5) implies the theorem.□

The following theorem, which is a direct consequence of Theorem 6.2.1, provides a test for the hypothesis (6.1.3).

THEOREM 6.2.2. *Let Conditions* 6.1(i–iii) *be satisfied. Then, for* $n \geq 2$, *the test with critical region*

$$Q_n^+ = \Big\{\mathbf{s}^n : \sum_{t=1}^{n-1} \beta_0^{t-1} \gamma_{tn}(\mathbf{s}^n) > \text{const}\Big\} \tag{6.2.7}$$

is locally most powerful for the hypothesis $H_0 \colon \beta = \beta_0$ *against alternatives* $H_1^+ : \beta > \beta_0$ *among tests based on* $\mathbf{S}^n(\beta_0)$.

With the reverse inequality, (6.2.7) provides a locally most powerful test for H_0 against left-sided alternatives. For two-sided alternatives $H_1 \colon \beta \neq \beta_0$ we suggest using the two-sided version of (6.2.7),

$$Q_n = \Big\{\mathbf{s}^n : \Big|\sum_{t=1}^{n-1} \beta_0^{t-1} \gamma_{tn}(\mathbf{s}^n)\Big| > \text{const}\Big\}. \tag{6.2.8}$$

For $S_k(\theta)$, $k = 1, \ldots, n$, as defined by (6.1.4) let

$$\Gamma_{tn}(\theta) = \sum_{k=t+1}^{n} S_{k-t}(\theta)\, S_k(\theta), \qquad t = 1, \ldots, n-1, \tag{6.2.9}$$

$$l_n^S(\theta) = \sum_{t=1}^{n-1} \theta^{t-1} \Gamma_{tn}(\theta). \tag{6.2.10}$$

Then $l_n^S(\beta_0)$ is the test statistic in (6.2.7) and (6.2.8).

Under the hypothesis H_0 the statistics $\Gamma_{tn}(\beta_0)$ become

$$\Gamma_{tn}(\beta_0) = \sum_{k=t+1}^{n} \operatorname{sign}(\varepsilon_{k-t}\varepsilon_k). \tag{6.2.11}$$

Subject to 6.1(i), they are distributed free of $G(x)$ for any n. Hence the null distribution of the test statistic $l_n^S(\beta_0)$ is also free of $G(x)$ for any n. Although this distribution depends on β_0, it can be computed for any given n and β_0, say, by simulation. Hence the critical values in (6.2.7) and (6.2.8) can also be evaluated.

Some numerical results of this kind are presented in §6.8. In particular, we give a table of quantiles, $c_n^\alpha(\beta_0)$, of the normalized statistic

$$T_{n,S}(\beta_0) = n^{-1/2}\sqrt{1-\beta_0^2}\, l_n^S(\beta_0)$$

for $n = 50(50)200$, $\beta_0 = 0, \pm 0.1, \pm 0.2, \dots, \pm 0.9$ and $\alpha = 0.05, 0.95$. (As follows from the relations (6.2.14) below, the statistic $T_{n,S}(\beta_0)$ has asymptotically zero mean and unit variance.) Further, for each n, the set

$$D_{n\alpha} = \{\theta \colon c_n^\alpha(\theta) \le T_{n,S}(\theta) \le c_n^{1-\alpha}(\theta)\} \tag{6.2.12}$$

is a confidence set for β with confidence level at least $1-2\alpha$. The quantile functions $c_n^\alpha(\theta)$ and $c_n^{1-\alpha}(\theta)$ can be evaluated with any desirable accuracy on a grid of values of θ, and the set $D_{n\alpha}$ can then be obtained by plotting the three curves $c_n^\alpha(\theta)$, $c_n^{1-\alpha}(\theta)$ and $T_{n,S}(\theta)$, $|\theta| < 1$. This will be done in examples to be given in §6.8

In the important case of the independence hypothesis $H_0\colon \beta = 0$, the test statistic $l_n^S(0)$ becomes

$$\Gamma_{1n}(0) = \sum_{k=2}^{n} \operatorname{sign}(u_{k-1}u_k). \tag{6.2.13}$$

Subject to Condition 6.1(i), the null distribution of this statistic reduces to the binomial distribution because the random variable

$$\frac{1}{2}\sum_{k=2}^{n}\bigl(1+\operatorname{sign}(\varepsilon_{k-1}\varepsilon_k)\bigr) = \frac{1}{2}\bigl(n-1+\Gamma_{1n}(0)\bigr)$$

has the binomial distribution $\mathrm{Bi}(n-1, 1/2)$. To see this, one can check (by induction in n) that under Condition 6.1(i) the random variables $\operatorname{sign}(\varepsilon_1\varepsilon_2), \operatorname{sign}(\varepsilon_2\varepsilon_3), \dots,$ $\operatorname{sign}(\varepsilon_{n-1}\varepsilon_n)$ are independent and

$$\mathbf{P}\{\operatorname{sign}(\varepsilon_{k-1}\varepsilon_k) = -1\} = \mathbf{P}\{\operatorname{sign}(\varepsilon_{k-1}\varepsilon_k) = 1\} = 1/2.$$

The sign test for independence with test statistic (6.2.13) appeared in the literature before. In particular, this test was proposed in Dufour [**23**] where its null distribution was given.

Subject to Condition 6.1(i), the null distributions of $\Gamma_{tn}(\beta_0)$ as in (6.2.11) for $t = 1, 2, \dots, n-1$ also reduce to the binomial distribution, namely,

$$\frac{1}{2}\bigl(n-t+\Gamma_{tn}(\beta_0)\bigr) \sim \mathrm{Bi}(n-t, 1/2).$$

Moreover, it is easily verified that

$$\mathbf{E}_{\beta_0}\Gamma_{tn}(\beta_0) = 0, \qquad \mathbf{E}_{\beta_0}\Gamma_{tn}(\beta_0)\Gamma_{rn}(\beta_0) = (n-t)\delta_{tr}, \tag{6.2.14}$$

with δ_{tr} denoting Kronecker's delta.

These properties imply the following theorem on the asymptotic null distribution of the test statistic.

THEOREM 6.2.3. *Let Condition* 6.1(i) *be satisfied. Then under* $H_0\colon \beta = \beta_0$

$$n^{-1/2} l_n^S(\beta_0) \xrightarrow{d_{\beta_0}} N\big(0, (1-\beta_0^2)^{-1}\big).$$

PROOF. Under the assumptions of the theorem, one has for $1 \le k < n-1$

$$l_n^S(\beta_0) = n^{-1/2} \sum_{t=1}^{n-1} \beta_0^{t-1} \Gamma_{tn}(\beta_0) = Z_{kn} + X_{kn},$$

where

$$Z_{kn} = n^{-1/2} \sum_{t=1}^{k} \beta_0^{t-1} \Gamma_{tn}(\beta_0),$$

$$X_{kn} = n^{-1/2} \sum_{t=k+1}^{n-1} \beta_0^{t-1} \Gamma_{tn}(\beta_0).$$

Rewrite Z_{kn} as

$$Z_{kn} = n^{-1/2} \sum_{j=1}^{n} a_{jk} + \delta_{kn},$$

where

$$a_{jk} = \operatorname{sign}(\varepsilon_j \varepsilon_{j+1}) + \beta_0 \operatorname{sign}(\varepsilon_j \varepsilon_{j+2}) + \cdots + \beta_0^{k-1} \operatorname{sign}(\varepsilon_j \varepsilon_{j+k}),$$

and $|\delta_{kn}| \le (1 + 2 + \cdots + k) n^{-1/2}$. The variables $\{a_{jk},\ j \in \mathbb{Z}\}$ form a k-dependent strictly stationary process, which in view of (6.2.14) implies

$$Z_{kn} \xrightarrow{d_{\beta_0}} Z_k \sim N\left(0, \frac{1-\beta_0^{2k}}{1-\beta_0^2}\right)$$

by Theorem 7.7.5 in Anderson [**2**]. Obviously,

$$Z_k \xrightarrow{d_{\beta_0}} N\big(0, (1-\beta_0^2)^{-1}\big) \qquad \text{as} \quad k \to \infty.$$

It remains to notice that, by virtue of (6.2.14), for any $\varepsilon > 0$

$$\sup_n \mathbf{E}_{\beta_0} X_{kn}^2 \le \sum_{t \ge k+1} \beta_0^{2(t-1)} < \varepsilon,$$

provided k is large enough. The proof is completed by applying Lemma 5.3.1. □

This theorem allows us to use the normal approximation for large n to determine the critical values in (6.2.7), (6.2.8). Moreover, the set

$$A_{n\alpha} = \left\{\theta : \left|n^{-1/2}(1-\theta^2)^{1/2} l_n^S(\theta)\right| < \xi_{1-\alpha}\right\} \tag{6.2.15}$$

is a confidence region for parameter β of asymptotic level $1 - 2\alpha$, where $\xi_{1-\alpha}$ denotes the $(1-\alpha)$-quantile of the standard normal distribution.

To conclude this section, we point out a simple but important fact: if $\{\varepsilon_i\}$ are nonidentically distributed with

$$\mathbf{P}\{\varepsilon_i < 0\} = \mathbf{P}\{\varepsilon_i > 0\} = 1/2, \qquad i \in \mathbb{Z}, \tag{6.2.16}$$

then $\mathbf{S}^n(\beta_0)$, and hence $\Gamma_{tn}(\beta_0)$ and $l_n^S(\beta_0)$ have the same distributions under H_0 as for identically distributed $\{\varepsilon_i\}$, so that they remain distributed free of the actual distributions of ε_i's. Therefore, under condition (6.2.16) the following statements hold.

- The set $D_{n\alpha}$ as in (6.2.12) remains a confidence set for β of level at least $1 - 2\alpha$.
- The assertion of Theorem 6.2.3 on the limiting distribution of $n^{-1/2} l_n^S(\beta_0)$ remains valid.
- The set $A_{n\alpha}$ as in (6.2.15) remains a confidence set for β of asymptotic level $1 - 2\alpha$.

The results of this section have been published in Boldin and Tyurin [**16**]. Theorem 6.2.1 was obtained by a different method by Tyurin [**89**].

6.3. Sign tests in a nonstationary autoregression

In this section we consider the autoregression model from §5.4,

$$u_i = \beta u_{i-1} + \varepsilon_i, \qquad i = 1, 2, \dots, \qquad u_0 = 0, \tag{6.3.1}$$

where $\{\varepsilon_i\}$ are i.i.d. random variables with an unknown distribution fuction $G(x)$ and the unknown parameter $\beta \in \mathbb{R}^1$.

It follows from (6.3.1) that

$$u_i = \sum_{j=0}^{i-1} \beta^j \varepsilon_{i-j}, \qquad i = 1, 2, \dots. \tag{6.3.2}$$

Let observations $u_1, \dots, u_n$ be generated by the relation (6.3.1). Our object in this section is to construct locally optimal sign tests for testing the hypothesis $H_0\colon \beta = \beta_0$ in the model (6.3.1) and to study their properties.

It is remarkable that this can be done for any $\beta_0 \in \mathbb{R}^1$. The results for finite n are quite similar to those in §6.2. The asymptotic results are similar to the results presented in §5.4.

So, let $\mathbf{S}^n(\beta_0)$ be the vector of signs defined by (6.1.4) and (6.1.5) and let $\mathbf{s}^n = (s_1, \dots, s_n)$, with $s_n = \pm 1$, denote possible realizations of $\mathbf{S}^n(\beta_0)$. Let the variables $\gamma_{tn}(\mathbf{s}^n)$ be defined by (6.2.2). The following theorem shows that the likelihood function of the vector $\mathbf{S}^n(\beta_0)$ in the model (6.3.1) with an arbitrary $\beta_0 \in \mathbb{R}^1$ admits the same expansion as in the model (6.1.1) for $|\beta_0| < 1$.

THEOREM 6.3.1. *Assume that Conditions* 6.1(i–iii) *are satisfied. Then for any* $\beta_0 \in \mathbb{R}^1$ *and* $n \geq 2$ *in the model* (6.3.1)

$$\mathbf{P}_\beta\{\mathbf{S}^n(\beta_0) = \mathbf{s}^n\} = \left(\frac{1}{2}\right)^n \left(1 + 2g(0)\mathbf{E}|\varepsilon_1| \sum_{t=1}^{n-1} \beta_0^{t-1} \gamma_{tn}(\mathbf{s}^n)(\beta - \beta_0)\right) + o(\beta - \beta_0) \qquad as \quad \beta \to \beta_0.$$

We omit the proof. It is obtained by using (6.3.2) in the same way as the proof of Theorem 6.2.1 is obtained with the aid of (6.1.2).

The following theorem immediately follows from Theorem 6.3.1.

THEOREM 6.3.2. *Assume that Conditions* 6.1(i–iii) *are satisfied. Then for any* $\beta_0 \in \mathbb{R}^1$ *and* $n \geq 2$ *in the model* (6.3.1) *the test with critical region*

$$Q_n^+ = \left\{ \mathbf{s}^n : \sum_{t=1}^{n-1} \beta_0^{t-1} \gamma_{tn}(\mathbf{s}^n) > \text{const} \right\} \tag{6.3.3}$$

is locally most powerful for the hypothesis $H_0\colon \beta = \beta_0$ *against* $H_1^+ : \beta > \beta_0$ *among the tests based on the signs* $\mathbf{S}^n(\beta_0)$.

Of course, (6.3.3) with reverse inequality provides the LMP test for H_0 against the left-sided alternative $H_0^- : \beta < \beta_0$.

Let the statistic $l_n^S(\theta)$, $\theta \in \mathbb{R}^1$, be defined by (6.2.10) and (6.2.9). Then $l_n^S(\beta_0)$ may be used as a test statistic for (6.3.3). As in the previous section, its null distribution is free of $G(x)$ for any $\beta_0 \in \mathbb{R}^1$ under the sole Condition 6.1(i).

Denote by $\widetilde{c}_n^{\alpha}(\beta_0)$ the α-quantile of $l_n^S(\beta_0)$. It can be computed by the Monte Carlo method for any $\beta_0 \in \mathbb{R}^1$. The set

$$\widetilde{D}_{n\alpha} = \{\theta\colon \widetilde{c}_n^{\alpha}(\beta_0) \leq l_n^S(\theta) \leq \widetilde{c}_n^{1-\alpha}(\beta_0)\}$$

is a confidence set for β of level at least $1 - 2\alpha$. One can explicitly obtain $\widetilde{D}_{n\alpha}$ in exactly the same way as in the stationary case (see §6.8).

Thus for finite n the sign tests can be treated quite analogously to the stationary case. A more interesting question concerns the asymptotic null distribution of $l_n^S(\beta_0)$ under Condition 6.1(i). First of all notice that if $I_n(\beta, \theta)$ denotes the Fisher information about β contained in $\mathbf{S}^n(\theta)$, then by Theorem 6.3.1

$$I_n(\beta, \beta) = \left(2g(0)\mathbf{E}|\varepsilon_1|\right)^2 \mathbf{E}_\beta(l_n^S(\beta))^2.$$

By using (6.2.14) we find

$$\mathbf{E}_\beta(l_n^S(\beta))^2 = \sum_{t=1}^{n-1} \beta^{2(t-1)}(n-t) = \begin{cases} \frac{1}{1-\beta^2}\left(n - \frac{1-\beta^{2n}}{1-\beta^2}\right), & |\beta| \neq 1, \\ \frac{1}{2}n(n-1), & |\beta| = 1. \end{cases}$$

Therefore, $I_n(\beta, \beta) \sim \left(2g(0)\mathbf{E}|\varepsilon_1|\right)^2 d_n^2(\beta)$, where $d_n^2(\beta)$ is defined by (5.4.4), i.e.,

$$d_n^2(\beta) = \begin{cases} \frac{n}{1-\beta^2}, & |\beta| < 1, \\ \frac{n^2}{2}, & |\beta| = 1, \\ \frac{\beta^{2n}}{(1-\beta^2)^2}, & |\beta| > 1. \end{cases}$$

The comparison of the sign information $I_n(\beta, \beta)$ with the Fisher information $J_n(\beta)$ dealt with in §5.4 (see (5.4.3)–(5.4.4)) shows that for each n they differ by the factor $\left(2g(0)\mathbf{E}|\varepsilon_1|\right)^2$.

Denote by $R(\beta)$ the distribution of $\beta\frac{1}{\sqrt{2}}(\zeta^2 - 1)$, where $\zeta \sim N(0, 1)$. Moreover, let $L(\beta)$ for $|\beta| > 1$ denote the distribution of the random variable $\frac{\beta^2-1}{\beta^2}\xi\eta$, where ξ and η are independent identically distributed random variables,

$$\xi = \sum_{i \geq 1} \beta^{-(i-1)} a_i,$$

with a_i, $i = 1, 2, \ldots$, being i.i.d. random variables such that $\mathbf{P}\{a_i = 1\} = \mathbf{P}\{a_i = -1\} = \frac{1}{2}$.

Theorem 6.3.3 describes the asymptotic null distribution of the test statistic.

THEOREM 6.3.3. *Let Condition* 6.1(i) *be satisfied. Then under the hypothesis* $H_0\colon \beta = \beta_0$ *in the model* (6.3.1)

$$d_n^{-1}(\beta_0) l_n^S(\beta_0) \xrightarrow{d_{\beta_0}} \begin{cases} N(0,1), & |\beta_0| < 1, \\ R(\beta_0), & |\beta_0| = 1, \\ L(\beta_0), & |\beta_0| > 1. \end{cases}$$

Theorem 6.3.3 is analogous to Theorem 5.4.1. In particular, $d_n^{-1}(\beta_0)\, l_n^S(\beta_0)$ for $|\beta_0| > 1$ is asymptotically distributed as a product of two independent identically distributed random variables with a finite support. Similarly, the nominator U_n of the LSE (see the proof of Theorem 5.4.1) was represented as a product of two i.i.d. Gaussian random variables.

Of course, Theorem 6.3.3 continues to hold for nonidentically distributed $\{\varepsilon_i\}$ satisfying (6.2.16).

PROOF. Obviously, Theorem 6.3.3 for $|\beta_0| < 1$ follows from Theorem 6.2.3.

Let $\beta_0 = 1$, the case $\beta_0 = -1$ is treated similarly. Put, for brevity, $a_i = \operatorname{sign} \varepsilon_i$. Then under H_0

$$l_n^S(1) = \sum_{t=1}^{n-1} \Gamma_{tn}(1) = \sum_{i=2}^{n} \sum_{j=1}^{i-1} a_j a_i.$$

Since under Condition 6.1(i)

$$\left(n^{-1/2} \sum_{i=1}^{n} a_i\right)^2 = 1 + 2n^{-1} \sum_{i=2}^{n} \sum_{j=1}^{i-1} a_j a_i \quad \text{a.s.},$$

we have with probability 1

$$\frac{1}{\sqrt{2}} \left[\left(n^{-1/2} \sum_{i=1}^{n} a_i\right)^2 - 1 \right] = \frac{\sqrt{2}}{n}\, l_n^S(1) = d_n^{-1}(1) l_n^S(1).$$

One has $n^{-1/2} \sum_{i=1}^{n} a_i \xrightarrow{d} N(0,1)$; therefore, $d_n^{-1}(1) l_n^S(1) \xrightarrow{d_1} R(1)$.

Now let $\beta_0 > 1$, the case of $\beta_0 < -1$ is handled similarly. Then

$$\begin{aligned} d_n^{-1}(\beta_0) l_n^S(\beta_0) &= \frac{\beta_0^2 - 1}{\beta_0^n} \sum_{t=1}^{n-1} \beta_0^{t-1} \Gamma_{tn} \\ &= \frac{\beta_0^2 - 1}{\beta_0^2} \Big(\Gamma_{n-1,n}(\beta_0) + \alpha \Gamma_{n-2,n}(\beta_0) + \dots, \alpha^{n-2} \Gamma_{1,n}(\beta_0) \Big), \end{aligned}$$

where $\alpha = \beta_0^{-1}$. Let $n + 1 > 2k$ for a fixed k. Then

$$S_n := \Gamma_{n-1,n}(\beta_0) + \alpha \Gamma_{n-2,n}(\beta_0) + \dots + \alpha^{n-2} \Gamma_{1n}(\beta_0) = Z_{kn} + X_{kn},$$

where

$$\begin{aligned} Z_{kn} &= \Gamma_{n-1,n}(\beta_0) + \dots + \alpha^{k-1} \Gamma_{n-k,n}(\beta_0), \\ X_{kn} &= \alpha^k \Gamma_{n-k-1,n}(\beta_0) + \dots + \alpha^{n-2} \Gamma_{1n}(\beta_0). \end{aligned}$$

By (6.2.14) under H_0

(6.3.4) $$\sup_n \mathbf{E}_{\beta_0} X_{kn}^2 \le \sum_{i\ge k} (i+1)\alpha^{2i} \to 0 \qquad \text{as} \quad k\to\infty.$$

Consider Z_{kn}. Under H_0 one has

$$Z_{kn} = a_1 a_n + \alpha(a_1 a_{n-1} + a_2 a_n) + \cdots + \alpha^{k-1}(a_1 a_{n-k+1} + a_2 a_{n-k+2} + \cdots + a_k a_n).$$

Let $\{b_i,\, i=1,2,\dots\}$ be a sequence of i.i.d. random variables independent of $\{a_i\}$, with $\mathbf{P}\{\, b_i = 1\} = \mathbf{P}\{\, b_i = -1\} = \frac{1}{2}$. Since $a_1,\dots,a_k$ and $a_{n-k+1},\dots,a_n$ are mutually independent for $n+1 > 2k$, the distribution of Z_{kn} coincides with that of

$$Z_k := a_1 b_1 + \alpha(a_1 b_2 + a_2 b_1) + \cdots + \alpha^{k-1}(a_1 b_k + a_2 b_{k-1} + \cdots + a_k b_1).$$

Therefore, for a fixed k

(6.3.5) $$Z_{kn} \xrightarrow{d_{\beta_0}} Z_k.$$

It is seen from the definition that

(6.3.6) $$Z_k \xrightarrow{d_{\beta_0}} (a_1 + \alpha a_2 + \dots)(b_1 + \alpha b_2 + \dots) \qquad \text{as} \quad k\to\infty.$$

The relations (6.3.4)–(6.3.6) ensure the validity of the conditions of Lemma 5.3.1, which implies that

(6.3.7) $$S_n \xrightarrow{d_{\beta_0}} (a_1 + \alpha a_2 + \dots)(b_1 + \alpha b_2 + \dots) = \xi\eta.$$

Further, one has $d_n^{-1}(\beta_0) l_n^S(\beta_0) = \frac{\beta_0^2-1}{\beta_0^2} S_n$, so that the conclusion of the theorem follows from (6.3.7) and the definition of $L(\beta_0)$. □

6.4. Uniform stochastic expansion: The power of sign tests under local alternatives

Now we consider the model (6.1.1) and study the asymptotic power of the test based on $n^{-1/2} l_n^S(\beta_0)$ (see (6.2.10), (6.2.9) for the definition) under the alternatives

$$H_{1n}(a)\colon \beta = \beta_n := \beta_0 + an^{-1/2} + o(n^{-1/2})$$

for some constant a (see (6.1.6)). The usual way of finding the limiting distributions of test statistics under contiguous alternatives is to employ Le Cam's Third Lemma. Instead of this, we derive the corresponding result for $n^{-1/2} l_n^S(\beta_0)$ from the following Theorem 6.4.1, which will be also used in §6.6 for the analysis of sign estimators. This theorem describes a linear stochastic expansion for the process $n^{-1/2} l_n^S(\beta_n + n^{-1/2}\theta)$ under $H_{1n}(a)$ which holds uniformly in $|\theta| \le \Theta_n$, where $\Theta_n \to \infty$ at a polynomial rate.

Let

(6.4.1) $$\lambda_S(\beta) = -2g(0)\,\mathbf{E}|\varepsilon_1|\,(1-\beta^2)^{-1}.$$

THEOREM 6.4.1. *Let Conditions* 6.1(i–iii) *be satisfied. Then, under the alternative* $H_{1n}(a)$,

$$\sup_{|\theta|\le\Theta_n} \left|n^{-1/2}l_n^S(\beta_n+n^{-1/2}\theta)-n^{-1/2}l_n^S(\beta_n)-\lambda_S(\beta_0)\theta\right| = o_p(1), \tag{6.4.2}$$

where $\Theta_n=\Theta n^\alpha$ *with an arbitrary* $0<\Theta<\infty$ *and* $\alpha<\frac{\delta}{2(1+2\delta)}$, $\delta=\min(\Delta,1)$.

The proof of this theorem is rather tedious, and we defer it to §6.9. This theorem for $\alpha=0$, i.e., for $\Theta_n\equiv\Theta$, was given in Boldin and Tyurin [**16**]. The strengthening to $\alpha>0$ is essential for the sequel; it will be particularly useful in §6.6.

The proof (namely, the equality (6.9.4), Lemma 6.9.2, and the proof of Lemma 6.9.6) shows that, under the assumptions of Theorem 6.4.1,

$$\sup_{|\theta|\le\Theta_n} \left|n^{-1/2}\Gamma_{tn}(\beta_n+n^{-1/2}\theta)-n^{-1/2}\Gamma_{tn}(\beta_n)-\lambda_S(0)\beta_0^{t-1}\theta\right| = o_p(1) \tag{6.4.3}$$

for any $t\in\mathbb{N}$. The comparison of (6.4.3) with the definitions of $l_n^S(\theta)$ and $\lambda_S(\beta)$ (see (6.2.10) and (6.4.1)) explains the second term of the expansion in (6.4.2).

We also formulate here a consequence of Theorem 6.4.1.

COROLLARY 6.4.1. *Let Conditions* 6.1(i–iii) *be satisfied. Then under* $H_{1n}(a)$

$$n^{-1/2}l_n^S(\widehat{\beta}_n)=n^{-1/2}l_n^S(\beta_n)+\lambda_S(\beta_0)\sqrt{n}(\widehat{\beta}_n-\beta_n)+o_p(1)$$

for any sequence of random variables $\widehat{\beta}_n$ *such that* $n^{1/2-\alpha}(\widehat{\beta}_n-\beta_n)=O_p(1)$ *with* $\alpha<\frac{\delta}{2(1+2\delta)}$, $\delta=\min(\Delta,1)$.

PROOF. One has for any $\varepsilon>0$ and $0<\Theta<\infty$, $\Theta_n=\Theta n^\alpha$

$$\begin{aligned}
&\mathbf{P}_{\beta_n}\{|n^{-1/2}l_n^S(\widehat{\beta}_n)-n^{-1/2}l_n^S(\beta_n)-\lambda_S(\beta_0)\sqrt{n}(\widehat{\beta}_n-\beta_n)|>\varepsilon\}\\
&\quad\le\mathbf{P}_{\beta_n}\{|n^{-1/2}l_n^S(\widehat{\beta}_n)-n^{-1/2}l_n^S(\beta_n)-\lambda_S(\beta_0)\sqrt{n}(\widehat{\beta}_n-\beta_n)|>\varepsilon,\\
&\qquad\qquad\qquad\qquad\qquad\qquad |\sqrt{n}(\widehat{\beta}_n-\beta_n)|\le\Theta_n\}\\
&\qquad+\mathbf{P}_{\beta_n}\{|\sqrt{n}(\widehat{\beta}_n-\beta_n)|>\Theta_n\}\\
&\quad\le\mathbf{P}_{\beta_n}\{\sup_{|\theta|\le\Theta_n}|n^{-1/2}l_n^S(\beta_n+n^{-1/2}\theta)-n^{-1/2}l_n^S(\beta_n)-\lambda_S(\beta_0)\theta|>\varepsilon\}\\
&\qquad+\mathbf{P}_{\beta_n}\{|\sqrt{n}(\widehat{\beta}_n-\beta_n)|>\Theta_n\}.
\end{aligned}$$

Both probabilities in the right-hand side can be made arbitrarily small for all sufficiently large n and for Θ large enough by Theorem 6.4.1 and by the condition $n^{1/2-\alpha}(\widehat{\beta}_n-\beta_n)=O_p(1)$. □

It is easily seen that, subject to Condition 6.1(i), $n^{-1/2}l_n^S(\beta_n)$ has asymptotically the same distribution under $H_{1n}(a)$ given by (6.1.6) as $n^{-1/2}l_n^S(\beta_0)$ under H_0 given by (6.1.3). Therefore, Corollary 6.4.1 (with $\widehat{\beta}_n=\beta_0$ and $\alpha=0$) and Theorem 6.2.3 imply

THEOREM 6.4.2. *Let Conditions* 6.1(i–iii) *be satisfied. Then, under the alternative* $H_{1n}(a)$,

$$n^{-1/2}l_n^S(\beta_0) = n^{-1/2}l_n^S(\beta_n) - \lambda_S(\beta_0)a + o_p(1)$$

and therefore

$$n^{-1/2}l_n^S(\beta_0) \xrightarrow{d_{\beta_n}} N\big(-\lambda_S(\beta_0)a,\ (1-\beta_0^2)^{-1}\big).$$

Thus, the asymptotic shift of the normalized statistic

$$T_{n,S}(\beta_0) = n^{-1/2}\sqrt{1-\beta_0^2}\, l_n^S(\beta_0)$$

under $H_{1n}(a)$ equals

$$\delta_S(\beta_0) = 2g(0)\mathbf{E}|\varepsilon_1|(1-\beta_0^2)^{-1/2}a. \tag{6.4.4}$$

Hence the asymptotic power of the corresponding level α test under

$$H_{1n}^+(a)\colon \beta = \beta_0 + n^{-1/2}a + o(n^{-1/2}), \qquad a > 0,$$

equals

$$\Phi\big(\delta_S(\beta_0) - \xi_{1-\alpha}\big).$$

Obviously, (6.4.3) implies that for any $t \in \mathbb{N}$ the statistic $n^{-1/2}\Gamma_{tn}(\beta_0)$ under the assumptions of Theorem 6.4.2 is asymptotically normal $N(-\lambda_S(0)\beta_0^{t-1}a, 1)$. Hence $n^{-1/2}\Gamma_{tn}(\beta_0)$ is asymptotically shifted by

$$\delta_{tS} = -\lambda_S(0)\beta_0^{t-1}a = 2g(0)\mathbf{E}|\varepsilon_1|\,\beta_0^{t-1}a.$$

Let $I_n(\beta,\theta)$ denote the Fisher information about β contained in $\mathbf{S}^n(\theta)$. By Theorem 6.2.1

$$\frac{\partial \log \mathbf{P}_\beta\{\mathbf{S}^n(\beta_0) = \mathbf{s}^n\}}{\partial\beta}\bigg|_{\beta=\beta_0} = 2g(0)\,\mathbf{E}|\varepsilon_1| \sum_{t=1}^{n-1} \beta_0^{t-1}\gamma_{tn}(\mathbf{s}^n). \tag{6.4.5}$$

It follows from (6.4.5) and the definition (6.2.10) of $l_n^S(\beta_0)$ that

$$I_n(\beta_0,\beta_0) = \big(2g(0)\mathbf{E}|\varepsilon_1|\big)^2 \mathbf{E}_{\beta_0}\big(l_n^S(\beta_0)\big)^2.$$

Note in passing that the sign information $I_n(\beta,\beta)$ in the stationary model (6.1.1) for $|\beta| < 1$ is the same as in the nonstationary model (6.3.1) (see §6.3).

Making use of (6.2.14), one easily deduces that

$$n^{-1}I_n(\beta_0,\beta_0) \sim \big(2g(0)\,\mathbf{E}|\varepsilon_1|\big)^2(1-\beta_0^2)^{-1}. \tag{6.4.6}$$

Hence the asymptotic shift $\delta_S(\beta_0)$ as given by (6.4.4) has the form

$$\delta_S(\beta_0) = \lim_{n\to\infty} \big(n^{-1}I_n(\beta_0,\beta_0)\big)^{1/2}a.$$

This representation is analogous to the well-known expression for the shift of a locally most powerful test statistic in the case of independent observations satisfying standard regularity conditions (see, for example, Cox and Hinkley [**20**], §9.3).

In the next section we will obtain asymptotic relative efficiencies of the sign test with test statistic $T_{n,S}(\beta_0)$ constructed here with respect to the tests based on the least squares and least absolute deviations estimators, as well as the rank and median tests.

6.5. Sign tests: Comparison with other nonparametric tests

In this section we consider the stationary model (6.1.1) and compare the sign tests, which have been constructed and studied in §§6.2 and 6.4, with tests based on widely used least squares and least absolute deviations estimators, $\widehat{\beta}_{n,LS}$ and $\widehat{\beta}_{n,LD}$, as well as with rank and median tests to be briefly discussed below.

We begin with the LSE $\widehat{\beta}_{n,LS}$. It was pointed out in 5.3.2 (see (5.3.12)) that if $\mathbf{E}\varepsilon_1 = 0$ and $0 < \mathbf{E}\varepsilon_1^2 < \infty$ then under the alternative

$$H_{1n}(a)\colon \beta = \beta_n := \beta_0 + an^{-1/2} + o(n^{-1/2}),$$

$\widehat{\beta}_{n,LS}$ is asymptotically normal; namely,

$$T_{n,LS}(\beta_0) = (1-\beta_0^2)^{-1/2}\sqrt{n}(\widehat{\beta}_{n,LS} - \beta_0) \xrightarrow{d_{\beta_n}} N\big(a(1-\beta_0^2)^{-1/2}, 1\big). \tag{6.5.1}$$

On account of (6.5.1) and (6.4.4) the asymptotic relative efficiency (ARE) of the test with test statistic

$$T_{n,S}(\beta_0) = n^{-1/2}\sqrt{1-\beta_0^2}\, l_n^S(\beta_0)$$

with respect to the one based on $T_{n,LS}(\beta_0)$ equals

$$e_{S,LS} = \big(2g(0)\,\mathbf{E}|\varepsilon_1|\big)^2. \tag{6.5.2}$$

This ARE does not depend on β_0 and on the scale parameter of $G(x)$. In particular, for $G(x)$ normal $e_{S,LS} = (2/\pi)^2 \approx 0.41$; for the Laplace d.f. $G(x)$ $e_{S,LS} = 1$; for $G(x)$ the logistic d.f. $e_{S,LS} = (\log 2)^2 \approx 0.48$. For a d.f. $G(x)$ with heavy tails, $e_{S,LS}$ may take arbitrarily large values. For example, for Tukey's contamination model $G(x) = (1-\varepsilon)\Phi(x) + \varepsilon\Phi(x/\tau)$,

$$e_{S,LS} = \left\{\frac{2}{\pi}\Big(1 - \frac{\tau-1}{\tau}\varepsilon\Big)\big(1 + (\tau-1)\varepsilon\big)\right\}^2 \to \infty \qquad \text{as} \quad \tau\to\infty.$$

In the class $\mathfrak{G}$ of distribution functions having an even density function with maximum at zero,

$$e_{S,LS} = \big(2g(0)\,\mathbf{E}|\varepsilon_1|\big)^2 \tag{6.5.3}$$

is minimized by a uniform distribution, in which case $e_{S,LS} = 1/4$. Thus the ARE is bounded from below on a natural class $\mathfrak{G}$. The proof of this fact is quite similar to the proof of Theorem 3.3 in Lehmann [**61**], Chapter 5, §3.

For the LAD estimator $\widehat{\beta}_{n,LD}$, subject to the conditions of Theorem 5.5.4, we have under $H_{1n}(a)$

$$\sqrt{n}(\widehat{\beta}_{n,LD} - \beta_0) \xrightarrow{d_{\beta_n}} N\Big(a, (1-\beta_0^2)\big((2g(0))^2\mathbf{E}\varepsilon_1^2\big)^{-1}\Big). \tag{6.5.4}$$

By (6.4.4) and (6.5.4), the ARE $e_{S,LD}$ of the test based on $T_{n,S}(\beta_0)$ with respect to $\sqrt{n}(\widehat{\beta}_{n,LD} - \beta_0)$ equals

$$e_{S,LD} = \big(\mathbf{E}|\varepsilon_1|\big)^2/\mathbf{E}\varepsilon_1^2. \tag{6.5.5}$$

In order to apply the test statistic $\sqrt{n}(\widehat{\beta}_{n,LD} - \beta_0)$ one has to know its limiting variance, which (both under H_0 and $H_{1n}(a)$) is equal to

$$\sigma^2_{LD}(\beta_0) = (1 - \beta_0^2)\big((2g(0))^2\mathbf{E}\varepsilon_1^2\big)^{-1},$$

or to use a consistent estimate for $\sigma^2_{LD}(\beta_0)$. For this reason, a more convenient test statistic is $T_{n,LD}(\beta_0)$ (see 5.5.2),

$$T_{n,LD}(\beta_0) = s_n^{-1}n^{-1/2}\sum_{k=1}^{n} u_{k-1}\,\mathrm{sign}(u_k - \beta_0 u_{k-1})$$

with

$$s_n^2 = n^{-1}\sum_{k=1}^{n} u_k^2.$$

This statistic is determined by the leading term of the expansion for $\sqrt{n}(\widehat{\beta}_{n,LD} - \beta)$ (see Theorem 5.5.2). By Theorem 5.5.4 under $H_{1n}(a)$

$$T_{n,LD}(\beta_0) \xrightarrow{d_{\beta_n}} N\Big((1 - \beta_0^2)^{-1/2}2g(0)(\mathbf{E}\varepsilon_1^2)^{1/2}a, 1\Big).$$

Hence, the ARE of the sign test with respect to the one based on $T_{n,LD}(\beta_0)$ is also equal to $e_{S,LD}$ as in (6.5.5) and does not depend on β_0 and the scale parameter of $G(x)$. It is seen that $e_{S,LD} < 1$. Recall, however, that the asymptotic normality of $\widehat{\beta}_{n,LD}$ under $H_{1n}(a)$ requires the moment condition $\mathbf{E}|\varepsilon_1|^{2+\Delta} < \infty$, whereas for the sign statistic only $\mathbf{E}|\varepsilon_1|^{1+\Delta} < \infty$ is needed.

Now we turn to the rank tests. Rank and signed-rank test statistics have been used in the models of autoregression–moving average and linear regression with autocorrelated errors for a long time, and this subject matter has a large bibliography. Mostly, consideration is focused on models with a finite variance satisfying the LAN condition. The asymptotic analysis employs the theory of U-statistics, the power of tests is obtained by means of Le Cam's Third Lemma, and the results are of a parametric nature (see Hallin, Ingenbleek, and Puri [**31, 32**], Hallin and Puri [**33, 34**], Ferretti, Kelmansky, and Yohai [**28**], Tyurin [**89**], and references therein).

Another approach, closely related to the one adopted in this book, is applied by Boldin [**15**], where rank estimators in an autoregression model with possibly infinite variance are treated. This approach is applicable for hypothesis testing as well.

Namely, let

$$\varepsilon_k(\theta) = u_k - \theta u_{k-1}, \qquad k = 1, \dots, n,$$

and let $R_k(\theta)$ be the rank of $\varepsilon_k(\theta)$ among $\varepsilon_1(\theta), \dots, \varepsilon_n(\theta)$.

It is well known (see, for example, Hallin, Ingenbleek, and Puri [**31**] and Tyurin [**89**]) that in testing $H_0\colon \beta = \beta_0$ against one-sided alternatives, an LMP test based on ranks $R_1(\beta_0), \dots, R_n(\beta_0)$, under natural regularity conditions has test statistic

$$\sum_{t=1}^{n-1} \beta_0^{t-1} \sum_{k=t+1}^{n} a_n\big(R_{k-t}(\beta_0), R_k(\beta_0)\big), \tag{6.5.6}$$

where $a_n(t,s) = \mathbf{E}\,\varepsilon_{(t)}\,l(\varepsilon_{(s)})$ with $l(x) = g'(x)/g(x)$. The test statistic (6.5.6) depends on $g(x)$, and the LMP rank test is a parametric one. It is seen that the scores $a_n(t,s)$ are approximately equal to $G^{-1}(t/(n+1))\,l\big(G^{-1}(s/(n+1))\big)$. The form of these approximate scores suggests a natural class of nonparametric test statistics

$$\sum_{t=1}^{n-1}\beta_0^{t-1}\sum_{k=t+1}^{n}\varphi_1\big(R_{k-t}(\beta_0)/(n+1)\big)\varphi_2\big(R_k(\beta_0)/(n+1)\big), \tag{6.5.7}$$

where functions $\varphi_i(u)$, $u \in [0,1]$, need not depend on $G(x)$.

To explore the statistic (6.5.7), consider the process

$$l_n^R(\theta) = \sum_{t=1}^{n-1}\theta^{t-1}\Gamma_{tn}^R(\theta) \tag{6.5.8}$$

with

$$\Gamma_{tn}^R(\theta) = \sum_{k=t+1}^{n}\varphi_1\big(R_{k-t}(\theta)/(n+1)\big)\varphi_2\big(R_k(\theta)/(n+1)\big).$$

We impose the following conditions on the functions $\varphi_i(u)$, $i = 1,2$, and $G(x)$.

CONDITION 6.5(i). $\mathbf{E}\varphi_i\big(G(\varepsilon_1)\big) = 0$, *$\varphi_i(u)$ are differentiable and $\varphi_i'(u)$ satisfy the Lipschitz condition.*

CONDITION 6.5(ii). $\mathbf{E}\varepsilon_1 = 0$, $\mathbf{E}|\varepsilon_1|^{1+\Delta} < \infty$ *for some* $\Delta > 0$.

CONDITION 6.5(iii). $\sup_x g(x) < \infty$, *$g(x)$ satisfies the Lipschitz condition.*

CONDITION 6.5(iv).

$$\lambda_R(\beta) := -\mathbf{E}\{\varepsilon_1\varphi_1(G(\varepsilon_1))\}\mathbf{E}\{\varphi_2'(G(\varepsilon_1))g(\varepsilon_1)\}(1-\beta^2)^{-1} \neq 0.$$

Assuming Conditions 6.5(i–iv), one can show along the lines of Boldin [**15**] that under $H_{1n}(a)$

$$\sup_{|\theta|\le\Theta\log n}\big|n^{-1/2}l_n^R(\beta_n + n^{-1/2}\theta) - n^{-1/2}l_n^R(\beta_n) - \lambda_R(\beta_0)\theta\big| = o_p(1)$$

for any $0 < \Theta < \infty$. This stochastic expansion implies quite similarly to Theorem 6.4.2 that under $H_{1n}(a)$

$$n^{-1/2}l_n^R(\beta_0) = n^{-1/2}l_n^R(\beta_n) - \lambda_R(\beta_0)a + o_p(1),$$

which yields

$$n^{-1/2}l_n^R(\beta_0) \xrightarrow{d_{\beta_n}} N\Big(-\lambda_R(\beta_0)a, \mathbf{E}\varphi_1^2\big(G(\varepsilon_1)\big)\mathbf{E}\varphi_2^2\big(G(\varepsilon_2)\big)(1-\beta_0^2)^{-1}\Big). \tag{6.5.9}$$

For continuous $G(x)$ the null distribution of $l_n^R(\beta_0)$ is free of $G(x)$ for finite n. Consider the Spearman scores $\varphi_i(u) = u - 1/2$. Let

$$T_{n,R}(\beta_0) = 12\sqrt{1-\beta_0^2}\,n^{-1/2}l_n^R(\beta_0).$$

By (6.5.9) the statistic $T_{n,R}(\beta_0)$ under H_0 has asymptotically the standard normal distribution. Under $H_{1n}(a)$

$$(6.5.10) \qquad T_{n,R}(\beta_0) \xrightarrow{d_{\beta_n}} N\Big(12\Big(\int g^2(x)dx \int xG(x)\,g(x)dx\Big)(1-\beta_0^2)^{-1/2}a, 1\Big).$$

It follows from (6.4.4) and (6.5.10) that the ARE, $e_{S,R}$, of the LMP sign test based on $T_{n,S}(\beta_0)$ with respect to the test based on $T_{n,R}(\beta_0)$ with the Spearman scores is equal to

$$(6.5.11) \qquad e_{S,R} = \frac{\big(2g(0)\mathbf{E}|\varepsilon_1|\big)^2}{144\Big(\int g^2(x)dx \int xG(x)\,g(x)dx\Big)^2}.$$

This ARE does not depend on β_0 and the scale parameter of $G(x)$. In particular, for $G(x)$ normal $e_{S,R} \approx 0.44$; for $G(x)$ the logistic d.f. $e_{S,R} \approx 0.48$; for the Laplace d.f. $G(x)$ $e_{S,R} \approx 0.82$.

In all these examples $e_{S,R} < 1$. As it was to be expected, the rank test performs better than the sign test. Recall, however, that the sign statistic $n^{-1/2}l_n^S(\beta_0)$ is advantageous in that it retains its finite sample distributions and the limiting normal distribution for nonidentically distributed $\{\varepsilon_i\}$, provided that (6.2.16) is fulfilled, whereas the corresponding distributional properties of rank statistics hold only for identically distributed $\{\varepsilon_i\}$.

To conclude this section, we discuss median tests related to the median estimator considered in 5.5.3. These tests were introduced in Boldin [**11**]. They are obtained in the following way.

Let $\widehat{\beta}_{n,M}$ be the median of the array

$$\Big\{\frac{u_k}{u_{k-1}},\ k = 1,\dots,n\Big\}.$$

Under Conditions 6.1(i–iii) one has

$$(6.5.12) \qquad \sqrt{n}(\widehat{\beta}_{n,M} - \beta) = -\lambda_M^{-1}(\beta)\,n^{-1/2}l_n^M(\beta) + o_p(1),$$

where

$$\lambda_M(\beta) = -2g(0)\,\mathbf{E}_\beta|u_1|, \qquad l_n^M(\theta) = \sum_{k=1}^n \operatorname{sign} u_{k-1}\operatorname{sign}(u_k - \theta u_{k-1}).$$

The test statistic for H_0 is determined by the leading term of the expansion (6.5.12) and hence is taken to be $T_{n,M}(\beta_0) = n^{-1/2}l_n^M(\beta_0)$. Assuming Conditions 6.1(i–iii), one has under $H_{1n}(a)$ (see Boldin [**11**])

$$(6.5.13) \qquad T_{n,M}(\beta_0) \xrightarrow{d_{\beta_n}} N(-\lambda_M(\beta_0)a, 1).$$

Recall (see 5.5.3) that under conditions $\mathbf{E}\log^+|\varepsilon_1| < \infty$, $\mathbf{P}(\varepsilon_1 < 0) = \mathbf{P}(\varepsilon_1 > 0) = 1/2$ the null distribution of $l_n^M(\beta_0)$ is free of $G(x)$, namely

$$(l_n^M(\beta_0) + n)/2 \sim \mathrm{Bi}(n, 1/2).$$

The relations (6.5.13) and (6.4.4) imply that the ARE $e_{S,M}(\beta_0)$ of the sign test with respect to the test based on $T_{n,M}(\beta_0)$ is equal to

$$e_{S,M}(\beta_0) = (1-\beta_0^2)^{-1}\big(\mathbf{E}|\varepsilon_1|/\mathbf{E}_{\beta_0}|u_1|\big)^2. \tag{6.5.14}$$

It follows from (6.1.1) that

$$\mathbf{E}|\varepsilon_1|(1+|\beta|)^{-1} \le \mathbf{E}_\beta|u_1| \le \mathbf{E}|\varepsilon_1|(1-|\beta|)^{-1},$$

which, combined with (6.5.14), entails

$$\frac{1-|\beta_0|}{1+|\beta_0|} \le e_{S,M}(\beta_0) \le \frac{1+|\beta_0|}{1-|\beta_0|}.$$

When $G(x)$ is normal, $e_{S,M}(\beta_0) = 1$ for all β_0, and in case $\beta_0 = 0$, $e_{S,M}(0) = 1$ for all $G(x)$ subject to Conditions 6.1(i–iii).

Finally, we particularly emphasize that in contrast to the commonly used tests based on the estimators $\widehat{\beta}_{n,LS}$ and $\widehat{\beta}_{n,LD}$, which are distributed free of $G(x)$ only in an asymptotic sense, the sign tests as well as rank and median tests are distribution free for all finite n.

6.6. Sign estimators

In this section we construct nonparametric sign estimators for β in the model (6.1.1) based on the observations $u_0, u_1, \dots, u_n$ satisfying (6.1.2). The construction relies on the results of §§6.2 and 6.4.

6.6.1. Sign estimator $\widehat{\beta}_{n,S}$. Define the variables

$$p_t(\theta) = 2P_\beta\big\{(u_1-\theta u_0)(u_{1+t}-\theta u_t) > 0\big\} - 1, \qquad t = 1,2,\dots. \tag{6.6.1}$$

For any θ one has for $\Gamma_{tn}(\theta)$ as defined by (6.2.9)

$$\mathbf{E}_\beta n^{-1}\Gamma_{tn}(\theta) = (1-t/n)\,p_t(\theta), \qquad t = 1,2,\dots,n-1. \tag{6.6.2}$$

By (6.6.2), taking into account the definition (6.2.10) of $l_n^S(\theta)$ and inequality $|p_t(\theta)| \le 1$, one obtains for $|\theta| < 1$

$$\mathbf{E}_\beta n^{-1} l_n^S(\theta) = \sum_{t=1}^{n-1} \theta^{t-1}(1-t/n)\,p_t(\theta) \to \Lambda_S(\theta),$$

where

$$\Lambda_S(\theta) = \sum_{t=1}^{\infty} \theta^{t-1} p_t(\theta). \tag{6.6.3}$$

The following theorem establishes the uniform convergence of $n^{-1}l_n^S(\theta)$ to $\Lambda_S(\theta)$ in probability.

THEOREM 6.6.1. *Let $G(x)$ satisfy the Lipschitz condition and $\mathbf{E}|\varepsilon_1|^\Delta < \infty$ for some $\Delta > 0$. Then for any $0 < \delta < 1$*

$$\sup_{|\theta|\le 1-\delta} |n^{-1} l_n^S(\theta) - \Lambda_S(\theta)| = o_p(1),$$

and the function $\Lambda_S(\theta)$ is uniformly continuous for $|\theta| \le 1 - \delta$.

The proof of this theorem in regard to its methodology and technique is close to that of Theorem 6.4.1, and hence is omitted. The interested reader can find it in Boldin and Tyurin [**16**].

Notice that under Conditions 6.1(i–iii) the derivative of $\Lambda_S(\theta)$ at $\theta = \beta$ can be easily found to equal $\lambda_S(\beta)$ as given by (6.4.1). Thus Theorems 6.6.1 and 6.4.1 (with $\beta_n = \beta_0 = \beta$) state that

$$n^{-1} l_n^S(\theta) = \Lambda_S(\theta) + o_p(1), \tag{6.6.4}$$

$$n^{-1/2} l_n^S(\beta + n^{-1/2}\theta) = n^{-1/2} l_n^S(\beta) + \lambda_S(\beta)\,\theta + o_p(1), \tag{6.6.5}$$

with remainder terms in (6.6.4) and (6.6.5) vanishing uniformly over compact sets $\{|\theta| \le 1 - \delta\}$ and $\{|\theta| \le \Theta_n\}$, respectively.

Obviously, $\Lambda_S(\beta) = 0$ under Condition 6.1(i). Together with Theorem 6.6.1 this implies that, whenever $\Lambda_S(\theta)$ has a unique root in an interval containing β, the equation

$$l_n^S(\theta) \div 0 \tag{6.6.6}$$

with probability tending to one has a root $\widehat{\beta}_{n,S}$, which is a consistent estimator for β. The symbol $\div$ in (6.6.6) means, as before, that a solution of this equation is understood as a point θ where the function $l_n^S(\theta)$ with discontinuities only of the first kind crosses the zero level.

Therefore, it is natural to take a root, $\widehat{\beta}_{n,S}$, of (6.6.6) lying in $(-1, 1)$ as an estimator for β.

We mentioned in §5.6 the class of RA-estimators, which are defined as the solution of an equation

$$\sum_{i=1}^{n-1} \psi_i(\mathbf{U}_n, \theta) = 0, \tag{6.6.7}$$

where $\mathbf{U}_n = (u_0, \dots, u_n)$,

$$\begin{aligned} \psi_i(\mathbf{U}_n, \theta) = {} & \eta(u_i - \theta u_{i-1}, u_{i+1} - \theta u_i) \\ & + \theta \eta(u_i - \theta u_{i-1}, u_{i+2} - \theta u_{i+1}) + \dots \\ & + \theta^{n-1} \eta(u_i - \theta u_{i-1}, u_n - \theta u_{n-1}). \end{aligned} \tag{6.6.8}$$

This class of estimators was introduced by Bustos and Yohai [**19**] for a general model of autoregression–moving average.

The definition (6.2.10) of $l_n^S(\theta)$ implies that the sign estimator $\widehat{\beta}_{n,S}$ is also an RA-estimator with

$$\eta(\xi_1, \xi_2) = \mathrm{sign}(\xi_1 \xi_2).$$

However, RA-estimators have been explored only for smooth functions η. Thus we cannot use the available results and have to directly study the sign estimators. The key result for this study is Theorem 6.4.1.

So, we consider the equation (6.6.6). Unfortunately, very mild assumptions of Theorem 6.6.1 do not allow us to localize the root $\theta = \beta$ of $\Lambda_S(\theta)$. Henceforth in this section, when dealing with equation (6.6.6), we assume the stronger Conditions 6.1(i–iii), which ensure the existence as well as the asymptotic normality of the solution $\widehat{\beta}_{n,S}$ of equation (6.6.6). The following theorem was given in Boldin and Tyurin [**16**].

THEOREM 6.6.2. *Let Conditions* 6.1(i–iii) *be satisfied. Then, with probability tending to* 1, *there exists a solution* $\widehat{\beta}_{n,S}$ *of equation* (6.6.6) *such that*

$$\sqrt{n}(\widehat{\beta}_{n,S} - \beta) = -\lambda_S^{-1}(\beta) n^{-1/2} l_n^S(\beta) + o_p(1),$$

and

$$\sqrt{n}(\widehat{\beta}_{n,S} - \beta) \xrightarrow{d_\beta} N\big(0, \sigma_S^2(\beta)\big), \qquad \text{where} \quad \sigma_S^2(\beta) = (1-\beta^2)\big(2g(0)\,\mathbf{E}|\varepsilon_1|\big)^{-2}.$$

PROOF. It follows from Theorem 6.4.1 (with $\beta_n = \beta_0 = \beta$) that, for sufficiently large $A > 0$

$$n^{-1/2} l_n^S(\beta - An^{-1/2}) = n^{-1/2} l_n^S(\beta) - \lambda_S(\beta)A + o_p(1) > 0$$

with probability arbitrarily close to 1 uniformly for all sufficiently large n, because

$$n^{-1/2} l_n^S(\beta) \xrightarrow{d_\beta} N\big(0, (1-\beta^2)^{-1}\big)$$

by Theorem 6.2.3 and $\lambda_S(\beta) < 0$. Together with the similar inequality

$$l_n^S(\beta + An^{-1/2}) < 0$$

this implies that, with probability arbitrarily close to 1 for all sufficiently large n, there is a root $\widehat{\beta}_{n,S}$ of equation (6.6.6) in the interval $(\beta - An^{-1/2}, \beta + An^{-1/2})$, i.e., $\sqrt{n}(\widehat{\beta}_{n,S} - \beta) = O_p(1)$. The jumps of $l_n^S(\theta)$ do not exceed $2(1-\theta^2)^{-1/2}$ in absolute value, so that $n^{-1/2} l_n^S(\widehat{\beta}_{n,S}) = o_p(1)$. Applying Corollary 6.4.1 (with $\widehat{\beta}_n = \widehat{\beta}_{n,S}$, $\beta_n = \beta_0 = \beta$, $\alpha = 0$) once more, we obtain the representation

$$\sqrt{n}(\widehat{\beta}_{n,S} - \beta) = -\lambda_S^{-1}(\beta) n^{-1/2} l_n^S(\beta) + o_p(1),$$

which in view of Theorem 6.2.3 implies

$$\sqrt{n}(\widehat{\beta}_{n,S} - \beta) \xrightarrow{d_\beta} N\big(0, \sigma_S^2(\beta)\big)$$

with

$$\sigma_S^2(\beta) = \lambda_S^{-2}(\beta)(1-\beta^2)^{-1} = (1-\beta^2)\big(2g(0)\,\mathbf{E}|\varepsilon_1|\big)^{-2}.$$

This completes the proof. □

Theorem 6.6.2 enables us to obtain the ARE's of the sign estimator $\widehat{\beta}_{n,S}$ with respect to other estimators, in particular, with respect to the least squares, $\widehat{\beta}_{n,LS}$, least absolute deviations, $\widehat{\beta}_{n,LD}$, and median, $\widehat{\beta}_{n,M}$, estimators. These ARE's turn out to equal $e_{S,LS}$, $e_{S,LD}$ and $e_{S,M}$, which are defined by (6.5.2), (6.5.5), and (6.5.14), respectively. Hence the comments about the ARE's of the corresponding tests in §6.5 concern the estimators as well.

Let us compare the sign estimator $\widehat{\beta}_{n,S}$ with the rank estimator obtainable as a solution of equation

$$l_n^R(\theta) \div 0,$$

where $l_n^R(\theta)$ is defined by (6.5.8). It is shown in Boldin [**15**] that under Conditions 6.5(i–iv) this equation with probability tending to 1 has a solution $\widehat{\beta}_{n,R}$ such that

$$\sqrt{n}(\widehat{\beta}_{n,R} - \beta) \xrightarrow{d_\beta} N\big(0, \sigma_R^2(\beta)\big),$$

where

$$\sigma_R^2(\beta) = (1-\beta^2)\frac{\mathbf{E}\varphi_1^2\big(G(\varepsilon_1)\big)\mathbf{E}\varphi_2^2\big(G(\varepsilon_1)\big)}{\big(\mathbf{E}\{\varepsilon_1\varphi_1(G(\varepsilon_1))\}\mathbf{E}\{\varphi_2'(G(\varepsilon_1))g(\varepsilon_1)\}\big)^2}.$$

Now the ARE of the sign estimator $\widehat{\beta}_{n,S}$ with respect to the rank estimator $\widehat{\beta}_{n,R}$ can readily by found. For the Spearman scores it is equal to $e_{S,R}$ as in (6.5.11).

6.6.2. Sign estimator $\beta_{n,S}^*$. The computation of $\widehat{\beta}_{n,S}$ involves no difficulties when the equation (6.6.6) has a unique root for $|\theta| < 1$. It can be found with a prescribed accuracy, for example, by successive dichotomizing of the interval. Otherwise the problem of selecting the proper root arises. One can take for $\widehat{\beta}_{n,S}$ the root of (6.6.6) closest to some $\sqrt{n}$-consistent estimator $\widehat{\beta}_n$ for β. For example, if $G(x)$ is symmetric about zero and $\mathbf{E}\log^+|\varepsilon_1| < \infty$ then $\sqrt{n}(\widehat{\beta}_{n,LS} - \beta) = O_p(1)$ (see 5.3.2); under Conditions 6.1(i–iii) one can use the median estimator $\widehat{\beta}_{n,M}$ (see 5.5.3).

Alternatively, we can construct an estimator asymptotically equivalent to $\widehat{\beta}_{n,S}$.

THEOREM 6.6.3. *Assume Conditions* 6.1(i–iii). *Let $\widehat{\beta}_n$ be a $\sqrt{n}$-consistent estimator for β and let $\widehat{e}_n$ be a consistent estimator for $\lambda_S(\beta)$. Put*

$$\beta_{n,S}^* = \widehat{\beta}_n - (n\widehat{e}_n)^{-1} l_n^S(\widehat{\beta}_n).$$

Then

$$\sqrt{n}(\beta_{n,S}^* - \beta) = -\lambda_S^{-1}(\beta) n^{-1/2} l_n^S(\beta) + o_p(1),$$

$$\sqrt{n}(\beta_{n,S}^* - \beta) \xrightarrow{d_\beta} N\big(0, \sigma_S^2(\beta)\big), \qquad \sigma_S^2(\beta) = (1-\beta^2)\big(2g(0)\,\mathbf{E}|\varepsilon_1|\big)^{-2}.$$

The proof is obtained by an immediate application of Corollary 6.4.1 (with $\beta_n = \beta_0 = \beta, \alpha = 0$) and Theorem 6.2.3.

The estimator $\beta_{n,S}^*$ is asymptotically equivalent to $\widehat{\beta}_{n,S}$ as in Theorem 6.6.2 in the sense that it has the same asymptotic distribution.

We mentioned above some possible choices of $\widehat{\beta}_n$. There are also a number of options for $\widehat{e}_n$. In view of Corollary 6.4.1 (with $\beta_n = \beta_0 = \beta$) one can take, for example,

$$\widehat{e}_n = \big[l_n^S(\widehat{\beta}_n + hn^{-1/2}) - l_n^S(\widehat{\beta}_n)\big](n^{1/2}h)^{-1}$$

for a fixed $h \neq 0$. Alternatively, one can consistently estimate $\lambda_S(\beta)$ by separately estimating $g(0)$ and $\mathbf{E}|\varepsilon_1|$. The latter mean value is consistently estimated, for example, by

$$\widehat{m}_n = n^{-1} \sum_{k=1}^{n} |u_k - \widehat{\beta}_n u_{k-1}|, \tag{6.6.9}$$

while for $g(0)$ one can use a consistent density esimate, for example, the Parzen–Rosenblatt statistic

$$\widehat{g}_n = (nh_n)^{-1} \sum_{k=1}^{n} \varphi\left(\frac{u_k - \widehat{\beta}_n u_{k-1}}{h_n}\right), \tag{6.6.10}$$

with $\varphi(x)$ being the standard normal density and $h_n = n^{-\alpha}$, $0 < \alpha < 1/4$. Then $\widehat{e}_n = -2\widehat{g}_n \widehat{m}_n (1 - \widehat{\beta}_n^2)^{-1}$.

6.6.3. Sign estimator $\widetilde{\beta}_{n,S}$. Next we discuss another construction of the sign estimator. The idea is that instead of solving the equation (6.6.6) we minimize the function $l_n^S(\theta)$, yet on a restricted parameter set, namely, we minimize this function for θ running over an "asymptotically small" compact set

$$Q(\widehat{\beta}_n) := \{\theta \colon |\theta - \widehat{\beta}_n| \leq n^{-1/2} \log n\},$$

where $\widehat{\beta}_n$ is some preliminary $\sqrt{n}$-consistent estimator of β. In order to guarantee the existence of a solution of this problem, it is expedient to replace the objective function $l_n^S(\theta)$ by the piecewise-constant function

$$l_n^S(\widehat{\beta}_n, \theta) = \sum_{t=1}^{n-1} \widehat{\beta}_n^{t-1} \Gamma_{tn}(\theta)$$

having discontinuities at the points u_k/u_{k-1}, $k = 1, \dots, n$. This function asymptotically approaches $l_n^S(\theta)$.

Thus we define the estimator for β, to be denoted by $\widetilde{\beta}_{n,S}$, as any measurable solution of the problem

$$\left|l_n^S(\widehat{\beta}_n, \theta)\right| \Longrightarrow \inf_{\theta \in Q(\widehat{\beta}_n)} . \tag{6.6.11}$$

One can take $\widetilde{\beta}_{n,S}$ to be, for example, the midpoint of the leftmost interval on which the objective function in (6.6.11) is minimized.

It can be shown (see Boldin [**14**]) that under Conditions 6.1(i–iii) for $0 < \Theta < \infty$ and $\Theta_n = \Theta \log n$

$$\sup_{|\theta| \leq \Theta_n} \left|n^{-1/2} l_n^S(\widehat{\beta}_n, \theta) - n^{-1/2} l_n^S(\theta)\right| = o_p(1).$$

This relation and Theorem 6.4.1 (with $\beta_n = \beta_0 = \beta$ and $\Theta_n = \Theta \log n$) imply the stochastic expansion

$$\sup_{|\theta| \leq \Theta_n} \left|n^{-1/2} l_n^S(\widehat{\beta}_n, \beta + \theta n^{-1/2}) - n^{-1/2} l_n^S(\beta) - \lambda_S(\beta)\theta\right| = o_p(1).$$

This expansion, in its turn, implies (see the proof of Corollary 6.4.1) that for any estimator β_n^* such that

$$\sqrt{n}(\beta_n^* - \beta)/\log n = O_p(1),$$

one has

$$n^{-1/2} l_n^S(\widehat{\beta}_n, \beta_n^*) = n^{-1/2} l_n^S(\beta) + \lambda_S(\beta)\sqrt{n}(\beta_n^* - \beta) + o_p(1). \tag{6.6.12}$$

Consider now

$$\widetilde{\beta}_n = -\lambda_S^{-1}(\beta) n^{-1/2} l_n^S(\beta).$$

By Theorem 6.2.3, $\sqrt{n}(\widetilde{\beta}_n - \beta) = O_p(1)$. Substituting $\widetilde{\beta}_n$ for β_n^* in (6.6.12), we see that

$$n^{-1/2} l_n^S(\widehat{\beta}_n, \widetilde{\beta}_n) = o_p(1). \tag{6.6.13}$$

Since $\widetilde{\beta}_n \in Q(\widehat{\beta}_n)$ with probability tending to 1, any solution $\widehat{\beta}_{n,S}$ of the problem (6.6.11) in view of (6.6.13) all the more fulfills a similar relation

$$n^{-1/2} l_n^S(\widehat{\beta}_n, \widetilde{\beta}_{n,S}) = o_p(1). \tag{6.6.14}$$

Now (6.6.12) (with $\beta_n^* = \widetilde{\beta}_{n,S}$), (6.6.14), and Theorem 6.2.3 entail the following theorem.

THEOREM 6.6.4. *Assume Conditions* 6.1(i–iii) *and let* $\sqrt{n}(\widehat{\beta}_n - \beta) = O_p(1)$. *Then the solution* $\widetilde{\beta}_{n,S}$ *of the problem* (6.6.11) *fulfills the following relations*:

$$\sqrt{n}(\widetilde{\beta}_{n,S} - \beta) = \lambda_S^{-1}(\beta) l_n^S(\beta) + o_p(1),$$

$$\sqrt{n}(\widetilde{\beta}_{n,S} - \beta) \xrightarrow{d_\beta} N\big(0, \sigma_S^2(\beta)\big), \qquad \sigma_S^2(\beta) = (1-\beta^2)\big(2g(0)\,\mathbf{E}|\varepsilon_1|\big)^{-2}.$$

In the next section we will study the estimators introduced above in regard to their robustness against outliers in observations $u_0, \dots, u_n$.

6.7. Influence functionals of sign estimators

Suppose we have contaminated observations $\mathbf{Y}_n = (y_0, \dots, y_n)$, where

$$y_i = u_i + z_i^\gamma \xi_i, \qquad i \in \mathbb{Z}. \tag{6.7.1}$$

We assume that the variables $\{u_i\}$ in (6.7.1) satisfy (6.1.1), $\{z_i^\gamma\}$ form a Bernoulli sequence of i.i.d. random variables taking values 1 and 0 with probabilities γ and $1-\gamma$, $0 \le \gamma \le 1$, and $\{\xi_i\}$ are i.i.d. random variables with distribution μ_ξ; the sequences $\{u_i\}$, $\{z_i^\gamma\}$, $\{\xi_i\}$ are mutually independent. Thus we condider the model of independent (individual) outliers dealt with in §5.6.

6.7.1. Influence functional of the sign estimator $\widehat{\beta}_{n,S}$. The sign estimator $\widehat{\beta}_{n,S}$ defined by equation (6.6.6) for an uncontaminated sample is determined in the model (6.7.1) from the equation

$$l_n^S(\theta) = \sum_{i=1}^{n-1} \psi_i^S(\mathbf{Y}_n, \theta) \div 0, \tag{6.7.2}$$

where

$$\begin{aligned}\psi_i^S(\mathbf{Y}_n,\theta) = {} & \operatorname{sign}\big((y_i-\theta y_{i-1})(y_{i+1}-\theta y_i)\big)\\ & +\theta\operatorname{sign}\big((y_i-\theta y_{i-1})(y_{i+2}-\theta y_{i+1})\big)+\dots\\ & +\theta^{n-1}\operatorname{sign}\big((y_i-\theta y_{i-1})(y_n-\theta y_{n-1})\big).\end{aligned} \tag{6.7.3}$$

As we pointed out in the previous section (see (6.6.7), (6.6.8) and the related comment), the sign estimator $\widehat{\beta}_{n,S}$ is an RA-estimator. The influence functionals for such estimators have been obtained by Martin and Yohai [**66**]. However they dealt only with smooth functions η determining the estimator (see (6.6.7), (6.6.8)). We will directly derive the influence functionals of the sign estimators without using these results.

Similarly to Theorem 6.6.1 it can be shown under the assumptions of the following Lemma 6.7.1 that for any $0\le\gamma\le1$ and $0<\delta<1$

$$n^{-1}l_n^S(\theta)=n^{-1}\sum_{i=1}^n\psi_i^S(\mathbf{Y}_n,\theta)\xrightarrow{\mathbf{P}_\beta}\Lambda_S(\gamma,\theta) \tag{6.7.4}$$

uniformly over $|\theta|\le1-\delta$, where

$$\Lambda_S(\gamma,\theta)=\sum_{t\ge1}\theta^{t-1}\left(2\mathbf{P}_\beta\big\{(y_1-\theta y_0)(y_{1+t}-\theta y_t)>0\big\}-1\right). \tag{6.7.5}$$

In fact (6.7.4) holds under weaker conditions than those of Lemma 6.7.1, but we stick to the conditions of this lemma to be used for the proof of the subsequent Theorem 6.7.1.

It is easily seen that under Condition 6.1(i) $\Lambda_S(0,\beta)=\Lambda_S(\beta)=0$, where $\Lambda_S(\theta)$ is defined by (6.6.3).

LEMMA 6.7.1. *Assume that* $\mathbf{E}\varepsilon_1=0$, $G(0)=1/2$, $\mathbf{E}|\xi_1|<\infty$, *and let* $g(x)$ *be continuous, bounded, and* $g(0)>0$. *Then* $\Lambda_S(\gamma,\theta)$ *has derivatives* $\frac{\partial}{\partial\gamma}\Lambda_S(\gamma,\theta)$, $\frac{\partial}{\partial\theta}\Lambda_S(\gamma,\theta)$ *continuous in* (γ,θ), $0\le\gamma\le1$, $|\theta|<1$, *and*

$$\frac{\partial}{\partial\theta}\Lambda_S(0,\beta)=\lambda_S(\beta)=-2g(0)\,\mathbf{E}|\varepsilon_1|\,(1-\beta^2)^{-1}\neq0, \tag{6.7.6}$$

$$\frac{\partial}{\partial\gamma}\Lambda_S(0,\beta)=\mathbf{E}_\beta\big(1-2G(-\xi_1)\big)\big(1-2G(\beta\xi_1)\big). \tag{6.7.7}$$

Recall that $\lambda_S(\beta)$ was defined by (6.4.1).

PROOF. Letting

$$p_t(\gamma,\theta)=\mathbf{P}_\beta\big\{(y_1-\theta y_0)(y_{1+t}-\theta y_t)>0\big\}-\frac{1}{2},$$

rewrite (6.7.5) as

$$\Lambda_S(\gamma,\theta)=2\sum_{t=1}^\infty\theta^{t-1}p_t(\gamma,\theta). \tag{6.7.8}$$

We can write $p_t(\gamma,\theta) = p_{t1}(\gamma,\theta) + p_{t2}(\gamma,\theta) - 1/2$, where

$$p_{t1}(\gamma,\theta) = \mathbf{P}_\beta\{y_1 - \theta y_0 < 0,\ y_{1+t} - \theta y_t < 0\},$$
$$p_{t2}(\gamma,\theta) = \mathbf{P}_\beta\{y_1 - \theta y_0 > 0,\ y_{1+t} - \theta y_t > 0\}.$$

By (6.7.8), it suffices to show that $p_{ti}(\gamma,\theta)$ are continuously differentiable and to find their derivatives. Consider $p_{t1}(\gamma,\theta)$ for $t = 1$. The other cases are treated similarly.

Let Ω be the σ-algebra generated by $\{\varepsilon_i, i \le 1, z_0^\gamma,\ z_1^\gamma,\ z_2^\gamma,\ \xi_0,\ \xi_1,\ \xi_2\}$. Then

$$\begin{aligned} p_{11}(\gamma,\theta) &= \mathbf{E}_\beta\{I(y_1 - \theta y_0 < 0)\, I(y_2 - \theta y_1 < 0)\} \\ &= \mathbf{E}_\beta \mathbf{E}_\beta\{I\big(\varepsilon_1 - (\theta-\beta)u_0 + z_1^\gamma \xi_1 - \theta z_0^\gamma \xi_0 < 0\big) \\ &\qquad \times I\big(\varepsilon_2 - (\theta-\beta)u_1 + z_2^\gamma \xi_2 - \theta z_1^\gamma \xi_1 < 0\big) \mid \Omega\} \\ &= \mathbf{E}_\beta\{I\big(\varepsilon_1 - (\theta-\beta)u_0 + z_1^\gamma \xi_1 - \theta z_0^\gamma \xi_0 < 0\big) \\ &\qquad \times G\big((\theta-\beta)u_1 - z_2^\gamma \xi_2 + \theta z_1^\gamma \xi_1\big)\}. \end{aligned} \tag{6.7.9}$$

Let H_i, $i = 0,1,2,3$, denote the event that there are i nonzero variables among z_0^γ, z_1^γ, z_2^γ. Rewrite (6.7.9) by the formula for total expectation as

$$\begin{aligned} &\sum_{i=0}^{3} \mathbf{E}_\beta\{I\big(\varepsilon_1 < (\theta-\beta)u_0 - z_1^\gamma \xi_1 + \theta z_0^\gamma \xi_0\big) \\ &\qquad \times G\big((\theta-\beta)u_1 - z_2^\gamma \xi_2 + \theta z_1^\gamma \xi_1\big) \mid H_i\}\mathbf{P}(H_i). \end{aligned} \tag{6.7.10}$$

Here $\mathbf{P}(H_i)$ are some polynomials in γ, while the conditional expectations do not depend on γ and are continuously differentiable with respect to θ under the conditions of the lemma. Therefore, $p_{11}(\gamma,\theta)$ is continuously differentiable with respect to (γ,θ), $0 \le \gamma \le 1$, $\theta \in \mathbb{R}^1$. On account of (6.6.10) this property, which holds in a similar way for other functions $p_{ti}(\gamma,\theta)$, implies that $\Lambda_S(\gamma,\theta)$ is continuously differentiable for $0 \le \gamma \le 1$, $|\theta| < 1$.

The equality (6.7.6) can easily be obtained by using (6.7.8) and rewriting every $p_t(\gamma,\theta)$ in a form similar to (6.7.9). Finally, for the proof of (6.7.7) observe that $p_1(0,\beta) = 0$ and $p_t(\gamma,\beta) = 0$ for $t \ge 2$, while

$$\begin{aligned} p_1(\gamma,\beta) &= \tfrac{1}{2}\gamma \mathbf{E}_\beta \operatorname{sign}(\varepsilon_1 + \xi_1)(\varepsilon_2 - \beta\xi_1) + o(\gamma) \\ &= \tfrac{1}{2}\gamma \mathbf{E}_\beta\big(1 - 2G(-\xi_1)\big)\big(1 - 2G(\beta\xi_1)\big) + o(\gamma). \end{aligned}$$

Thus the proof is completed. □

Lemma 6.7.1 and condition $\Lambda_S(0,\beta) = 0$ imply that the equation

$$\Lambda_S(\gamma,\theta) = 0$$

in a neighborhood of the point $(0,\beta)$ determines a differentiable function $\theta_\gamma^S = \theta(\gamma)$ with $\theta_0^S = \beta$ and

$$\left.\frac{d\theta_\gamma^S}{d\gamma}\right|_{\gamma=0} = -\lambda_S^{-1}(\beta)\frac{\partial}{\partial\gamma}\Lambda_S(0,\beta), \tag{6.7.11}$$

where $\lambda_S(\beta)$ and $\frac{\partial}{\partial\gamma}\Lambda_S(0,\beta)$ are defined by (6.7.6), (6.4.1), and (6.7.7).

As a function of θ, $\Lambda_S(\gamma,\theta)$ is strictly decreasing in a neighborhood of β for sufficiently small γ, since $\frac{\partial}{\partial\theta}\Lambda_S(0,\beta) < 0$ by (6.7.6) and hence $\frac{\partial}{\partial\theta}\Lambda_S(\gamma,\theta) < 0$ in a neighborhood of $(0,\beta)$. This fact and the relation (6.7.4) imply that with probability tending to 1 there exists a solution $\widehat{\beta}_{n,S}$ of the equation (6.7.2) such that

$$\widehat{\beta}_{n,S} \xrightarrow{\mathbf{P}_\beta} \theta_\gamma^S. \tag{6.7.12}$$

The relations (6.7.11) and (6.7.12) and the definition of the influence functional (5.6.2) imply the following theorem.

THEOREM 6.7.1. *Assume the conditions of Lemma* 6.7.1 *to hold. Then with probability tending to* 1 *the equation* (6.7.2) *has a solution* $\widehat{\beta}_{n,S}$ *such that*

$$\widehat{\beta}_{n,S} \xrightarrow{\mathbf{P}_\beta} \theta_\gamma^S.$$

The influence functional of the estimator $\widehat{\beta}_{n,S}$ *equals*

$$\mathrm{IF}(\theta_\gamma^S,\mu_\xi) = \frac{1-\beta^2}{2g(0)\,\mathbf{E}|\varepsilon_1|}\mathbf{E}_\beta\big(1-2G(-\xi_1)\big)\big(1-2G(\beta\xi_1)\big). \tag{6.7.13}$$

For $\beta = 0$ the influence functional $\mathrm{IF}(\theta_\gamma^S,\mu_\xi)$ equals zero. If $\xi_i = \xi$ with a constant ξ, then $\mathrm{IF}(\theta_\gamma^S,\mu_\xi) = \mathrm{IF}(\theta_\gamma^S,\xi)$ and

$$\mathrm{IF}(\theta_\gamma^S,\xi) = \frac{1-\beta^2}{2g(0)\,\mathbf{E}|\varepsilon_1|}\big(1-2G(-\xi)\big)\big(1-2G(\beta\xi)\big), \tag{6.7.14}$$

which is a bounded continuous function of ξ.

Let $\mathfrak{M}_i$, $i = 1, 2$, be the class of distributions μ_ξ having a finite ith absolute moment. The expectation in (6.7.13) is no greater than 1 in absolute value and for a constant $\xi_1 = \xi$ its absolute value tends to 1 as $\xi \to \infty$ and $\beta \neq 0$. Therefore

$$\mathrm{GES}(\mathfrak{M}_1,\theta_\gamma^S) = \frac{1-\beta^2}{2g(0)\mathbf{E}|\varepsilon_1|} < \infty, \qquad \beta \neq 0, \tag{6.7.15}$$

so that outliers in data have little effect on the sign estimator $\widehat{\beta}_{n,S}$. This is one of its advantages over the least squares estimators and the least absolute deviations estimators, which have unbounded gross error sensitivity even over the narrower class $\mathfrak{M}_2$ (see (5.6.9)–(5.6.10) and (5.6.12)–(5.6.13)). It is of interest to compare (6.7.15) with the sensitivity of the median estimator. The latter is finite on $\mathfrak{M}_1$. In particular, for $G(x)$ Gaussian we find from (5.6.15) that $\mathrm{GES}(\mathfrak{M}_1,\theta_\gamma^M) = \frac{\pi}{2}(1-\beta^2)^{1/2}$. Hence in this case

$$\frac{\mathrm{GES}(\mathfrak{M}_1,\theta_\gamma^S)}{\mathrm{GES}(\mathfrak{M}_1,\theta_\gamma^M)} = (1-\beta^2)^{1/2} < 1,$$

i.e., the sign estimator is preferable in this respect to the median estimator.

6.7.2. Influence functional of the sign estimator $\widetilde{\beta}_{n,S}$. Consider now the sign estimator $\widetilde{\beta}_{n,S}$ defined by (6.6.11) in the no contamination case. For contaminated observations (6.7.1) $\widetilde{\beta}_{n,S}$ is defined as the solution of the problem

$$|l_n^S(\widehat{\beta}_n, \theta)| \Longrightarrow \inf_{\theta \in Q(\widehat{\beta}_n)}, \tag{6.7.16}$$

where $Q(\widehat{\beta}_n) = \{\theta \colon |\theta - \widehat{\beta}_n| \le n^{-1/2} \log n\}$ and the preliminary estimator $\widehat{\beta}_n$ as well as $l_n^S(\widehat{\beta}_n, \theta)$ are based on $\mathbf{Y}_n$.

CONDITION 6.7(i). $\widehat{\beta}_n \xrightarrow{\mathbf{P}_\beta} \theta_\gamma$, $\theta_0 = \beta$, *and influence functional* $\mathrm{IF}(\theta_\gamma, \mu_\xi)$ *exists.*

By definition, $\widetilde{\beta}_{n,S}$ has the same limit

$$\widetilde{\beta}_{n,S} \xrightarrow{\mathbf{P}_\beta} \widetilde{\theta}_\gamma^S = \theta_\gamma,$$

and thus

$$\mathrm{IF}\left(\widetilde{\theta}_\gamma^S, \mu_\xi\right) = \mathrm{IF}\left(\theta_\gamma, \mu_\xi\right).$$

The last equality means that the sign estimator $\widetilde{\beta}_{n,S}$ inherits the influence functional and GES of the preliminary estimator $\widehat{\beta}_n$. In particular, if $\widehat{\beta}_n$ is taken to be $\widehat{\beta}_{n,LS}$, $\widehat{\beta}_{n,LD}$, or $\widehat{\beta}_{n,M}$, the influence functional of $\widetilde{\beta}_{n,S}$ is given by (5.6.9), (5.6.12), or (5.6.15), respectively.

6.7.3. Influence functional of the sign estimator $\beta^*_{n,S}$. Finally, consider the sign estimator $\beta^*_{n,S}$ introduced in Theorem 6.6.3. This estimator is defined by

$$\beta^*_{n,S} = \widehat{\beta}_n - (n\widehat{e}_n)^{-1} l_n^S(\widehat{\beta}_n), \tag{6.7.17}$$

where the preliminary estimator $\widehat{\beta}_n$ as well as $\widehat{e}_n$ and $l_n^S(\widehat{\beta}_n)$ are based on $\mathbf{Y}_n$.

We assume here the conditions of Lemma 6.7.1, Condition 6.7(i) and

CONDITION 6.7(ii). $\widehat{e}_n \xrightarrow{\mathbf{P}_\beta} \theta_\gamma^e$, *and* $\theta_\gamma^e \to \theta_0^e = \lambda_S(\beta)$ *as* $\gamma \to 0$.

Condition 6.7(ii) is satisfied by the estimator $\widehat{e}_n = -2\widehat{g}_n \widehat{m}_n (1 - \widehat{\beta}_n^2)^{-1}$, constructed from observations $\mathbf{Y}_n$ similarly to (6.6.9)–(6.6.10), provided that $\sqrt{n}(\widehat{\beta}_n - \theta_\gamma) = O_p(1)$.

For small γ one has $|\theta_\gamma| < 1$, and $\Lambda_S(\gamma, \theta)$ defined by (6.7.5) is continuous as a function of θ at $\theta = \theta_\gamma$ by Lemma 6.7.1. One readily infers from this and (6.7.4) that

$$n^{-1} l_n^S(\widehat{\beta}_n) \xrightarrow{\mathbf{P}_\beta} \Lambda_S(\gamma, \theta_\gamma). \tag{6.7.18}$$

Indeed, let $\delta > 0$ be such that $|\beta| < 1 - \delta$ and γ is small enough for $|\theta_\gamma| < 1 - \delta$. Then for any $\varepsilon > 0$

$$\begin{aligned}
&\mathbf{P}_\beta\{|n^{-1} l_n^S(\widehat{\beta}_n) - \Lambda_S(\gamma, \theta_\gamma)| > \varepsilon\} \\
&\quad \le \mathbf{P}_\beta\{|n^{-1} l_n^S(\widehat{\beta}_n) - \Lambda_S(\gamma, \widehat{\beta}_n)| + |\Lambda_S(\gamma, \widehat{\beta}_n) - \Lambda_S(\gamma, \theta_\gamma)| > \varepsilon, |\widehat{\beta}_n| \le 1 - \delta\} \\
&\qquad + \mathbf{P}_\beta\{|\widehat{\beta}_n| > 1 - \delta\} \\
&\quad \le \mathbf{P}_\beta\{\sup_{|\theta| \le 1 - \delta} |n^{-1} l_n^S(\theta)| > \varepsilon/2\} + \mathbf{P}_\beta\{|\Lambda_S(\gamma, \widehat{\beta}_n) - \Lambda_S(\gamma, \theta_\gamma)| > \varepsilon/2\} \\
&\qquad + \mathbf{P}_\beta\{|\widehat{\beta}_n - \theta_\gamma| + |\theta_\gamma| > 1 - \delta\}.
\end{aligned}$$

The probabilities in the right-hand side become arbitrarily small for sufficiently large n: the first one by (6.7.4) and the last two by virtue of continuity of $\Lambda_S(\gamma,\theta)$ with respect to θ and convergence of $\widehat{\beta}_n$ to θ_γ. This proves (6.7.18).

Now the definition (6.7.17) of $\beta^*_{n,S}$, relation (6.7.18) and Conditions 6.7(i, ii) imply the convergence

$$\beta^*_{n,S} \xrightarrow{\mathbf{P}_\beta} \theta^{*S}_\gamma := \theta_\gamma - (\theta^e_\gamma)^{-1}\Lambda_S(\gamma,\theta_\gamma). \tag{6.7.19}$$

Since $\Lambda_S(\gamma,\theta)$ is continuously differentiable, (6.7.19), Conditions 6.7(i, ii), and equalities

$$\Lambda_S(0,\beta)=0, \quad \mathrm{IF}(\theta^S_\gamma,\mu_\xi) = -\lambda_S^{-1}(\beta)\frac{\partial}{\partial\gamma}\Lambda_S(0,\beta)$$

imply

$$\begin{aligned}\theta^{*S}_\gamma &= \theta_\gamma - \left[\lambda_S^{-1}(\beta)+o(1)\right]\left[\Lambda_S(0,\beta)+\frac{\partial\Lambda_S(0,\beta)}{\partial\gamma}\gamma+\frac{\partial\Lambda_S(0,\beta)}{\partial\theta}(\theta_\gamma-\beta)+o(\gamma)\right]\\ &= \theta_\gamma + \mathrm{IF}(\theta^S_\gamma,\mu_\xi)\gamma - \mathrm{IF}(\theta_\gamma,\mu_\xi)\gamma + o(\gamma).\end{aligned}$$

The last relation and the definition of the influence functional (5.6.2) yield

$$\lim_{\gamma\to 0}\frac{\theta^{*S}_\gamma-\beta}{\gamma} := \mathrm{IF}(\theta^{*S}_\gamma,\mu_\xi) = \mathrm{IF}(\theta^S_\gamma,\mu_\xi),$$

where $\mathrm{IF}(\theta^S_\gamma,\mu_\xi)$ was defined in Theorem 6.7.1.

Thus the estimators $\widehat{\beta}_{n,S}$ and $\beta^*_{n,S}$ have equal influence functionals and gross error sensitivities, which are given by (6.7.13)–(6.7.15). Therefore outliers in data have little effect on $\beta^*_{n,S}$.

6.8. Simulation results: Evaluation of quantiles, confidence sets, and contaminated samples

In this section we consider the stationary model (6.1.1) and present the results of simulation experiments on computation of quantiles $c^\alpha_n(\beta_0)$ from §6.2, confidence sets for β in the model (6.1.1) for moderate and large n, as well as on robustness of sign estimator $\widehat{\beta}_{n,S}$ to outliers in observations.

6.8.1. Evaluation of quantiles. We define the α-quantile ζ_α, $0<\alpha<1$, of a discrete random variable with distribution function $F(x)$ as follows: if the equation $F(x)=\alpha$ has no solution, then $\zeta_\alpha = \sup\{x\colon F(x)<\alpha\}$; otherwise, ζ_α is the midpoint of the interval formed by the solutions of equation $F(x)=\alpha$.

On account of the results of §6.2, the sign test statistic $l^S_n(\beta_0)$ defined by (6.2.10), (6.2.9), subject to Condition 6.1(i), is distributed free of $G(x)$ under the null hypothesis $H_0\colon \beta=\beta_0$. Of course, the same property holds for the normalized statistic $T_{n,S}(\beta_0) = n^{-1/2}\sqrt{1-\beta_0^2}\, l^S_n(\beta_0)$, whose null distribution converges to the standard normal one. For finite n the null distribution of $T_{n,S}(\beta_0)$ coincides with the distribution of the random variable

$$\nu_n(\beta_0) = \sqrt{\frac{1-\beta_0^2}{n}}\sum_{t=1}^{n-1}\beta_0^{t-1}\sum_{k=t+1}^{n}\xi_{k-t}\xi_k,$$

where $\xi_1, \dots, \xi_n$ are i.i.d. random variables with $\mathbf{P}\{\xi_i = +1\} = \mathbf{P}\{\xi_i = -1\} = 1/2$. This distribution as well as the α-quantiles $c_n^\alpha(\beta_0)$ of $\nu_n(\beta_0)$ can be computed for any β_0 by simulation.

For this purpose we present $\nu_n(\beta_0)$ in a special form. Namely, we introduce the variables $\eta_k = \xi_k \xi_{k+1}$, $k = 1, \dots, n-1$, which, like ξ_k's, are i.i.d. with $\mathbf{P}\{\eta_k = +1\} = \mathbf{P}\{\eta_k = -1\} = 1/2$. Since $\eta_k^2 = 1$, the statistic $\nu_n(\beta_0)$ can be written in the following form:

$$\nu_n(\beta_0) = \nu_n(\beta_0, \eta) = \sqrt{\frac{1-\beta_0^2}{n}}\, l_n^S(\beta_0, \eta),$$

where $\eta = (\eta_1, \eta_2, \dots, \eta_{n-1})$ and

$$\begin{aligned}
l_n^S(\beta, \eta) &= \sum_{t=1}^{n-1} \beta^{t-1} \sum_{k=t+1}^{n} \xi_k \xi_{k-t} && (6.8.1)\\
&= \eta_1 + \eta_2 + \eta_3 + \dots + \eta_{n-1}\\
&\quad + \beta\eta_1\eta_2 + \beta\eta_2\eta_3 + \dots + \beta\,\eta_{n-2}\eta_{n-1}\\
&\quad + \beta^2\eta_1\eta_2\eta_3 + \dots + \beta^2\eta_{n-3}\eta_{n-2}\eta_{n-1}\\
&\quad + \dots + \beta^{n-2}\eta_1\eta_2\dots\eta_{n-1}\\
&= \eta_1 + \eta_2(1+\beta\eta_1) + \eta_3(1+\beta\eta_2(1+\beta\eta_1)) + \dots\\
&\quad + \eta_{n-1}(1+\beta\,\eta_{n-2}(1+\beta\,\eta_{n-3}+\dots))\\
&= \sum_{k=1}^{n-1} v_k \eta_k, && (6.8.2)
\end{aligned}$$

with $v_1 = 1$, $v_k = 1 + \beta\eta_{k-1}v_{k-1}$, $k = 2, \dots, n-1$.

In this way the computation of $\nu_n(\beta)$ reduces to generating random numbers $\eta_1, \dots, \eta_{n-1}$ taking values ± 1 with probabilities $1/2$, recursive computation of the coefficients v_k, and computation of the linear combination of η_k with these coefficients.

Now we show that the quantiles $c_n^{1-\alpha}(\beta)$ possess the following property of symmetry about zero:

$$c_n^{1-\alpha}(\beta) = -c_n^\alpha(-\beta). \tag{6.8.3}$$

Let $F_n(x, \beta) = \mathbf{P}_\beta\{\nu_n(\beta) < x\}$ be the distribution function of $\nu_n(\beta)$. We will show that

$$F_n(x+0, \beta) + F_n(-x, -\beta) = 1 \tag{6.8.4}$$

for any $x \in \mathbb{R}^1$ and $\beta \in (-1, 1)$, which directly implies (6.8.3). The statistic $l_n^S(\beta, \eta)$ as in (6.8.1), (6.8.2) can be represented as a polynomial in β,

$$l_n^S(\beta, \eta) = \sum_{t=0}^{n-2} a_t \beta^t,$$

with coefficients $a_t = \sum_{k=t+1}^{n-1} \prod_{r=k-t}^{k} \eta_r$. Combining the terms with even and odd powers of β, rewrite $\nu_n(\beta, \eta)$ as

$$\nu_n(\beta,\eta) = \nu_{n1}(\beta,\eta) + \nu_{n2}(\beta,\eta),$$

where

$$\nu_{n1}(\beta,\eta) = \sqrt{\frac{1-\beta^2}{n}} \sum_{t=0}^{[(n-2)/2]} \beta^{2t} \sum_{k=2t+1}^{n-1} \prod_{r=k-2t}^{k} \eta_r, \tag{6.8.5}$$

$$\nu_{n2}(\beta,\eta) = \sqrt{\frac{1-\beta^2}{n}} \sum_{t=0}^{[(n-3)/2]} \beta^{2t+1} \sum_{k=2t+2}^{n-1} \prod_{r=k-2t-1}^{k} \eta_r. \tag{6.8.6}$$

Then

$$\begin{aligned} F_n(x{+}0,\beta) &+ F_n(-x,-\beta) \\ &= \mathbf{P}_\beta\{\nu_{n1}(\beta,\eta) + \nu_{n2}(\beta,\eta) \le x\} \\ &\quad + \mathbf{P}_\beta\{\nu_{n1}(-\beta,\eta) + \nu_{n2}(-\beta,\eta) < -x\} \\ &= \mathbf{P}_\beta\{\nu_{n1}(\beta,\eta) + \nu_{n2}(\beta,\eta) \le x\} \\ &\quad + \mathbf{P}_\beta\{-\nu_{n1}(\beta,\eta) + \nu_{n2}(\beta,\eta) > x\} \\ &= \frac{1}{2^{n-1}} \sum_{k=1}^{2^{n-1}} \{I(\nu_{n1}(\beta,\eta^{(k)}) + \nu_{n2}(\beta,\eta^{(k)}) \le x) \\ &\qquad + I(-\nu_{n1}(\beta,\eta^{(k)}) + \nu_{n2}(\beta,\eta^{(k)}) > x)\}, \end{aligned} \tag{6.8.7}$$

where $\eta^{(k)} = (\eta_1^{(k)}, \eta_2^{(k)}, \dots, \eta_{n-1}^{(k)})$, $k = 1, 2, \dots, 2^{n-1}$, are vectors of ± 1's running over all possible values of the vector $\eta = (\eta_1, \eta_2, \dots, \eta_{n-1})$. Now (6.8.7) is identically equal to

$$\begin{aligned} \frac{1}{2}\frac{1}{2^{n-1}} \sum_{k=1}^{2^{n-1}} \Big\{ & I(\nu_{n1}(\beta,\eta^{(k)}) + \nu_{n2}(\beta,\eta^{(k)}) \le x) \\ & + I(-\nu_{n1}(\beta,\eta^{(k)}) + \nu_{n2}(\beta,\eta^{(k)}) > x) \\ & + I(\nu_{n1}(\beta,-\eta^{(k)}) + \nu_{n2}(\beta,-\eta^{(k)}) \le x) \\ & + I(-\nu_{n1}(\beta,-\eta^{(k)}) + \nu_{n2}(\beta,-\eta^{(k)}) > x)\Big\}. \end{aligned} \tag{6.8.8}$$

The polynomial (6.8.5) includes only even powers of β, and its coefficients contain products of different components of η with an odd number of factors. Likewise, the coefficients of the odd polynomial (6.8.6) contain products with an even number of factors. Therefore $\nu_{n1}(\beta,-\eta) = -\nu_{n1}(\beta,\eta)$, $\nu_{n2}(\beta,-\eta) = \nu_{n2}(\beta,\eta)$. Hence (6.8.8) is equal to

$$\begin{aligned} \frac{1}{2}\frac{1}{2^{n-1}} \sum_{k=1}^{2^{n-1}} \Big\{ & I(\nu_{n1}(\beta,\eta^{(k)}) + \nu_{n2}(\beta,\eta^{(k)}) \le x) \\ & + I(-\nu_{n1}(\beta,\eta^{(k)}) + \nu_{n2}(\beta,\eta^{(k)}) > x) \\ & + I(-\nu_{n1}(\beta,\eta^{(k)}) + \nu_{n2}(\beta,\eta^{(k)}) \le x) \\ & + I(\nu_{n1}(\beta,\eta^{(k)}) + \nu_{n2}(\beta,\eta^{(k)}) > x)\Big\} = 1. \end{aligned}$$

TABLE 6.8.1. ($\alpha = 0.05$)

$\beta_0 \backslash n$	50	100	150	200
−0.9	−1.85	−1.85	−1.84	−1.82
−0.8	−1.83	−1.81	−1.79	−1.77
−0.7	−1.81	−1.78	−1.75	−1.75
−0.6	−1.78	−1.75	−1.74	−1.73
−0.5	−1.75	−1.73	−1.71	−1.71
−0.4	−1.73	−1.71	−1.70	−1.69
−0.3	−1.70	−1.69	−1.68	−1.68
−0.2	−1.68	−1.67	−1.67	−1.66
−0.1	−1.65	−1.65	−1.65	−1.65
0.0	−1.56	−1.70	−1.71	−1.63
0.1	−1.60	−1.62	−1.63	−1.63
0.2	−1.57	−1.60	−1.61	−1.62
0.3	−1.55	−1.58	−1.60	−1.60
0.4	−1.52	−1.56	−1.57	−1.58
0.5	−1.47	−1.53	−1.55	−1.57
0.6	−1.42	−1.50	−1.53	−1.54
0.7	−1.35	−1.45	−1.49	−1.51
0.8	−1.26	−1.37	−1.43	−1.46
0.9	−1.07	−1.23	−1.31	−1.35

Thus (6.8.4) and hence (6.8.3) is proven. Due to (6.8.3) it is enough to compute $c_n^\alpha(\beta)$ for $\beta \geq 0$, obtaining the values for $\beta < 0$ by symmetry. Likewise, having found $c_n^\alpha(\beta)$ one immediately obtains $c_n^{1-\alpha}(\beta)$, $\beta \in (-1,1)$. In Table 6.8.1 the values of $c_n^\alpha(\beta_0)$ are presented for $n = 50(50), 200$, $\beta_0 = 0, \pm 0.1, \pm 0.2, \ldots, \pm 0.9$ and $\alpha = 0.05$. The computation is based on 60,000 replications of $\nu_n(\beta_0)$ for each β_0.

It is seen from Table 6.8.1 that the quantiles of $\nu_n(\beta_0)$ converge to the corresponding quantile of the standard normal distribution (which is equal here to 1.65) at a rate which drastically depends on β_0, the convergence being slower for β_0 close to ± 1. It can be shown that the convergence is uniform over

$$|\beta_0| \leq 1 - \delta, \quad 0 < \delta < 1.$$

The shape of the distribution of $\nu_n(\beta_0)$ and its rate of convergence to normality for various β_0 are illustrated also by Figures 6.8.1a and 6.8.1b. Figure 6.8.1a shows the graphs of the distribution function $F_n(x, \beta_0)$ of $\nu_n(\beta_0)$ for $\beta_0 = \pm 0.5$ and $n = 100$. Figure 6.8.1b shows the same graphs on the normal probability paper, with the straight line corresponding to the normal distribution. (The distribution functions have also been computed from 60,000 replications of $\nu_n(\beta_0)$.) The deviation from the normal distribution is very well visible.

Although these numerical results show the validity of the normal approximation for the sign test statistics, some hundreds of observations are needed to achieve a satisfactory accuracy of this approximation. This is an additional evidence in favor of the use (whenever possible, of course) of exactly valid nonparametric procedures for hypothesis testing and estimation. Thus we turn to exact confidence sets for β.

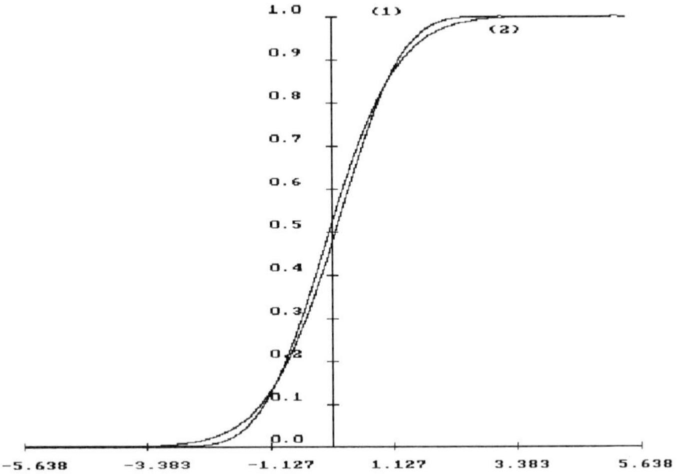

FIGURE 6.8.1a. Distribution functions of $\nu_n(\beta_0)$ $n = 100$; (1) $\beta_0 = -0.5$; (2) $\beta_0 = 0.5$

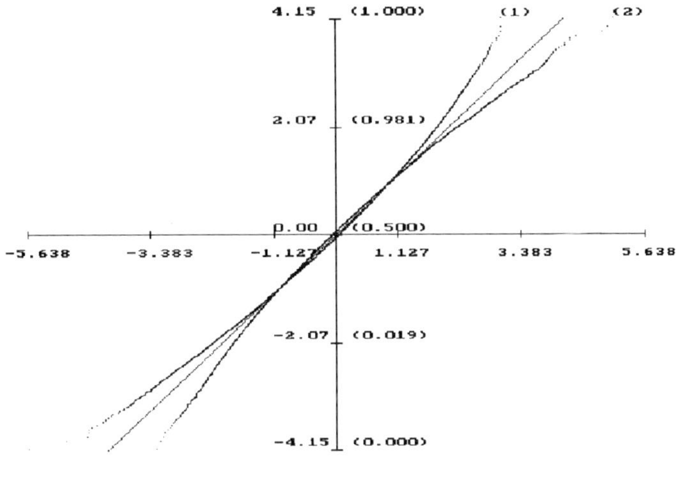

FIGURE 6.8.1b. Distribution functions of $\nu_n(\beta_0)$ on the normal probability paper $n = 100$; (1) $\beta_0 = -0.5$; (2) $\beta_0 = 0.5$

6.8.2. Confidence estimation of β. Let $T_{n,S}(\theta) = n^{-1/2}\sqrt{1-\theta^2}\, l_n^S(\theta)$, $|\theta| < 1$. Under the sole Condition 6.1(i), the set

$$D_{n\alpha} = \left\{ \theta\colon c_n^{\alpha}(\theta) \le T_{n,S}(\theta) \le c_n^{1-\alpha}(\theta) \right\},$$

is a confidence set for β of condidence level at least $1 - 2\alpha$.

As it was described in 6.8.1, the quantiles $c_n^{\alpha}(\theta)$ and $c_n^{1-\alpha}(\theta)$ can be evaluated on a grid of values of θ, $|\theta| < 1$, and the set $D_{n\alpha}$ can then be obtained by plotting the three curves $c_n^{\alpha}(\theta)$, $c_n^{1-\alpha}(\theta)$, and $T_{n,S}(\theta)$.

The simulation results for $n = 100$, 200, and 400 for two error distributions, normal and Cauchy, are presented in Figures 6.8.2–6.8.7. Each figure shows three independent trajectories of the random process $T_{n,S}(\theta)$ based on observations

TABLE 6.8.2

Trajectory	Point estimate	90%-confidence interval
1 (*)	0.04	$(-0.18, 0.54)$
2 (+)	0.19	$(-0.13, 0.45)$
3 (×)	0.15	$(-0.13, 0.56)$

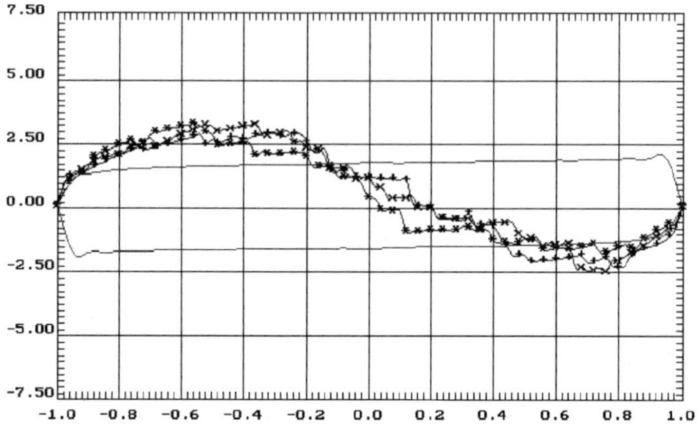

FIGURE 6.8.2. Normal $N(0,1)$ error distribution; $n = 100$, $\beta = 0.2$

$u_0, u_1, \ldots, u_n$ for $\beta = 0.2$ as well as quantile curves $c_n^{\alpha}(\theta)$, $c_n^{1-\alpha}(\theta)$ for $\alpha = 0.05$. The confidence sets for β are formed by those values of θ for which $T_{n,S}(\theta)$ lies between $c_n^{\alpha}(\theta)$ and $c_n^{1-\alpha}(\theta)$. All graphs are equally scaled in order to facilitate the comparison of the results obtained for different sample sizes and distributions. Each figure is accompanied by a table (Tables 6.8.2–6.8.7) showing the numerical values of the corresponding point and interval estimates for β. The point estimate $\widehat{\beta}_{n,S}$ is obtained from the equation $l_n^S(\theta) \div 0$, $|\theta| < 1$.

We can draw the following conclusions from the simulation results.

First, in order to estimate the autoregression parameter with reasonable accuracy one needs several hundreds of observations, much more than is usually needed in the case of i.i.d. observations. For instance, one of the trajectories in Figure 6.8.5 for $n = 100$ has a tail falling between $(c_n^{\alpha}(\theta)$ and $c_n^{1-\alpha}(\theta))$, which gives rise to an additional interval in the confidence set for β. There are no effects like this for $n = 400$.

Secondly, the estimation accuracy increases for the heavy-tailed error distribution. It is seen from the comparison between Figures 6.8.2 and 6.8.5, Figures 6.8.3 and 6.8.6, or Figures 6.8.4 and 6.8.7 that in the case of Cauchy errors the curve $T_{n,S}(\theta)$ in a vicinity of the true value of β falls off more sharply than for normal errors. This means that for equal sample sizes confidence intervals corresponding to Cauchy distributed errors are typically narrower than for normally distributed errors. Note that there is no way to construct (asymptotic) confidence intervals based on the commonly used least squares estimate when the errors $\{\varepsilon_i\}$ have the Cauchy distribution. At the same time our sign procedures are applicable and perform reasonably well.

TABLE 6.8.3

Trajectory	Point estimate	90%-confidence interval
1 (*)	0.00	(−0.21, 0.32)
2 (+)	0.11	(−0.06, 0.30)
3 (×)	0.14	(−0.06, 0.33)

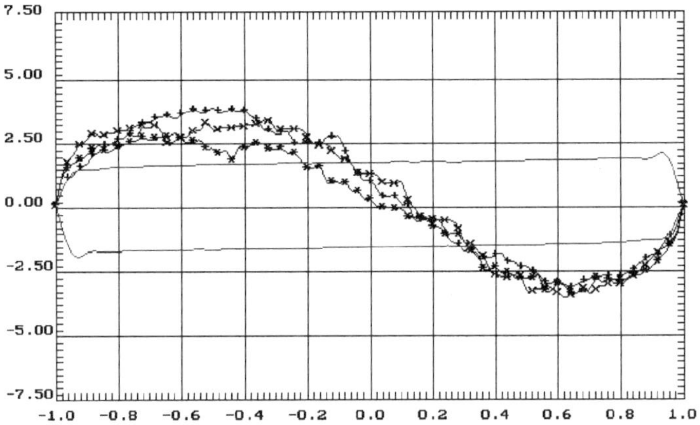

FIGURE 6.8.3. Normal $N(0,1)$ error distribution; $n = 200$, $\beta = 0.2$

TABLE 6.8.4

Trajectory	Point estimate	90%-confidence interval
1 (*)	0.18	(0.09, 0.25)
2 (+)	0.19	(0.08, 0.30)
3 (×)	0.19	(0.10, 0.32)

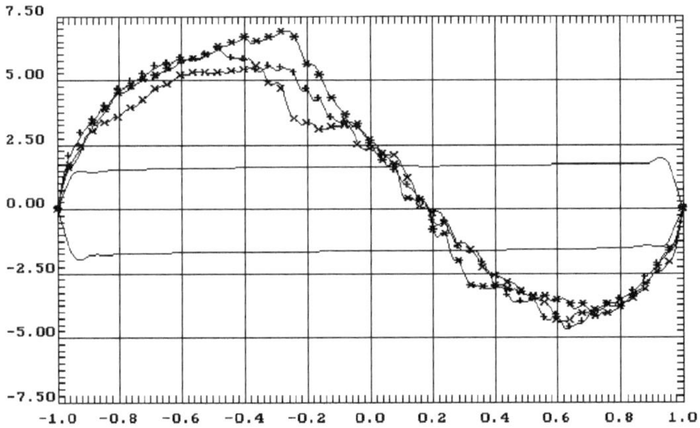

FIGURE 6.8.4. Normal $N(0,1)$ error distribution; $n = 400$, $\beta = 0.2$

TABLE 6.8.5

Trajectory	Point estimate	90%-confidence interval
1 (∗)	0.21	$(0.15, 0.44)$
2 (+)	0.19	$(0.00, 0.46) \cup (0.64, 1.0)$
3 (×)	0.15	$(0.09, 0.35)$

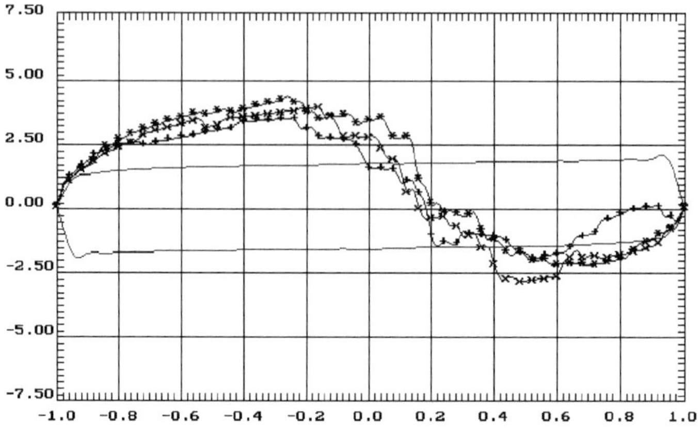

FIGURE 6.8.5. Cauchy (0,1) error distribution $n = 100$, $\beta = 0.2$

TABLE 6.8.6

Trajectory	Point estimate	90%-confidence interval
1 (∗)	0.18	$(0.12, 0.28)$
2 (+)	0.20	$(0.07, 0.26)$
3 (×)	0.22	$(0.15, 0.39)$

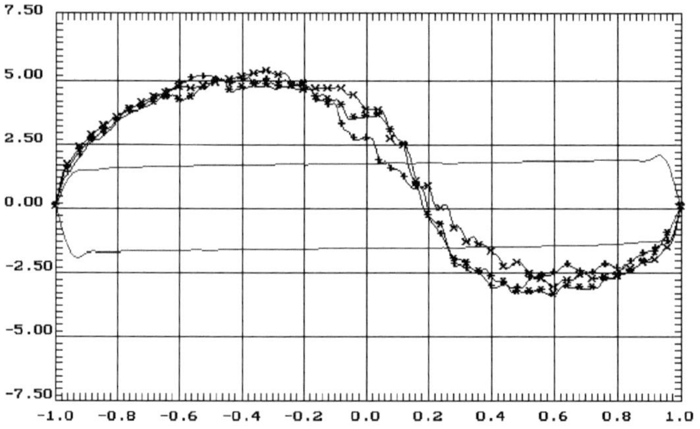

FIGURE 6.8.6. Cauchy (0,1) error distribution $n = 200$, $\beta = 0.2$

TABLE 6.8.7

Trajectory	Point estimate	90%-confidence interval
1 (*)	0.20	$(0.16, 0.32)$
2 (+)	0.20	$(0.18, 0.28)$
3 (×)	0.18	$(0.14, 0.20)$

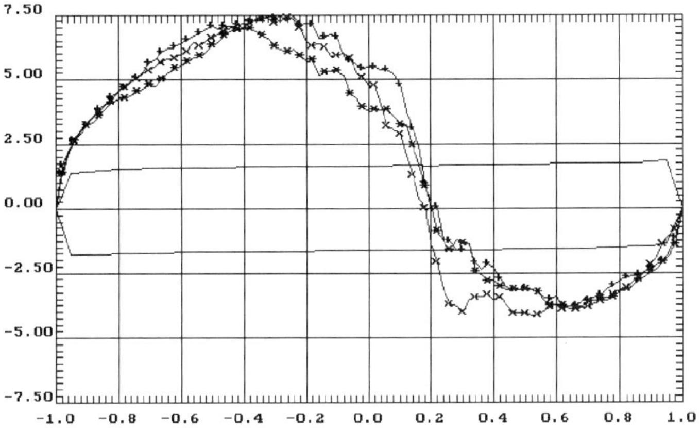

FIGURE 6.8.7. Cauchy (0,1) error distribution $n = 400$, $\beta = 0.2$

6.8.3. Sign estimation from contaminated samples. Here we compare the performance of the sign estimator $\widehat{\beta}_{n,S}$ from 6.6.1 and the least squares estimator (LSE) from 5.3.1 for contaminated observations. Suppose that instead of $\{u_i\}$ satisfying (6.1.1) we observe the process $\{y_i\}$, $i \in \mathbb{Z}$, as given by (6.7.1). In order to see the effect of outliers, consider the estimators based on a simulated process $\{y_i\}$ with $\beta = 0.2$, $\gamma = 0.1$, $\varepsilon_i \sim N(0,1)$, $\xi_i \sim N(0,25)$.

Figure 6.8.8 (on the next page) depicts three independent trajectories of the process $T_{n,S}(\theta, Y_n) := n^{-1/2}\sqrt{1-\theta^2}\, l_n^S(\theta, Y_n)$ for $n = 400$. The process $l_n^S(\theta, Y_n)$ is constructed from $Y_n = (y_0, \ldots, y_n)$ in the same way as $l_n^S(\theta)$ from $u_0, \ldots, u_n$. Table 6.8.8 (on the next page) contains the values of the sign estimator $\widehat{\beta}_{n,S}$ and the LSE based on these contaminated data as well as on the corresponding non-contaminated data (with $\gamma = 0$). We present also the 90%-confidence intervals; for the LSE these are asymptotic confidence intervals obtained according to (5.3.9). Of course, the standard procedures are not valid for contaminated observations because the quantiles of $T_{n,S}(\theta, Y_n)$ differ then from $c_n^\alpha(\theta)$ and the LSE is no longer $\sqrt{n}$-asymptotically normal. Nevertheless we applied the standard procedures in order to see the effect of contamination on them. These numerical results show that the sign estimators are very robust to outliers. They also confirm the fact that the LSE is very sensitive to contamination.

In §§5.6 and 5.7 we discussed how the robustness of estimators is characterized by the influence function and gross error sensitivity. These characteristics for the sign estimator $\widehat{\beta}_{n,S}$ were obtained in 6.7.1 (see (6.7.13)–(6.7.15)). Figure 6.8.9 (on the next page) shows the graphs of the influence function $\mathrm{IF}(\theta_\gamma^S, \xi)$ of the estimator $\widehat{\beta}_{n,S}$ for three values of β for standard normal $G(x)$. It is seen that these influence functions are bounded, in accordance with the finite sensitivity of the sign estimator $\widehat{\beta}_{n,S}$.

TABLE 6.8.8

Method of estimation	Point estimate	90%-confidence interval
LSE 1 (∗)	0.06	$(-0.02, 0.14)$
LSE 1 ($\gamma = 0$) (∗)	0.21	$(0.13, 0.29)$
sign 1 (∗)	0.20	$(0.11, 0.29)$
sign 1 ($\gamma = 0$) (∗)	0.26	$(0.11, 0.41)$
LSE 2 (+)	−0.03	$(-0.11, 0.05)$
LSE 2 ($\gamma = 0$) (+)	0.20	$(0.12, 0.28)$
sign 2 (+)	0.16	$(0.08, 0.28)$
sign 2 ($\gamma = 0$) (+)	0.16	$(0.13, 0.25)$
LSE 3 (×)	0.10	$(0.01, 0.18)$
LSE 3 ($\gamma = 0$) (×)	0.23	$(0.14, 0.31)$
sign 3 (×)	0.15	$(0.04, 0.28)$
sign 3 ($\gamma = 0$) (×)	0.23	$(0.10, 0.33)$

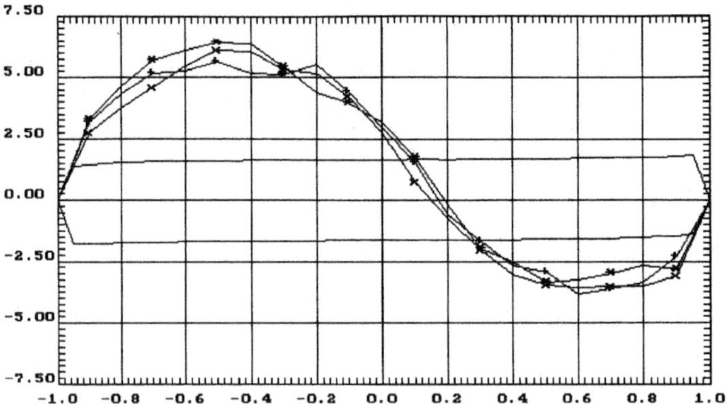

FIGURE 6.8.8. Contaminated observations; $\gamma = 0.1$, $\varepsilon_i \sim N(0,1)$, $\xi_i \sim N(0,25)$, $n = 400$, $\beta = 0.2$

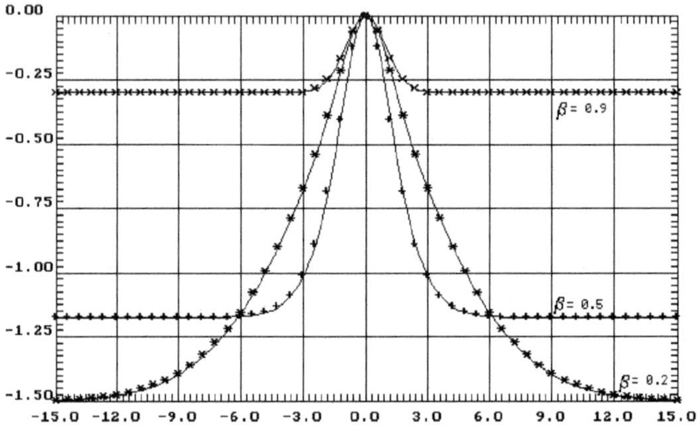

FIGURE 6.8.9. Influence functions of sign estimator for independent ouliers ∗: $\beta = 0.2$; +: $\beta = 0.5$; ×: $\beta = 0.9$

6.9. Proof of Theorem 6.4.1

It follows from the definition (6.2.10) that

$$n^{-1/2}l_n^S(\beta_n+n^{-1/2}\theta)-n^{-1/2}l_n^S(\beta_n)$$

$$=n^{-1/2}\sum_{t=1}^{n-1}(\beta_n+n^{-1/2}\theta)^{t-1} \tag{6.9.1}$$

$$\times\left(\Gamma_{tn}(\beta_n+n^{-1/2}\theta)-\Gamma_{tn}(\beta_n)\right)$$

$$+n^{-1/2}\sum_{t=1}^{n-1}\left((\beta_n+n^{-1/2}\theta)^{t-1}-\beta_n^{t-1}\right)\Gamma_{tn}(\beta_n). \tag{6.9.2}$$

Consider first (6.9.2). Throughout this section we suppress the index β_n of the probability and expectation signs.

LEMMA 6.9.1. *Assume Condition* 6.1(i) *to hold and let $H_{1n}(a)$ be true. Then the supremum of* (6.9.2) *over $|\theta|\le\Theta n^r$ for any $0<\Theta<\infty$, $r<1/2$, is $o_p(1)$.*

PROOF. Let us abbreviate $\Theta_n=\Theta n^r$. For any $\varepsilon>0$ and b such that $|\beta_0|<b<1$ one has for all sufficiently large n

$$\mathbf{P}\Big\{\sup_{|\theta|\le\Theta_n}\Big|n^{-1/2}\sum_{t=1}^{n-1}\left((\beta_n+n^{-1/2}\theta)^{t-1}-\beta_n^{t-1}\right)\Gamma_{tn}(\beta_n)\Big|\ge\varepsilon\Big\}$$

$$\le\mathbf{P}\Big\{\sup_{|\theta|\le\Theta_n}\Theta_n n^{-1}\sum_{t=2}^{n-1}\sum_{s=0}^{t-2}\left|(\beta_n+n^{-1/2}\theta)\right|^{t-2-s}|\beta_n|^s|\Gamma_{tn}(\beta_n)|\ge\varepsilon\Big\}$$

$$\le\mathbf{P}\Big\{\Theta_n n^{-1}\sum_{t=2}^{n-1}(t-1)b^{t-2}|\Gamma_{tn}(\beta_n)|\ge\varepsilon\Big\}.$$

By Chebyshev's inequality the last probability does not exceed

$$\varepsilon^{-1}\Theta_n n^{-1}\sum_{t=2}^{n-1}(t-1)b^{t-2}\mathbf{E}|\Gamma_{tn}(\beta_n)|=O(n^{-1/2+r})=o(1),$$

since $n^{-1/2}\mathbf{E}|\Gamma_{tn}(\beta_n)|\le n^{-1/2}\left(\mathbf{E}\Gamma_{tn}^2(\beta_n)\right)^{1/2}=\left((n-t)/n\right)^{1/2}\le1$ by (6.2.14). This proves the lemma. □

Consider now (6.9.1). Let

$$\Delta_k(x)=\begin{cases}1, & \varepsilon_k<x\\ 1/2, & \varepsilon_k=x\\ 0, & \varepsilon_k>x.\end{cases} \tag{6.9.3}$$

For brevity, put $\eta_n=n^{-1/2}\theta$. Then

$$S_k(\beta_n+n^{-1/2}\theta)=\operatorname{sign}\left(u_k-(\beta_n+n^{-1/2}\theta)u_{k-1}\right)$$
$$=\operatorname{sign}(\varepsilon_k-\eta_n u_{k-1})=1-2\Delta_k(\eta_n u_{k-1}).$$

Together with the definition of $\Gamma_{tn}(\theta)$ in (6.2.9) this implies the representation

$$(6.9.4)\quad n^{-1/2}\Gamma_{tn}(\beta_n+n^{-1/2}\theta)-n^{-1/2}\Gamma_{tn}(\beta_n)=4x_{tn}(\eta_n)+4y_{tn}(\eta_n)-2u_{tn}(\eta_n),$$

where

$$\begin{aligned}
x_{tn}(\eta_n)=n^{-1/2}\sum_{k=t+1}^{n}\big(&\Delta_{k-t}(\eta_n u_{k-t-1})\Delta_k(\eta_n u_{k-1})\\
&-\Delta_{k-t}(\eta_n u_{k-t-1})G(\eta_n u_{k-1})\\
&-\Delta_{k-t}(0)\Delta_k(0)+\Delta_{k-t}(0)G(0)\big),\\
y_{tn}(\eta_n)=n^{-1/2}\sum_{k=t+1}^{n}\big(&\Delta_{k-t}(\eta_n u_{k-t-1})G(\eta_n u_{k-1})-\Delta_{k-t}(0)G(0)\big),\\
u_{tn}(\eta_n)=n^{-1/2}\sum_{k=t+1}^{n}\big(&\Delta_{k-t}(\eta_n u_{k-t-1})-\Delta_{k-t}(0)\big)\\
&+n^{-1/2}\sum_{k=t+1}^{n}\big(\Delta_k(\eta_n u_{k-1})-\Delta_k(0)\big).
\end{aligned}$$

With the aid of (6.9.4) rewrite (6.9.1) in the form

$$(6.9.5)\qquad \sum_{t=1}^{n-1}(\beta_n+\eta_n)^{t-1}\big(4x_{tn}(\eta_n)+4y_{tn}(\eta_n)-2u_{tn}(\eta_n)\big).$$

LEMMA 6.9.2. *Assume Conditions* 6.1(ii, iii) *to hold. Let* $r>1/4$, $\Theta_n=\Theta n^{-r}$, *and* $N_n\sim\log_b n^{-1}$, *where* $|\beta_0|<b<1$. *Then under* $H_{1n}(a)$:

1°. $\sup_{t\le N_n}\sup_{|\eta_n|\le\Theta_n}|x_{tn}(\eta_n)|=o_p(1)$;

2°. $\sup_{|\eta_n|\le\Theta_n}\left|n^{-1/2}\sum_{k=1}^{n}(\Delta_k(\eta_n u_{k-1})-\Delta_k(0))\right|=o_p(1).$

PROOF. Both assertions are proved along the same lines, so we present only the proof of assertion 1°.

Divide the interval $[-\Theta_n,\Theta_n]$ into 3^{m_n} subintervals by the points

$$\eta_{sn}=-\Theta_n+2\Theta_n 3^{-m_n}s,\qquad s=0,1,\dots,3^{m_n}.$$

Choose r_1 such that $0<r_1<r$ and $r+r_1>1/2$ and let $3^{m_n}\sim n^{r_1}$. Define the variables

$$(6.9.6)\qquad \widehat{u}_{ks}=u_k\big(1-2\Theta_n 3^{-m_n}\eta_{sn}^{-1}I(u_k\le 0)\big),$$

$$(6.9.7)\qquad \widetilde{u}_{ks}=u_k\big(1-2\Theta_n 3^{-m_n}\eta_{sn}^{-1}I(u_k\ge 0)\big).$$

For an arbitrary $|\eta_n|\le\Theta_n$ select the point η_{ln} from among the points $\{\eta_{sn}\}$ such that

$$(6.9.8)\qquad 0\le\eta_{ln}-\eta_n\le\Theta_n 3^{-m_n}.$$

It follows from the definitions (6.9.6)–(6.9.7) and the inequality (6.9.8) that

$$\eta_{ln}\widetilde{u}_{kl} \le \eta_n u_k \le \eta_{ln}\widehat{u}_{kl}, \qquad k = 0, 1, \dots, n-1. \tag{6.9.9}$$

Let

$$\widehat{\mathbf{U}}_{sn} = (\widehat{u}_{0s}, \dots, \widehat{u}_{n-1,s}), \qquad \widetilde{\mathbf{U}}_{sn} = (\widetilde{u}_{0s}, \dots, \widetilde{u}_{n-1,s}), \qquad \mathbf{U}_n = (u_0, \dots, u_{n-1}).$$

To indicate the dependence of $x_{tn}(\eta_n)$ on the components of $\mathbf{U}_n$, we will write it as

$$x_{tn}(\eta_n) := x_{tn}(\eta_n, \mathbf{U}_n).$$

We will also use the variables $x_{tn}(\eta_{sn}, \widehat{\mathbf{U}}_{sn})$ and $x_{tn}(\eta_{sn}, \widetilde{\mathbf{U}}_{sn})$ which are defined by the substitution of $\widehat{\mathbf{U}}_{sn}$ and $\widetilde{\mathbf{U}}_{sn}$ for $\mathbf{U}_n$ and η_{sn} for η_n in $x_{tn}(\eta_n, \mathbf{U}_n)$. It follows from (6.9.9) and monotonicity of functions $\Delta_k(y)$ and $G(y)$ that

$$\begin{aligned} x_{tn}(\eta_n, \mathbf{U}_n) \le x_{tn}(\eta_{ln}, \widehat{\mathbf{U}}_{ln}) &\\ + n^{-1/2} \sum_{k=t+1}^{n} \big(&\Delta_{k-t}(\eta_{ln}\widehat{u}_{k-t-1,l})G(\eta_{ln}\widehat{u}_{k-1,l}) \\ &- \Delta_{k-t}(\eta_{ln}\widetilde{u}_{k-t-1,l})G(\eta_{ln}\widetilde{u}_{k-1,l})\big), \\ x_{tn}(\eta_n, \mathbf{U}_n) \ge x_{tn}(\eta_{ln}, \widetilde{\mathbf{U}}_{ln}) &\\ - n^{-1/2} \sum_{k=t+1}^{n} \big(&\Delta_{k-t}(\eta_{ln}\widehat{u}_{k-t-1,l})G(\eta_{ln}\widehat{u}_{k-1,l}) \\ &- \Delta_{k-t}(\eta_{ln}\widetilde{u}_{k-t-1,l})G(\eta_{ln}\widetilde{u}_{k-1,l})\big). \end{aligned}$$

These inequalities imply

$$\begin{aligned} &\sup_{t \le N_n} \sup_{|\eta_n| \le \Theta_n} |x_{tn}(\eta_n, \mathbf{U}_n)| \\ &\quad \le \sup_{t \le N_n} \sup_{s \le 3^{m_n}} \left\{ |x_{tn}(\eta_{sn}, \widehat{\mathbf{U}}_{sn})| + |x_{tn}(\eta_{sn}, \widetilde{\mathbf{U}}_{sn})| \right\} && (6.9.10) \\ &\quad + \sup_{t \le N_n} \sup_{s \le 3^{m_n}} n^{-1/2} \sum_{k=t+1}^{n} \big(\Delta_{k-t}(\eta_{sn}\widehat{u}_{k-t-1,s}) && (6.9.11) \\ &\qquad \times G(\eta_{sn}\widehat{u}_{k-1}, s) \\ &\qquad - \Delta_{k-t}(\eta_{sn}\widetilde{u}_{k-t-1,s}) \\ &\qquad \times G(\eta_{sn}\widetilde{u}_{k-1,s})\big). \end{aligned}$$

Consider the first term, $x_{tn}(\eta_{sn}\widehat{\mathbf{U}}_{sn})$, in (6.9.10). Put

$$\begin{aligned} \xi_k &= \xi_k(n, s, t) \\ &= \Delta_{k-t}(\eta_{sn}\widehat{u}_{k-t-1,s})\Delta_k(\eta_{sn}\widehat{u}_{k-1,s}) \\ &\quad - \Delta_{k-t}(\eta_{sn}\widehat{u}_{k-t-1,s})G(\eta_{sn}\widehat{u}_{k-1,s}) - \Delta_{k-t}(0)\Delta_k(0) + \Delta_{k-t}(0)G(0). \end{aligned} \tag{6.9.12}$$

Then

$$x_{tn}(\eta_{sn}, \widehat{\mathbf{U}}_{sn}) = n^{-1/2} \sum_{k=t+1}^{n} \xi_k.$$

Let Ω_j be the σ-algebra generated by the random variables $\{\varepsilon_i,\, i \le j\}$.

Then $\mathbf{E}\{\xi_k|\Omega_{k-1}\} = 0$ a.s. (i.e., the sequence $\{\xi_k, \Omega_k\}$, $k = 1, 2, \dots$, forms a martingale difference). Hence $\mathbf{E}\xi_k = \mathbf{E}\{\mathbf{E}(\xi_k|\Omega_{k-1})\} = 0$. Similarly, for $k < j$

$$\mathbf{E}\xi_k\xi_j = \mathbf{E}\{\mathbf{E}(\xi_k\xi_j|\Omega_{j-1})\} = \mathbf{E}\{\xi_k\mathbf{E}(\xi_j|\Omega_{j-1})\} = 0.$$

By a simple calculation one obtains

$$\mathbf{E}\xi_k^2 \le 2\sup_x g(x)\{\mathbf{E}|\eta_{sn}\widehat{u}_{k-1,s}| + \mathbf{E}|\eta_{sn}\widehat{u}_{k-t-1,s}|\} = O(n^{-r})$$

uniformly in k, s, t since $|\eta_{sn}| \le \Theta_n = O(n^{-r})$, $|\widehat{u}_{k-1,s}| \le 3|u_{k-1}|$. Therefore

$$\sup_{t \le N_n} \sup_{s \le 3^{m_n}} \mathbf{E}x_{tn}^2(\eta_{sn}, \widehat{\mathbf{U}}_{sn}) = O(n^{-r}). \tag{6.9.13}$$

By the Chebyshev inequality, (6.9.13) implies that for any $\varepsilon > 0$

$$\mathbf{P}\{\sup_{t \le N_n} \sup_{s \le 3^{m_n}} |x_{tn}(\eta_{sn}, \widehat{\mathbf{U}}_{sn})| > \varepsilon\} \le \sum_{t=1}^{N_n} \sum_{s=0}^{3^{m_n}} \mathbf{P}\{|x_{tn}(\eta_{sn}, \widehat{\mathbf{U}}_{sn})| > \varepsilon\}$$
$$\le \varepsilon^{-2} \sum_{t=1}^{N_n} \sum_{s=0}^{3^{m_n}} \mathbf{E}x_{tn}^2(\eta_{sn}, \widehat{\mathbf{U}}_{sn}) = O(N_n 3^{m_n} n^{-r}) = O(n^{r_1 - r} \log n) = o(1).$$

By a similar argument one shows that

$$\sup_{t \le N_n} \sup_{s \le 3^{m_n}} |x_{tn}(\eta_{sn}, \widetilde{\mathbf{U}}_{sn})| = o_p(1).$$

Thus (6.9.10) is $o_p(1)$.

Consider now (6.9.11). Obviously, it is no greater than

$$\sup_{t \le N_n} \sup_{s \le 3^{m_n}} n^{-1/2} \sum_{k=t+1}^{n} (\Delta_{k-t}(\eta_{sn}\widehat{u}_{k-t-1,s}) - \Delta_{k-t}(\eta_{sn}\widetilde{u}_{k-t-1,s})) \times G(\eta_{sn}\widehat{u}_{k-1,s}) \tag{6.9.14}$$

$$+ \sup_{t \le N_n} \sup_{s \le 3^{m_n}} n^{-1/2} \sum_{k=t+1}^{n} (G(\eta_{sn}\widehat{u}_{k-1,s}) - G(\eta_{sn}\widetilde{u}_{k-1,s})) \times \Delta_{k-t}(\eta_{sn}\widetilde{u}_{k-t-1,s}). \tag{6.9.15}$$

Here (6.9.15) does not exceed

$$\begin{aligned} &\sup_{s \le 3^{m_n}} n^{-1/2} \sum_{k=1}^{n} (G(\eta_{sn}\widehat{u}_{k-1,s}) - G(\eta_{sn}\widetilde{u}_{k-1,s})) \\ &\le \sup_{s \le 3^{m_n}} \sup_x g(x) n^{-1/2} \sum_{k=1}^{n} (\eta_{sn}\widehat{u}_{k-1,s} - \eta_{sn}\widetilde{u}_{k-1,s}). \end{aligned} \tag{6.9.16}$$

It follows from the definitions (6.9.6) and (6.9.7) that the last expression is bounded by

$$2\Theta_n \sup_x g(x) 3^{-m_n} n^{-1/2} \sum_{k=1}^{n} |u_{k-1}| = O_p(n^{1/2-r_1-r}) = o_p(1).$$

It remains to consider (6.9.14). Obviously, it does not exceed

$$\sup_{s\le 3^{m_n}} n^{-1/2} \sum_{k=1}^{n} \big(\Delta_k(\eta_{sn}\widehat{u}_{k-1,s}) - \Delta_k(\eta_{sn}\widetilde{u}_{k-1,s})\big)$$

$$(6.9.17) \qquad \le \sup_{s\le 3^{m_n}} \Big| n^{-1/2} \sum_{k=1}^{n} \big(\Delta_k(\eta_{sn}\widehat{u}_{k-1,s}) - G(\eta_{sn}\widehat{u}_{k-1,s}) - \Delta_k(\eta_{sn}\widetilde{u}_{k-1,s}) + G(\eta_{sn}\widetilde{u}_{k-1,s})\big)\Big|$$

$$(6.9.18) \qquad + \sup_{s\le 3^{m_n}} n^{-1/2} \sum_{k=1}^{n} \big(G(\eta_{sn}\widehat{u}_{k-1,s}) - G(\eta_{sn}\widetilde{u}_{k-1,s})\big).$$

The expression (6.9.18) conincides with (6.8.16) which is $o_p(1)$. Consider (6.9.17). Put

$$\nu_k = \nu_k(n,s) = \Delta_k(\eta_{sn}\widehat{u}_{k-1,s}) - G(\eta_{sn}\widehat{u}_{k-1,s}) - \Delta_k(\eta_{sn}\widetilde{u}_{k-1,s}) + G(\eta_{sn}\widetilde{u}_{k-1,s}).$$

Then (6.9.17) becomes

$$\sup_{s\le 3^{m_n}} \Big| n^{-1/2} \sum_{k=1}^{n} \nu_k \Big|.$$

Similarly to ξ_k's defined by (6.9.12), the variables ν_k's are centered, uncorrelated, and $\sup_{s\le 3^{m_n}} \mathbf{E}\nu_k^2 = O(n^{-r}3^{-m_n})$ uniformly in k. Therefore, by the Chebyshev inequality for any $\varepsilon > 0$

$$\mathbf{P}\Big\{ \sup_{t\le N_n} \sup_{s\le 3^{m_n}} \Big| n^{-1/2} \sum_{k=1}^{n} \nu_k \Big| \ge \varepsilon \Big\} \le \varepsilon^{-2} \sum_{t=1}^{N_n} \sum_{s=0}^{3^{m_n}} \mathbf{E}\Big\{ n^{-1/2} \sum_{k=1}^{n} \nu_k \Big\}^2 = O(N_n n^{-r}) = o(1).$$

Thus (6.9.17) is $o_p(1)$, which completes the proof of the lemma. □

LEMMA 6.9.3. *Assume Conditions* 6.1(ii, iii) *to hold. Let* $N_n \sim \log_b n^{-1}$, *where* $|\beta_0| < b < 1$, *let* $\delta = \min(\Delta, 1)$ *and* $\Theta_n = \Theta n^{-r}$ *for* $r > 0$ *and* $0 < \Theta < \infty$. *Then under* $H_{1n}(a)$

$$\mathbf{E} \sup_{t\le N_n} \sup_{|\eta_n|\le \Theta_n} \Big| n^{-1} \sum_{k=t+1}^{n} \big(\Delta_{k-t}(\eta_n u_{k-t-1}) - \Delta_{k-t}(0)\big) u_{k-1} \Big| = O(n^{-r\delta/(1+\delta)} \log^2 n).$$

PROOF. Divide the interval $[-\Theta_n, \Theta_n]$ into 3^{m_n} parts by the points $\eta_{sn} = -\Theta_n + 2\Theta_n 3^{-m_n}$, $s = 0, 1, \dots, 3^{m_n}$, and let $3^{m_n} \sim N_n$. Define the random variables $\widehat{u}_{ks}$ and $\widetilde{u}_{ks}$ by (6.9.6) and (6.9.7), respectively. For an arbitrary $|\eta_n| \leq \Theta_n$ select the point η_{ln} from among the points $\{\eta_{sn}\}$ such that

$$0 \leq \eta_{ln} - \eta_n \leq 2\Theta_n 3^{-m_n}.$$

Then (6.9.9) is fulfilled. Due to monotonicity of $\Delta_k(y)$ this inequality implies

$$\begin{aligned}\Delta_{k-t}(\eta_n u_{k-t-1}) - \Delta_{k-t}(0) &\leq \Delta_{k-t}(\eta_{ln}\widehat{u}_{k-t-1,l}) - \Delta_{k-t}(0)\\ &\leq |\Delta_{k-t}(\eta_l\widehat{u}_{k-t-1,l}) - \Delta_{k-t}(0)| + |\Delta_{k-t}(\eta_l\widetilde{u}_{k-t-1,l}) - \Delta_{k-t}(0)|,\end{aligned}$$

and similarly

$$\begin{aligned}\Delta_{k-t}(\eta_n u_{k-t-1}) - \Delta_{k-t}(0) &\geq \Delta_{k-t}(\eta_l\widetilde{u}_{k-t-1,l}) - \Delta_{k-t}(0)\\ &\geq -|\Delta_{k-t}(\eta_l\widehat{u}_{k-t-1,l}) - \Delta_{k-t}(0)| - |\Delta_{k-t}(\eta_l\widetilde{u}_{k-t-1,l}) - \Delta_{k-t}(0)|.\end{aligned}$$

Therefore

$$\begin{aligned}&\sup_{|\eta_n|\leq\Theta_n} |\Delta_{k-t}(\eta_n u_{k-t-1}) - \Delta_{k-t}(0)|\\ &\qquad\leq \sup_{s\leq 3^{m_n}} |\Delta_{k-t}(\eta_s\widehat{u}_{k-t-1,s}) - \Delta_{k-t}(0)| \qquad (6.9.19)\\ &\qquad\quad+ \sup_{s\leq 3^{m_n}} |\Delta_{k-t}(\eta_s\widetilde{u}_{k-t-1,s}) - \Delta_{k-t}(0)|. \qquad (6.9.20)\end{aligned}$$

Consider (6.9.19). Obviously, the mean value of this quantity satisfies the inequality

$$\begin{aligned}&\mathbf{E}\sup_{s\leq 3^{m_n}} |\Delta_{k-t}(\eta_s\widehat{u}_{k-t-1,s}) - \Delta_{k-t}(0)|\\ &\qquad\leq \sum_{s=0}^{3^{m_n}} \mathbf{E}|\Delta_{k-t}(\eta_s\widehat{u}_{k-t-1,s}) - \Delta_{k-t}(0)|\\ &\qquad\leq \sum_{s=0}^{3^{m_n}} \Big(\mathbf{P}\{0 < \varepsilon_{k-t} < |\eta_s\widehat{u}_{k-t-1,s}|\}\\ &\qquad\qquad + \mathbf{P}\{-|\eta_s\widehat{u}_{k-t-1,s}| < \varepsilon_{k-t} < 0\}\Big).\end{aligned} \tag{6.9.21}$$

In each term of the sum in (6.9.21) the first probability is bounded by

$$\mathbf{E}\{G(|\eta_s\widehat{u}_{k-t-1,s}|) - G(0)\} \leq \sup_x g(x)\mathbf{E}|\eta_s\widehat{u}_{k-t-1,s}| = O(n^{-r})$$

uniformly in s, k, t. The second probablity is estimated in the same way, so that (6.9.19) is $O(n^{-r}\log n)$.

The mean value of (6.9.20) is treated similarly to yield

$$\mathbf{E}\sup_{|\eta_n|\leq\Theta_n} |\Delta_{k-t}(\eta_n u_{k-t-1}) - \Delta_{k-t}(0)| = O(n^{-r}\log n) \tag{6.9.22}$$

uniformly in k and t. Hence, applying Hölder's inequality we obtain

$$\begin{aligned}
&\mathbf{E}\sup_{t\le N_n}\sup_{|\eta_n|\le\Theta_n}\left|n^{-1}\sum_{k=t+1}^{n}\big(\Delta_{k-t}(\eta_n u_{k-t-1})-\Delta_{k-t}(0)\big)u_{k-1}\right| \\
&\le \sum_{t=1}^{N_n}\left\{n^{-1}\sum_{k=t+1}^{n}\mathbf{E}\sup_{|\eta_n|\le\Theta_n}|\Delta_{k-t}(\eta_n u_{k-t-1})-\Delta_{k-t}(0)||u_{k-1}|\right\} \\
&\le N_n n^{-1}\sum_{k=t+1}^{n}\left\{\mathbf{E}\{\sup_{|\eta_n|\le\Theta_n}|\Delta_{k-t}(\eta_n u_{k-t-1})-\Delta_{k-t}(0)|\}^{(1+\delta)/\delta}\right\}^{\delta/(1+\delta)} \\
&\qquad\qquad\times\left\{\mathbf{E}|u_{k-1}|^{1+\delta}\right\}^{1/(1+\delta)} \\
&= O\big(N_n(n^{-r}\log n)^{\delta/(1+\delta)}\big) = O\big(n^{-r\delta/(1+\delta)}\log^2 n\big)
\end{aligned}$$

by (6.9.22). This completes the proof of Lemma 6.9.3. □

LEMMA 6.9.4. *Assume Conditions* 6.1(ii, iii) *to hold and let* $\Theta_n = \Theta n^{-r}$ *with* $r > 1/4$, $0 < \Theta < \infty$. *Then under* $H_{1n}(a)$:

1°. $$\sup_{|\eta_n|\le\Theta_n}\left|\sum_{t=1}^{n-1}(\beta_n+\eta_n)^{t-1}x_{tn}(\eta_n)\right| = o_p(1),$$

2°. $$\sup_{|\eta_n|\le\Theta_n}\left|\sum_{t=1}^{n-1}(\beta_n+\eta_n)^{t-1}u_{tn}(\eta_n)\right| = o_p(1).$$

PROOF. The proofs of assertions 1° and 2° of this lemma are similar. They use assertions 1° and 2° of Lemma 6.9.2, respectively. We will prove assertion 1°. Obviously, $|x_{tn}(\eta_n)| \le 4n^{1/2}$ for any $t = 1,\dots,n-1$. Hence for an arbitrary $\varepsilon > 0$, and for b such that $|\beta_0| < b < 1$, one has, taking $N_n \sim \log_b n^{-1}$,

$$\begin{aligned}
&\mathbf{P}\left\{\sup_{|\eta_n|\le\Theta_n}\left|\sum_{t=N_n+1}^{n-1}(\beta_n+\eta_n)^{t-1}x_{tn}(\eta_n)\right|\ge\varepsilon\right\} \\
&\qquad\le \mathbf{P}\left\{4n^{1/2}\sum_{t\ge N_n+1}b^{t-1}>\varepsilon\right\} = 0
\end{aligned}$$

for all sufficiently large n because $n^{1/2}\sum_{t\ge N_n+1}b^{t-1} = O(n^{1/2}b^{N_n}) = o(1)$. Moreover, for all sufficiently large n

$$\sup_{|\eta_n|\le\Theta_n}\left|\sum_{t=1}^{N_n}(\beta_n+\eta_n)^{t-1}x_{tn}(\eta_n)\right| \le \sup_{t\le N_n}\sup_{|\eta_n|\le\Theta_n}|x_{tn}(\eta_n)|\sum_{t\ge1}b^{t-1} = o_p(1)$$

on account of assertion 1° of Lemma 6.9.2. □

LEMMA 6.9.5. *Assume that Condition* 6.1(ii) *holds. Then under* $H_{1n}(a)$

$$n^{-1}\sum_{k=t+1}^{n}\Delta_{k-t}(0)u_{k-1} = -\frac{1}{2}\mathbf{E}|\varepsilon_1|\beta_0^{t-1}+\delta_{tn},$$

where, for $N_n \sim \log_b n^{-1}$ *with* $|\beta_0| < b < 1$,

$$\sup_{t\le N_n}\mathbf{E}|\delta_{tn}| = O(n^{-\delta/(1+\delta)}\log n), \qquad \delta = \min(\Delta, 1).$$

PROOF. We will consider the more difficult case $0 < \Delta < 1$. First we state bounds for a sum of i.i.d. random variables and its expectation to be used in the proof. Let $\xi_1, \xi_2, \dots, \xi_n$ be i.i.d. random variables with $\mathbf{E}\xi_1 = 0$ and $\mathbf{E}|\xi_1|^{1+\Delta} < \infty$. Then

$$n^{-1/(1+\Delta)} \sum_{i=1}^{n} \xi_i \xrightarrow{\text{a.s.}} 0. \tag{6.9.23}$$

Further, since for $0 < \Delta < 1$

$$\mathbf{E}|\xi_1 + \cdots + \xi_n|^{1+\Delta} \le 2^{1+\Delta} n \mathbf{E}|\xi_1|^{1+\Delta}$$

(see von Bahr and Esséen [**3**]), one has

$$\mathbf{E}\left| n^{-1/(1+\Delta)} \sum_{i=1}^{n} \xi_i \right|^{1+\Delta} \le 2^{1+\Delta} \mathbf{E}|\xi_1|^{1+\Delta},$$

hence the sequence $n^{-1/(1+\Delta)} \sum_{i=1}^{n} \xi_i$ is uniformly integrable. Therefore

$$\mathbf{E}\left| n^{-1/(1+\Delta)} \sum_{i=1}^{n} \xi_i \right| \to 0,$$

or

$$\mathbf{E}\left| n^{-1} \sum_{i=1}^{n} \xi_i \right| = o(n^{-\Delta/(1+\Delta)}). \tag{6.9.24}$$

Now we write by (6.1.1)

$$u_{k-1} = \sum_{s=0}^{t-2} \beta_n^s \varepsilon_{k-1-s} + \beta_n^{t-1} \varepsilon_{k-t} + \beta_n^t u_{k-t-1},$$

so that

$$n^{-1} \sum_{k=t+1}^{n} \Delta_{k-t}(0) u_{k-1}$$

$$= n^{-1} \sum_{k=t+1}^{n} \sum_{s=0}^{t-2} \beta_n^s \Delta_{k-t}(0) \varepsilon_{k-1-s} \tag{6.9.25}$$

$$+ \beta_n^t n^{-1} \sum_{k=t+1}^{n} \Delta_{k-t}(0) \varepsilon_{k-t} \tag{6.9.26}$$

$$+ \beta_n^t n^{-1} \sum_{k=t+1}^{n} \Delta_{k-t}(0) u_{k-t-1}. \tag{6.9.27}$$

Consider (6.9.25). The expectation of its absolute value is bounded by

$$\sum_{s=0}^{t-2} \beta_n^s \mathbf{E}\left| n^{-1} \sum_{k=1}^{n-t} \Delta_k(0) \varepsilon_{k+t-s-1} \right|. \tag{6.9.28}$$

For $N_n \sim \log_b n^{-1}$ one has

$$\begin{aligned}
&\sup_{t\le N_n}\sup_{1\le m\le N_n}\mathbf{E}\left|n^{-1}\sum_{k=1}^{n-t}\Delta_k(0)\varepsilon_{k+m}\right| \\
&= \sup_{1\le m\le N_n}\mathbf{E}\left|n^{-1}\sum_{k=1}^{n}\Delta_k(0)\varepsilon_{k+m}\right| + \delta_{1n} \\
&= \sup_{1\le m\le N_n}\mathbf{E}\left|n^{-1}\sum_{j=1}^{m}\sum_{i=0}^{n/m}\Delta_{j+(m+1)i}\varepsilon_{j+(m+1)i+m}\right| \\
&\quad + \delta_{1n} + \delta_{2n},
\end{aligned} \tag{6.9.29}$$

with

$$\mathbf{E}|\delta_{jn}| \le N_n^{-1}n^{-1} = O(n^{-1}\log n), \qquad j = 1,\, 2.$$

The expectation in (6.9.29) is no greater than

$$m^{-1}\sum_{j=1}^{m}\mathbf{E}\left|(mn^{-1})\sum_{i=0}^{n/m}\Delta_{j+(m+1)i}\varepsilon_{j+(m+1)i+m}\right|. \tag{6.9.30}$$

The terms of the inner sum in (6.9.30) are i.i.d. random variables with zero mean and finite absolute moment of order $1+\Delta$. Hence, by (6.9.24), the quantity in (6.9.30) is $o\big((n/m)^{-\Delta/(1+\Delta)}\big)$. It readily follows from this that (6.9.29) is $O\big(n^{-\Delta/(1+\Delta)}\log n\big)$, and hence (6.9.28) is also $O\big(n^{-\Delta/(1+\Delta)}\log n\big)$. Thus (6.9.25) is δ_{3tn}, where

$$\sup_{t\le N_n}\mathbf{E}|\delta_{3tn}| = O\big(n^{-\Delta/(1+\Delta)}\log n\big).$$

Further, (6.9.26) equals $-\frac{1}{2}\beta_0^{t-1}\mathbf{E}|\varepsilon_1| + \delta_{4tn}$, where

$$\sup_{t\le N_n}\mathbf{E}|\delta_{4tn}| = O\big(n^{-1}\log n + n^{-\Delta/(1+\Delta)}\big) = O\big(n^{-\Delta/(1+\Delta)}\big).$$

Finally, making use of the representation

$$u_{k-t-1} = \sum_{s\ge 0}\beta_n^s\varepsilon_{k-t-1-s},$$

and arguing as in the analysis of (6.9.25), we obtain that (6.9.27) is δ_{5tn} with

$$\sup_{t\le N_n}\mathbf{E}|\delta_{5tn}| = O\big(n^{-\Delta/(1+\Delta)}\log n\big).$$

This completes the proof of Lemma 6.9.5 □

LEMMA 6.9.6. *Assume Conditions* 6.1(ii, iii) *to hold. Let* $\delta = \min(\Delta, 1)$ *and let* $\Theta_n = \Theta n^{-r}$ *with* $r > (1+\delta)/[2(1+2\delta)]$, $0 < \Theta < \infty$. *Then under* $H_{1n}(a)$

$$\sup_{|\eta_n|\le\Theta_n}\left|\sum_{t=1}^{n-1}(\beta_n+\eta_n)^{t-1}y_{tn}(\eta_n) + \frac{1}{2}g(0)\,\mathbf{E}|\varepsilon_1|\,(1-\beta_0^2)^{-1}\theta\right| = o_p(1).$$

PROOF. Let $N_n \sim \log_b n^{-1}$ with $|\beta_0| < b < 1$. As in the proof of Lemma 6.9.4, one verifies that

$$\sup_{|\eta_n|\le\Theta_n}\left|\sum_{t=N_n+1}^{n-1}(\beta_n+\eta_n)^{t-1}y_{tn}(\eta_n)\right| = o_p(1).$$

By Taylor's formula

$$\sum_{t=1}^{N_n}(\beta_n+\eta_n)^{t-1}y_{tn}(\eta_n)$$

$$= G(0)\sum_{t=1}^{N_n}(\beta_n+\eta_n)^{t-1} \times\left\{n^{-1/2}\sum_{k=t+1}^{n}\big(\Delta_{k-t}(\eta_n u_{k-t-1})-\Delta_{k-t}(0)\big)\right\} \tag{6.9.31}$$

$$+\sum_{t=1}^{N_n}(\beta_n+\eta_n)^{t-1}\left\{n^{-1/2}\sum_{k=t+1}^{n}\Delta_{k-t}(\eta_n u_{k-t-1}) \times g(\theta_k\eta_n u_{k-1})\eta_n u_{k-1}\right\}, \tag{6.9.32}$$

where $\theta_k \in (0,1)$. By assertion 2° of Lemma 6.9.2 the absolute value of (6.9.31) is $o_p(1)$ uniformly over $|\eta_n| \le \Theta_n$ since $r > (1+\delta)/[2(1+2\delta)] > \frac{1}{4}$.

By the condition

$$|g(x)-g(0)| \le h|x|^\delta, \qquad \delta = \min(\Delta, 1)$$

(see the proof of Theorem 6.2.1), we can rewrite (6.9.32) in the form

$$g(0)\eta_n\sum_{t=1}^{N_n}(\beta_n+\eta_n)^{t-1}\left\{n^{-1/2}\sum_{k=t+1}^{n}\Delta_{k-t}(\eta_n u_{k-t+1})u_{k-1}\right\}+\varepsilon_n, \tag{6.9.33}$$

where, for $|\eta_n| \le \Theta_n$, $|\beta| < b < 1$, and all sufficiently large n,

$$|\varepsilon_n| \le hn^{1/2}|\eta_n|^{1+\delta}\sum_{t=1}^{N_n}b^{t-1}\left\{n^{-1}\sum_{k=1}^{n}|u_{k-1}|^{1+\delta}\right\} = o_p(1)$$

for $r > (1+\delta)/2(1+2\delta)$. The main term of (6.9.33) can be represented as

$$g(0)\eta_n n^{1/2}\sum_{t=1}^{N_n}(\beta_n+\eta_n)^{t-1} \times\left\{n^{-1}\sum_{k=t+1}^{n}\big(\Delta_{k-t}(\eta_n u_{k-t-1})-\Delta_{k-t}(0)\big)u_{k-1}\right\} \tag{6.9.34}$$

$$+\,g(0)\eta_n n^{1/2}\sum_{t=1}^{N_n}(\beta_n+\eta_n)^{t-1} \times\left\{n^{-1}\sum_{k=t+1}^{n}\Delta_{k-t}(0)u_{k-1}\right\}. \tag{6.9.35}$$

By Lemma 6.9.3 for $r > (1+\delta)/[2(1+2\delta)]$ the expected value of the supremum over $|\eta_n| \leq \Theta_n$ of the absolute value of (6.9.34) is $o(1)$.

Finally, (6.9.35) is

$$-\frac{1}{2}g(0)\mathbf{E}|\varepsilon_1|(1-\beta_0^2)^{-1}\theta + \varepsilon_{1n},$$

where $\sup_{|\eta_n|\leq\Theta_n} |\varepsilon_{1n}| = o_p(1)$ by Lemma 6.9.5. The proof of Lemma 6.9.6 is completed. □

Now (6.9.5), Lemmas 6.9.4 and 6.9.6, and the definition $\eta_n = n^{-1/2}\theta$ immediately imply that (6.9.1) is equal to

$$\lambda_S(\beta_0)\,\theta + \varepsilon_{2n}(\theta) \qquad \text{with} \qquad \sup_{|\theta|\leq\Theta n^r} |\varepsilon_{2n}(\theta)| = o_p(1) \tag{6.9.36}$$

for $r < \frac{\delta}{2(1+2\delta)}$. Lemma 6.9.1 and (6.9.36) imply Theorem 6.4.1.

CHAPTER 7

Sign-based Analysis of the Multiparameter Autoregression

7.1. Introduction

In this chapter we consider the autoregression model

$$u_i = \beta_1 u_{i-1} + \cdots + \beta_q u_{i-q} + \varepsilon_i, \qquad i \in \mathbb{Z}. \tag{7.1.1}$$

The order q is assumed to be known; $\beta_1, \ldots, \beta_q$ are unknown nonrandom coefficients such that the roots of the characteristic equation

$$x^q = \beta_1 x^{q-1} + \cdots + \beta_q \tag{7.1.2}$$

lie inside the unit disk; $\{\varepsilon_i\}$ are i.i.d. random variables with an unknown (nondegenerate) distribution function $G(x)$. These conditions will be assumed throughout this chapter.

Let $\boldsymbol{\beta} = (\beta_1, \ldots, \beta_q)^T$. Define the sequence $\{\delta_t = \delta_t(\boldsymbol{\beta}),\, t = 1-q, 2-q, \ldots\}$ by the recurrent relation

$$\delta_t = \beta_1 \delta_{t-1} + \cdots + \beta_q \delta_{t-q}, \qquad t = 1, 2, \ldots, \tag{7.1.3}$$

with initial conditions $\delta_{1-q} = \cdots = \delta_{-1} = 0$, $\delta_0 = 1$.

It is well known (see Anderson [**2**], Section 5.2.1) that if the roots of (7.1.2) lie in the unit disk, δ_t decrease at an exponential rate, i.e.,

$$|\delta_t(\beta)| \le c\, b^t, \qquad t = 1, 2, \ldots, \tag{7.1.4}$$

for some $c > 0$ and $0 < b < 1$ independent of t.

Arguing as in the proof of Lemma 5.2.2, one readily infers that under condition $\mathbf{E} \log^+ |\varepsilon_1| < \infty$ the equation (7.1.1) a.s. has a unique strictly stationary solution, which is representable as

$$u_i = \sum_{t=0}^{\infty} \delta_t(\boldsymbol{\beta}) \varepsilon_{i-t}, \tag{7.1.5}$$

where the series absolutely converges a.s. If $\mathbf{E}|\varepsilon_1|^{1+\Delta} < \infty$ then the series in (7.1.5) converges also in $L^{1+\Delta}$, which can be shown similarly to the proof of Lemma 5.2.1.

Let $u_{1-q}, \ldots, u_n$ be observations satisfying (7.1.5). We will construct nonparametric sign tests based on these observations for testing hypotheses about the vector $\boldsymbol{\beta}$ as well as nonparametric sign esimators for $\boldsymbol{\beta}$, which extend the corresponding tests and estimators in Chapter 6. Our procedures will be based on the vector of signs

$$\mathbf{S}^n(\boldsymbol{\theta}) = (S_1(\boldsymbol{\theta}), \dots, S_n(\boldsymbol{\theta})), \qquad \boldsymbol{\theta} = (\theta_1, \dots, \theta_q)^T \in \mathbb{R}^q, \tag{7.1.6}$$

where

$$S_k(\boldsymbol{\theta}) = \operatorname{sign}(u_k - \theta_1 u_{k-1} - \dots - \theta_q u_{k-q}), \qquad k = 1, \dots, n. \tag{7.1.7}$$

In §7.2 we deal with testing the hypothesis

$$H_0 \colon \boldsymbol{\beta} = \boldsymbol{\beta}_0, \tag{7.1.8}$$

for a completely specified $\boldsymbol{\beta}_0 = (\beta_{10}, \dots, \beta_{q0})^T$. Let

$$\boldsymbol{\beta}_0^j = (\beta_{10}, \dots, \beta_{j-1,0}, \beta_j, \beta_{j+1,0}, \dots, \beta_{q0})^T$$

be a vector with the only variable component β_j. We will first consider alternatives to (7.1.8) of the form

$$H_j^+ \colon \boldsymbol{\beta} = \boldsymbol{\beta}_0^j, \qquad \beta_j > \beta_{j0}, \quad j = 1, \dots, q, \tag{7.1.9}$$

and similar left-sided alternatives H_j^-. Our test statistics $l_{jn}^S(\boldsymbol{\beta}_0)$ are based on the vector $\mathbf{S}^n(\boldsymbol{\beta}_0)$ and the corresponding tests are locally most powerful (LMP) among tests based on the vector of signs $\mathbf{S}^n(\boldsymbol{\beta}_0)$. Next we form a statistic $L_{n,S}(\boldsymbol{\beta}_0)$, which is a quadratic form of $l_{jn}^S(\boldsymbol{\beta}_0)$, $j = 1, \dots, q$, for testing H_0 against

$$H_1 \colon \boldsymbol{\beta} \neq \boldsymbol{\beta}_0. \tag{7.1.10}$$

The matrix of the quadratic form is proportional to $\mathbf{I}_n^{-1}(\boldsymbol{\beta}_0, \boldsymbol{\beta}_0)$, where $\mathbf{I}_n(\boldsymbol{\beta}, \boldsymbol{\theta})$ is Fisher's information about $\boldsymbol{\beta}$ contained in the vector $\mathbf{S}^n(\boldsymbol{\theta})$. It is shown that

$$\mathbf{I}_n(\boldsymbol{\beta}, \boldsymbol{\beta}) \sim n\big(2g(0)\mathbf{E}|\varepsilon_1|\big)^2 \mathbf{K}(\boldsymbol{\beta}),$$

where

$$\mathbf{K}(\boldsymbol{\beta}) = \big(k_{ij}(\boldsymbol{\beta})\big)_{i,j=1,\dots,q}, \qquad k_{ij}(\boldsymbol{\beta}) = \sum_{t=0}^{\infty} \delta_t(\boldsymbol{\beta})\delta_{t+|i-j|}(\boldsymbol{\beta}). \tag{7.1.11}$$

The matrix $\mathbf{K}(\boldsymbol{\beta})$ plays the key role in our analysis. Indeed, under usual regularity conditions on $G(x)$ (namely, when $\mathbf{E}\varepsilon_1^2 < \infty$, $G(x)$ has a density $g(x)$, and the Fisher information, $I(g)$, of $g(x)$ with respect to the location parameter is finite) the matrix of Fisher information about $\boldsymbol{\beta}$ contained in the observations $u_{1-q}, \dots, u_n$ is

$$\mathbf{J}_n(\boldsymbol{\beta}) \sim n\, I(g)\mathbf{E}\varepsilon_1^2 \mathbf{K}(\boldsymbol{\beta})$$

(see, for example, Cox and Hinkley [**20**], Section 9.2.3, Example 9.11). Note also that $\det\big(\mathbf{K}(\boldsymbol{\beta})\big) \neq 0$ (see Anderson [**2**], Lemma 5.5.5). Moreover, when $\mathbf{E}\varepsilon_1 = 0$, $0 < \mathbf{E}\varepsilon_1^2 < \infty$, the least squares estimator (LSE) $\widehat{\boldsymbol{\beta}}_{n,LS}$ is asymptotically normal,

$$\sqrt{n}(\widehat{\boldsymbol{\beta}}_{n,LS} - \boldsymbol{\beta}) \xrightarrow{d_{\boldsymbol{\beta}}} N\big(\mathbf{0}, \boldsymbol{\Sigma}_{LS}(\boldsymbol{\beta})\big),$$

with $\boldsymbol{\Sigma}_{LS}(\boldsymbol{\beta}) = \mathbf{K}^{-1}(\boldsymbol{\beta})$ (Anderson [**2**], Theorem 5.5.7).

In §7.3 we explore the asymptotic behavior of test statistics introduced in §7.2 under local alternatives

$$H_{1n}(\mathbf{a})\colon \boldsymbol{\beta} = \boldsymbol{\beta}_n := \boldsymbol{\beta}_0 + \mathbf{a}n^{-1/2} + o(n^{-1/2}) \tag{7.1.12}$$

with a fixed vector $\mathbf{a}$. We show, in particular, that the asymptotic relative efficiency (ARE) of our sign test with respect to the test based on $\sqrt{n}(\widehat{\boldsymbol{\beta}}_{n,LS} - \boldsymbol{\beta}_0)$ is

$$e_{S,LS} = \big(2g(0)\mathbf{E}|\varepsilon_1|\big)^2,$$

which coincides with the ARE obtained in §6.5 for $q = 1$.

The key tool for obtaining this result is a stochastic expansion of the vector random field $\mathbf{l}_n^S(\boldsymbol{\beta}_n + n^{-1/2}\boldsymbol{\theta})$, where $\mathbf{l}_n^S(\boldsymbol{\theta}) = \big(l_{1n}^S(\boldsymbol{\theta}), \dots, l_{qn}^S(\boldsymbol{\theta})\big)^T$, which holds uniformly over $|\boldsymbol{\theta}| \leq \Theta_n$ with $\Theta_n \to \infty$ at a polynomial rate. Here and throughout, $|\cdot|$ denotes the Euclidean norm of a vector or a matrix.

In §7.4 we construct tests for linear hypotheses about $\boldsymbol{\beta}$. Namely, let $\boldsymbol{\beta}$ consist of two subvectors, $\boldsymbol{\beta}^T = (\boldsymbol{\beta}^{(1)T}, \boldsymbol{\beta}^{(2)T})$, of dimension m and $q - m$ respectively, $1 \leq m < q$. For a given vector $\boldsymbol{\beta}_0^{(2)}$ we test the hypothesis

$$H_0'\colon \boldsymbol{\beta}^{(2)} = \boldsymbol{\beta}_0^{(2)}, \tag{7.1.13}$$

with $\boldsymbol{\beta}^{(1)}$ being a nuisance parameter. When $\boldsymbol{\beta}_0^{(2)} = \mathbf{0}$, the hypothesis (7.1.13) states that the equation (7.1.1) is actually of order $m < q$.

We study the asymptotic distribution of the test statistic under local alternatives and show that the ARE of our sign test with respect to the test based on $\widehat{\boldsymbol{\beta}}_{n,LS}$ is equal to $e_{S,LS}$.

Next, from sign tests we derive in §7.5 nonparametric sign estimators and in §7.6 we study their influence functionals in the contamination model of independent outliers. We construct several such estimators, namely, $\widehat{\boldsymbol{\beta}}_{n,S}$, $\widetilde{\boldsymbol{\beta}}_{n,S}$, $\boldsymbol{\beta}^*_{n,S}$. They are asymptotically equivalent to each other in the sense of their asymptotic distributions. In particular, we show that in the model (7.1.1) (without contamination)

$$\sqrt{n}(\widetilde{\boldsymbol{\beta}}_{n,S} - \boldsymbol{\beta}) \xrightarrow{d_{\boldsymbol{\beta}}} N\big(\mathbf{0}, \boldsymbol{\Sigma}_S(\boldsymbol{\beta})\big),$$

where $\boldsymbol{\Sigma}_S(\boldsymbol{\beta}) = \big(2g(0)\mathbf{E}|\varepsilon_1|\big)^{-2}\mathbf{K}^{-1}(\boldsymbol{\beta})$. The covariance matrix $\boldsymbol{\Sigma}_S(\boldsymbol{\beta})$ differs by only a constant factor $e_{S,LS}^{-1}$ from the covariance matrix $\boldsymbol{\Sigma}_{LS}(\boldsymbol{\beta})$ of the LSE.

In §7.7 the empirical distribution function $\widehat{G}_n(x)$ based on $\varepsilon_1(\widehat{\beta}_n), \cdots, \varepsilon_n(\widehat{\beta}_n)$ is studied. These are "residuals", which estimate the unobservable errors $\varepsilon_1, \cdots, \varepsilon_n$ with the aid of a $\sqrt{n}$-consistent estimator $\widehat{\boldsymbol{\beta}} = (\widehat{\beta}_{1n}, \cdots, \widehat{\beta}_{qn})^T$ for $\widehat{\boldsymbol{\beta}}$:

$$\varepsilon_k(\widehat{\boldsymbol{\beta}}_n) := u_k - \widehat{\beta}_{1n}u_{k-1} - \cdots - \widehat{\beta}_{qn}u_{k-q}, \qquad k = 1, \dots, n.$$

Thus

$$\widehat{G}_n(x) := n^{-1}\sum_{k=1}^{n} I\big(\varepsilon_k(\widehat{\boldsymbol{\beta}}_n) < x\big),$$

with $I(\cdot)$ denoting the indicator of an event. Let $G_n(x)$ denote the (unobservable) empirical distribution function of the errors $\varepsilon_1, \cdots, \varepsilon_n$. Under conditions similar to 6.1(ii, iii) we show that

$$\sup_x \sqrt{n}|\widehat{G}_n(x) - G_n(x)| = o_p(1). \tag{7.1.14}$$

Making use of (7.1.14) the well-known asymptotic properties of $G_n(x)$ are carried over to $\widehat{G}_n(x)$. In particular, empirical processes obtained from $\widehat{G}_n(x)$ weakly converge in the Skorokhod space $D[0,1]$ to the same limiting processes as the empirical processes obtained from $G_n(x)$. This enables us to construct tests of Kolmogorov and omega-square type based on $\widehat{G}_n(x)$ for testing simple and composite hypotheses about $G(x)$.

7.2. Test statistics and their null distributions

Consider the model (7.1.1) where the observations $u_{-q+1}, \ldots, u_n$ satisfy (7.1.5) with $\delta_t(\boldsymbol{\beta})$ defined by (7.1.3), and let $\mathbf{S}^n(\boldsymbol{\theta})$ be defined by (7.1.6), (7.1.7). Our aim in this section is to construct LMP tests based on $\mathbf{S}^n(\boldsymbol{\beta}_0)$ for testing $H_0\colon \boldsymbol{\beta} = \boldsymbol{\beta}_0$ as in (7.1.8) against right-sided alternatives H_j^+, $j = 1, \ldots, q$, as in (7.1.9) and similar left-sided alternatives H_j^-, and to test H_0 against $H_1\colon \boldsymbol{\beta} \neq \boldsymbol{\beta}_0$ as in (7.1.10).

In the setting under consideration, the possible realizations of $\mathbf{S}^n(\boldsymbol{\beta}_0)$ are vectors $\mathbf{s}^n = (s_1, \ldots, s_n)$, $s_k = \pm 1$. Denote

$$\gamma_{tn}(\mathbf{s}^n) = \sum_{k=t+1}^{n} s_{k-t} s_k, \qquad t = 1, \ldots, n-1. \tag{7.2.1}$$

Our procedures rely on the following expansion for the likelihood function of $\mathbf{S}^n(\boldsymbol{\beta}_0)$ under $\boldsymbol{\beta} = \boldsymbol{\beta}_0^j$ as $\beta_j \to \beta_{j0}$.

THEOREM 7.2.1. *Let Conditions* 6.1(i–iii) *be satisfied. Then, for* $n \geq j+1$, $j = 1, \ldots, q$,

$$\begin{aligned}\mathbf{P}_{\boldsymbol{\beta}_0^j}&\{\mathbf{S}^n(\boldsymbol{\beta}_0) = \mathbf{s}^n\} \\ &= \left(\frac{1}{2}\right)^n \left(1 + 2g(0)\mathbf{E}|\varepsilon_1| \sum_{t=0}^{n-j-1} \delta_t(\boldsymbol{\beta}_0)\gamma_{t+j,n}(\mathbf{s}^n)(\beta_j - \beta_{j0})\right) + o(\beta_j - \beta_{j0})\end{aligned}$$

as $\beta_j \to \beta_{j0}$.

For $q = 1$ this theorem coincides with Theorem 6.2.1.

PROOF. One has for $\boldsymbol{\beta} = \boldsymbol{\beta}_0^j$, $j = 1, \ldots, q$, and $k = 1, \ldots, n$

$$u_k - \beta_{10}u_{k-1} - \cdots - \beta_{j0}u_{k-j} - \cdots - \beta_{q0}u_{k-q} = \varepsilon_k + (\beta_j - \beta_{j0})u_{k-j}. \tag{7.2.2}$$

We will establish a recurrent relation between the likelihood functions of $\mathbf{S}^n(\boldsymbol{\beta}_0)$ and $\mathbf{S}^{n-1}(\boldsymbol{\beta}_0)$. One has for $n \geq j+1$

$$\begin{aligned}
&\mathbf{P}_{\boldsymbol{\beta}_0^j}\{\mathbf{S}^n(\boldsymbol{\beta}_0) = \mathbf{s}^n\} \\
&\quad = \mathbf{E}_{\boldsymbol{\beta}_0^j}\Big\{I\big(\mathbf{S}^{n-1}(\boldsymbol{\beta}_0) = \mathbf{s}^{n-1}\big) \\
&\qquad\qquad \times \Big[I\big(S_n(\boldsymbol{\beta}_0) = 1\big)\frac{1+s_n}{2} + I\big(S_n(\boldsymbol{\beta}_0) = -1\big)\frac{1-s_n}{2}\Big]\Big\} \\
&\quad = \mathbf{E}_{\boldsymbol{\beta}_0^j}\Big\{I\big(\mathbf{S}^{n-1}(\boldsymbol{\beta}_0) = \mathbf{s}^{n-1}\big) \\
&\qquad\qquad \times \Big[\frac{1+s_n}{2} - s_n I\big(u_n - \beta_{10}u_{n-1} - \cdots - \beta_{q0}u_{n-q} < 0\big)\Big]\Big\} \qquad (7.2.3)\\
&\quad = \mathbf{E}_{\boldsymbol{\beta}_0^j}\Big\{I\big(\mathbf{S}^{n-1}(\boldsymbol{\beta}_0) = \mathbf{s}^{n-1}\big) \\
&\qquad\qquad \times \Big[\frac{1+s_n}{2} - s_n I\big(\varepsilon_n < (\beta_{j0} - \beta_j)u_{n-j}\big)\Big]\Big\}
\end{aligned}$$

by (7.2.2).

Let $\Omega_{\leq i}$ be the σ-algebra generated by $\{\varepsilon_r,\ r \leq i\}$. On account of representation (7.1.5), ε_n and $\Omega_{\leq n-1}$ are independent. Under Condition 6.1(iii), $g(x)$ satisfies Hölder's condition of any order $\delta \in (0,1]$. Let $\delta = \min(1, \Delta)$. Then by the formula for total probability (7.2.3) is equal to

$$\begin{aligned}
&\mathbf{E}_{\boldsymbol{\beta}_0^j}\Big\{\mathbf{E}_{\boldsymbol{\beta}_0^j}\Big\{I\big(\mathbf{S}^{n-1}(\boldsymbol{\beta}_0) = \mathbf{s}^{n-1}\big) \\
&\qquad\qquad \times \Big[\frac{1+s_n}{2} - s_n I\big(\varepsilon_n < (\beta_{j0} - \beta_j)u_{n-j}\big)\Big] \Big| \Omega_{n-1}\Big\}\Big\} \\
&\quad = \mathbf{E}_{\boldsymbol{\beta}_0^j}\Big\{I\big(\mathbf{S}^{n-1}(\boldsymbol{\beta}_0) = \mathbf{s}^{n-1}\big) \\
&\qquad\qquad \times \Big[\frac{1+s_n}{2} - s_n G\big((\beta_{j0} - \beta_j)u_{n-j}\big)\Big]\Big\} \\
&\quad = \mathbf{E}_{\boldsymbol{\beta}_0^j}\Big\{I\big(\mathbf{S}^{n-1}(\boldsymbol{\beta}_0) = \mathbf{s}^{n-1}\big) \\
&\qquad\qquad \times \Big[\frac{1+s_n}{2} - s_n\Big(\frac{1}{2} + g(0)(\beta_{j0} - \beta_j)u_{n-j}\Big)\Big]\Big\} \qquad (7.2.4)\\
&\qquad + O\big(|\beta_j - \beta_{j0}|^{1+\delta}\big) \\
&\quad = \mathbf{E}_{\boldsymbol{\beta}_0^j}\Big\{I\big(\mathbf{S}^{n-1}(\boldsymbol{\beta}_0) = \mathbf{s}^{n-1}\big) \\
&\qquad\qquad \times \Big[\frac{1}{2} + s_n g(0)u_{n-j}(\beta_j - \beta_{j0})\Big]\Big\} \\
&\qquad + o(\beta_j - \beta_{j0}) \\
&\quad = \frac{1}{2}\mathbf{P}_{\boldsymbol{\beta}_0^j}\{\mathbf{S}^{n-1}(\boldsymbol{\beta}_0) = \mathbf{s}^{n-1}\} + s_n g(0) \\
&\qquad\qquad \times \mathbf{E}_{\boldsymbol{\beta}_0^j}\{I\big(\mathbf{S}^{n-1}(\boldsymbol{\beta}_0) = \mathbf{s}^{n-1}\big)u_{n-j}\}(\beta_j - \beta_{j0}) \\
&\qquad + o(\beta_j - \beta_{j0}).
\end{aligned}$$

By (7.1.5)

$$u_{n-j} = \sum_{t=0}^{\infty} \delta_t(\boldsymbol{\beta}_0^j)\varepsilon_{n-j-t}.$$

Substitute this expression into (7.2.4), and apply the same argument utilizing the formula for total probability to evaluate the expectation

$$\mathbf{E}_{\boldsymbol{\beta}_0^j}\{I(\mathbf{S}^{n-1}(\boldsymbol{\beta}_0) = \mathbf{s}^{n-1})\varepsilon_{n-j-t}\}.$$

By a simple calculation (7.2.4) becomes

$$\left(\frac{1}{2}\right)^{n-1} g(0)\mathbf{E}|\varepsilon_1| \sum_{t=0}^{n-j-1} s_n s_{n-j-t}\delta_t(\boldsymbol{\beta}_0^j)(\beta_j - \beta_{j0}) + o(\beta_j - \beta_{j0}). \tag{7.2.5}$$

By (7.1.3) each $\delta_t(\boldsymbol{\beta})$ is a continuous function of $\boldsymbol{\beta}$. Hence (7.2.5) can be rewritten as

$$\left(\frac{1}{2}\right)^{n-1} g(0)\mathbf{E}|\varepsilon_1| \sum_{t=0}^{n-j-1} s_n s_{n-j-t}\delta_t(\boldsymbol{\beta}_0)(\beta_j - \beta_{j0}) + o(\beta_j - \beta_{j0}). \tag{7.2.6}$$

Now (7.2.6) implies

$$\begin{aligned}
&\mathbf{P}_{\boldsymbol{\beta}_0^j}\{\mathbf{S}^n(\boldsymbol{\beta}_0) = \mathbf{s}^n\} \\
&\quad = \frac{1}{2}\mathbf{P}_{\boldsymbol{\beta}_0^j}\{\mathbf{S}^{n-1}(\boldsymbol{\beta}_0) = \mathbf{s}^{n-1}\} + \left(\frac{1}{2}\right)^{n-1} g(0)\mathbf{E}|\varepsilon_1| \\
&\qquad \times \sum_{t=0}^{n-j-1} s_n s_{n-j-t}\delta_t(\boldsymbol{\beta}_0)(\beta_j - \beta_{j0}) + o(\beta_j - \beta_{j0}).
\end{aligned} \tag{7.2.7}$$

Similarly to (7.2.7) we obtain for $n = j$

$$\mathbf{P}_{\boldsymbol{\beta}_0^j}\{\mathbf{S}^j(\boldsymbol{\beta}_0) = \mathbf{s}^j\} = \left(\frac{1}{2}\right)^j + o(\beta_j - \beta_{j0}). \tag{7.2.8}$$

The recurrent relation (7.2.7) with initial condition (7.2.8) implies the theorem. □

The following theorem is a direct consequence of Theorem 7.2.1.

THEOREM 7.2.2. *Let Conditions* 6.1(i–iii) *be satisfied. Then, for* $n \geq j+1$, $j = 1, \dots, q$, *the test for* $H_0\colon \boldsymbol{\beta} = \boldsymbol{\beta}_0$ *against* H_j^+ *as in* (7.1.9) *with critical region*

$$Q_{jn}^+ = \left\{\mathbf{s}^n\colon \sum_{t=0}^{n-j-1} \delta_t(\boldsymbol{\beta}_0)\gamma_{t+j,n}(\mathbf{s}^n) > \text{const}\right\} \tag{7.2.9}$$

is locally most powerful among tests based on $\mathbf{S}^n(\boldsymbol{\beta}_0)$.

With the inequality sign reversed, (7.2.9) determines a locally most powerful test against left-sided alternatives $H_j^-\colon \boldsymbol{\beta} = \boldsymbol{\beta}_0^j$, $\beta_j < \beta_{j0}$. For two-sided alternatives $H_{1j}\colon \boldsymbol{\beta} = \boldsymbol{\beta}_0^j$, $\beta_j \neq \beta_{j0}$, one can use the two-sided version of (7.2.9)

$$Q_{jn} = \left\{\mathbf{s}^n\colon \left|\sum_{t=0}^{n-j-1} \delta_t(\boldsymbol{\beta}_0)\gamma_{t+j,n}(\mathbf{s}^n)\right| > \text{const}\right\}. \tag{7.2.10}$$

For $S_k(\boldsymbol{\beta}_0)$, $k = 1, \ldots, n$, as defined by (7.1.7), let

$$\Gamma_{tn}(\boldsymbol{\theta}) = \sum_{k=t+1}^{n} S_{k-t}(\boldsymbol{\theta}) S_k(\boldsymbol{\theta}), \qquad t = 1, \ldots, n-1, \tag{7.2.11}$$

$$l_{jn}^S(\boldsymbol{\theta}) = \sum_{t=0}^{n-j-1} \delta_t(\boldsymbol{\theta}) \Gamma_{t+j,n}(\boldsymbol{\theta}), \qquad j = 1, \ldots, q. \tag{7.2.12}$$

Then $l_{jn}^S(\boldsymbol{\beta}_0)$ is the test statistic in (7.2.9) and (7.2.10).

Under the hypothesis $H_0 \colon \boldsymbol{\beta} = \boldsymbol{\beta}_0$

$$\Gamma_{tn}(\boldsymbol{\beta}_0) = \sum_{k=t+1}^{n} \operatorname{sign}(\varepsilon_{k-t}\varepsilon_k). \tag{7.2.13}$$

Subject to Condition 6.1(i), this statistic is obviously distributed free of $G(x)$; as was pointed out in §6.2 (see (6.2.14)),

$$\mathbf{E}_{\boldsymbol{\beta}_0} \Gamma_{tn}(\boldsymbol{\beta}_0) = 0, \qquad \mathbf{E}_{\boldsymbol{\beta}_0} \Gamma_{tn}(\boldsymbol{\beta}_0) \Gamma_{rn}(\boldsymbol{\beta}_0) = (n-t)\delta_{tr}, \tag{7.2.14}$$

where δ_{tr} denotes Kronecker's delta, $t, r = 1, \ldots, n-1$, and

$$\frac{1}{2}\big(n - t + \Gamma_{tn}(\boldsymbol{\beta}_0)\big) \sim \operatorname{Bi}\left(n-t, \frac{1}{2}\right),$$

i.e., it has the binomial distribution with parameters $n-t$ and $1/2$.

Thus, subject to Condition 6.1(i), the statistics $\Gamma_{tn}(\boldsymbol{\beta}_0)$ are distributed free of $G(x)$ under H_0, and hence so are the statistics $l_{jn}^S(\boldsymbol{\beta}_0)$, $j = 1, \ldots, q$. Their distribution, depending on $\boldsymbol{\beta}_0$, can be computed by simulation similarly to §6.8. A large sample approximation for the null distribution of $l_{jn}^S(\boldsymbol{\beta}_0)$ will be given in Theorem 7.2.3.

Now we construct a statistic for testing $H_0 \colon \boldsymbol{\beta} = \boldsymbol{\beta}_0$ against $H_1 \colon \boldsymbol{\beta} \neq \boldsymbol{\beta}_0$. Let $n \geq q + 1$. Define

$$\mathbf{l}_n^S(\boldsymbol{\theta}) = \big(l_{1n}^S(\boldsymbol{\theta}), \ldots, l_{qn}^S(\boldsymbol{\theta})\big)^T \tag{7.2.15}$$

with $l_{jn}^S(\boldsymbol{\theta})$ given by (7.2.12). By Theorem 7.2.1

$$\left.\frac{\partial \log \mathbf{P}_{\boldsymbol{\beta}_0^j}\{\mathbf{S}^n(\boldsymbol{\beta}_0) = \mathbf{s}^n\}}{\partial \beta_j}\right|_{\beta_j = \beta_{j0}} = 2g(0)\mathbf{E}|\varepsilon_1| \sum_{t=0}^{n-j-1} \delta_t(\boldsymbol{\beta}_0) \gamma_{t+j,n}(\mathbf{s}^n) \tag{7.2.16}$$

for $j = 1, \ldots, q$.

This formula and the definitions (7.2.12), (7.2.15) suggest taking the quadratic form

$$L_{n,S}(\boldsymbol{\beta}_0) = \big(2g(0)\,\mathbf{E}|\varepsilon_1|\,\big)^2 \big(\mathbf{l}_n^S(\boldsymbol{\beta}_0)\big)^T \mathbf{I}_n^{-1}(\boldsymbol{\beta}_0, \boldsymbol{\beta}_0)\, \mathbf{l}_n^S(\boldsymbol{\beta}_0) \tag{7.2.17}$$

as a test statistic for testing H_0 against H_1, where $\mathbf{I}_n(\boldsymbol{\beta}, \boldsymbol{\theta})$ is the Fisher information about $\boldsymbol{\beta}$ contained in $\mathbf{S}^n(\boldsymbol{\theta})$.

One can readily write down $\mathbf{I}_n(\boldsymbol{\beta},\boldsymbol{\beta})$. Indeed, by (7.2.16), (7.2.12), (7.2.15), and by the definition of the Fisher information,

$$\mathbf{I}_n(\boldsymbol{\beta},\boldsymbol{\beta}) = \big(2g(0)\mathbf{E}|\varepsilon_1|\big)^2 \mathbf{E}_{\boldsymbol{\beta}} \mathbf{l}_n^S(\boldsymbol{\beta})\big(\mathbf{l}_n^S(\boldsymbol{\beta})\big)^T.$$

One easily infers from (7.2.14) that

$$\mathbf{E}_{\boldsymbol{\beta}} \mathbf{l}_n^S(\boldsymbol{\beta}) = \mathbf{0}, \qquad \mathbf{E}_{\boldsymbol{\beta}} \mathbf{l}_n^S(\boldsymbol{\beta})\big(\mathbf{l}_n^S(\boldsymbol{\beta})\big)^T = \mathbf{K}_n(\boldsymbol{\beta}), \tag{7.2.18}$$

where

$$\begin{aligned} \mathbf{K}_n(\boldsymbol{\beta}) &= (k_{ij}^n(\boldsymbol{\beta})), \qquad i,j = 1,\dots,q, \\ k_{ij}^n(\boldsymbol{\beta}) &= \sum_{t=0}^{n-\max(i,j)-1} \delta_t(\boldsymbol{\beta})\delta_{t+|j-i|}(\boldsymbol{\beta})(n-t-\max(i,j)). \end{aligned} \tag{7.2.19}$$

Therefore

$$\mathbf{I}_n(\boldsymbol{\beta},\boldsymbol{\beta}) = \big(2g(0)\mathbf{E}|\varepsilon_1|\big)^2 \mathbf{K}_n(\boldsymbol{\beta}). \tag{7.2.20}$$

Note that $\delta_t(\boldsymbol{\beta}) \to 0$ at an exponential rate (see (7.1.4)); hence

$$n^{-1}k_{ij}^n(\boldsymbol{\beta}) \to k_{ij}(\boldsymbol{\beta})$$

with $k_{ij}(\boldsymbol{\beta})$ defined by (7.1.11). Therefore

$$n^{-1}\mathbf{K}_n(\boldsymbol{\beta}) \to \mathbf{K}(\boldsymbol{\beta}), \tag{7.2.21}$$

with $\mathbf{K}(\boldsymbol{\beta})$ defined by (7.1.11), and

$$\mathbf{I}_n(\boldsymbol{\beta},\boldsymbol{\beta}) \sim n\big(2g(0)\mathbf{E}|\varepsilon_1|\big)^2 \mathbf{K}(\boldsymbol{\beta}). \tag{7.2.22}$$

Let $\mathbf{J}_n(\boldsymbol{\beta})$ be the Fisher information matrix of observations $u_{-q+1},\dots,u_n$. If $\mathbf{E}\varepsilon_1^2 < \infty$ then under usual regularity conditions

$$\mathbf{J}_n(\boldsymbol{\beta}) \sim n\, I(g)\mathbf{E}\varepsilon_1^2 \mathbf{K}(\boldsymbol{\beta}), \tag{7.2.23}$$

where $I(g)$ is the Fisher information of density $g(x)$ with respect to the location parameter.

It is easily verified that $I(g) \geq [2g(0)]^2$, which becomes an equality only for the Laplace distribution. Hence (7.2.22), (7.2.23) and inequality $\mathbf{E}\varepsilon_1^2 \geq \big(\mathbf{E}|\varepsilon_1|\big)^2$ imply that for any $g(x)$ a strict inequality

$$\mathbf{I}_n(\boldsymbol{\beta},\boldsymbol{\beta}) < \mathbf{J}_n(\boldsymbol{\beta})$$

holds (at least for sufficiently large n).

Making use of (7.2.20), rewrite (7.2.17) as

$$L_{n,S}(\boldsymbol{\beta}_0) = \big(\mathbf{l}_n^S(\boldsymbol{\beta}_0)\big)^T \mathbf{K}_n^{-1}(\boldsymbol{\beta}_0)\, \mathbf{l}_n^S(\boldsymbol{\beta}_0). \tag{7.2.24}$$

Under Condition 6.1(i) the null distribution of $L_{n,S}(\boldsymbol{\beta}_0)$ is free of $G(x)$. Its large sample approximation will be given in Theorem 7.2.4. In an important particular case of $\boldsymbol{\beta}_0 = \mathbf{0}$, which is the hypothesis of independence, (7.2.24) becomes

$$L_{n,S}(\mathbf{0}) = \sum_{t=1}^{q} (n-t)^{-1}\Gamma_{tn}^2(\mathbf{0}), \qquad \Gamma_{tn}(\mathbf{0}) = \sum_{k=t+1}^{n} \operatorname{sign}(u_{k-t}u_k).$$

The null distribution of $L_{n,S}(\mathbf{0})$ can be computed and tabulated for any q and n.

The following theorem gives the asymptotic null distribution of $\mathbf{l}_n^S(\boldsymbol{\beta}_0)$.

THEOREM 7.2.3. *Let Condition* 6.1(i) *be satisfied. Then under* $H_0\colon \boldsymbol{\beta} = \boldsymbol{\beta}_0$

$$n^{-1/2}\,\mathbf{l}_n^S(\boldsymbol{\beta}_0) \xrightarrow{d_{\boldsymbol{\beta}_0}} N\big(\mathbf{0}, \mathbf{K}(\boldsymbol{\beta}_0)\big).$$

PROOF. We briefly sketch the proof. It suffices to show the asymptotic normality of scalar variables,

$$n^{-1/2}\mathbf{c}^T\mathbf{l}_n^S(\boldsymbol{\beta}_0) \xrightarrow{d_{\boldsymbol{\beta}_0}} N\big(0, \mathbf{c}^T\mathbf{K}(\boldsymbol{\beta}_0)\mathbf{c}\big)$$

for any nonrandom vector $\mathbf{c} = (c_1, \ldots, c_q)^T$. It follows from (7.2.18) and (7.2.21) that

$$\begin{aligned} \mathbf{E}_{\boldsymbol{\beta}_0} n^{-1/2}\mathbf{c}^T\mathbf{l}_n^S(\boldsymbol{\beta}_0) &= 0, \\ \mathrm{Var}_{\boldsymbol{\beta}_0}\{n^{-1/2}\mathbf{c}^T\mathbf{l}_n^S(\boldsymbol{\beta}_0)\} &= n^{-1}\mathbf{c}^T\mathbf{E}_{\boldsymbol{\beta}_0}\mathbf{l}_n^S(\boldsymbol{\beta}_0)\big(\mathbf{l}_n^S(\boldsymbol{\beta}_0)\big)^T\mathbf{c} \\ &= n^{-1}\mathbf{c}^T\mathbf{K}_n(\boldsymbol{\beta}_0)\mathbf{c} \to \mathbf{c}^T\mathbf{K}(\boldsymbol{\beta}_0)\mathbf{c}. \end{aligned}$$

This establishes the convergence of the mean and variance of $n^{-1/2}\mathbf{c}^T\mathbf{l}_n^S(\boldsymbol{\beta}_0)$ to the mean and variance of the limiting normal distribution. The convergence of the distribution of $n^{-1/2}\mathbf{c}^T\mathbf{l}_n^S(\boldsymbol{\beta}_0)$ to normality is shown along the same lines as in the proof of Theorem 6.2.3. □

Theorem 7.2.3 implies that, subject to Condition 6.1(i), one has under H_0

$$n^{-1/2}l_{jn}(\boldsymbol{\beta}_0) \xrightarrow{d_{\boldsymbol{\beta}_0}} N\big(0, k_{jj}(\boldsymbol{\beta}_0)\big), \quad k_{jj}(\boldsymbol{\beta}_0) = \sum_{t=0}^{\infty}\delta_t^2(\boldsymbol{\beta}_0), \quad j = 1, \ldots, q. \tag{7.2.25}$$

Hence the normal approximation can be used to evaluate the critical constants in (7.2.9) and (7.2.10).

Let $\chi^2(q)$ denote the chi-square distribution with q degrees of freedom. Theorem 7.2.3 and (7.2.21) imply

THEOREM 7.2.4. *Let Condition* 6.1(i) *be satisfied. Then under the hypothesis* $H_0\colon \boldsymbol{\beta} = \boldsymbol{\beta}_0$

$$L_{n,S}(\boldsymbol{\beta}_0) \xrightarrow{d_{\boldsymbol{\beta}_0}} \chi^2(q).$$

This theorem implies the following construction of a confidence region for $\boldsymbol{\beta}$. Denote by $\chi^2_{\alpha,q}$ the α-quantile of $\chi^2(q)$, and let C be the set of those $\boldsymbol{\theta} = (\theta_1, \ldots, \theta_q)^T$ for which all the roots of equation $x^q = \theta_1 x^{q-1} + \cdots + \theta_q$ lie in the unit disk. Then the set

$$A_{n\alpha} = \{\boldsymbol{\theta}\colon \boldsymbol{\theta} \in C, L_{n,S}(\boldsymbol{\theta}) < \chi_{q,1-\alpha}\} \tag{7.2.26}$$

is a confidence set for $\boldsymbol{\theta}$ of asymptotic level $1 - \alpha$.

We conclude this section by the following useful observation similar to the one in §6.2. For nonidentically distributed $\{\varepsilon_i\}$ satisfying the condition

$$\mathbf{P}\{\varepsilon_i < 0\} = \mathbf{P}\{\varepsilon_i > 0\} = 1/2, \qquad i \in \mathbb{Z},$$

the following statements hold:

- subject to Condition 6.1(i), the statistics $l_{jn}^S(\boldsymbol{\beta}_0)$, $\mathbf{l}_n^S(\boldsymbol{\beta}_0)$, and $L_{n,S}(\boldsymbol{\beta}_0)$ remain distribution free under H_0; in fact, their distributions are the same as for identically distributed $\{\varepsilon_i\}$;
- the assertions of Theorems 7.2.3 and 7.2.4 remain valid;
- the set $A_{n\alpha}$ as in (7.2.26) remains a confidence set for $\boldsymbol{\beta}$ of asymptotic level $1 - \alpha$.

The main results of this section have been published in Boldin [**12**].

7.3. Uniform stochastic expansion: The power of sign tests under local alternatives

In this section we obtain the power of sign tests from §7.2 under local alternatives (7.1.12)

$$H_{1n}(\mathbf{a})\colon \boldsymbol{\beta} = \boldsymbol{\beta}_n := \boldsymbol{\beta}_0 + \boldsymbol{a}n^{-1/2} + o(n^{-1/2}).$$

Similarly to the one-parameter case (see §6.4), this result relies on Theorem 7.3.1 giving a stochastic expansion for the vector random field $n^{-1/2}\mathbf{l}_n^S(\boldsymbol{\beta}_n + n^{-1/2}\boldsymbol{\theta})$, which holds uniformly over $|\boldsymbol{\theta}| \le \Theta_n$, where $\Theta_n \to \infty$ at a polynomial rate. This expansion will also be used in §7.4 dealing with testing linear hypotheses and in §7.5 where sign estimators are treated.

Let the vector $\mathbf{l}_n^S(\boldsymbol{\theta})$ be defined by (7.2.15) and (7.2.11)–(7.2.12), and let the matrix $\mathbf{K}(\boldsymbol{\beta})$ be defined by (7.1.11); put

$$\boldsymbol{\lambda}_S(\beta) = -2g(0)\mathbf{E}|\varepsilon_1|\mathbf{K}(\boldsymbol{\beta}). \tag{7.3.1}$$

THEOREM 7.3.1. *Let Conditions* 6.1(i–iii) *be satisfied. Then, under* $H_{1n}(\mathbf{a})$ *as in* (7.1.12)

$$\sup_{|\boldsymbol{\theta}|\le\Theta_n} \left|n^{-1/2}\mathbf{l}_n^S(\boldsymbol{\beta}_n + n^{-1/2}\boldsymbol{\theta}) - n^{-1/2}\mathbf{l}_n^S(\boldsymbol{\beta}_n) - \boldsymbol{\lambda}_S(\boldsymbol{\beta}_0)\boldsymbol{\theta}\right| = o_p(1) \tag{7.3.2}$$

for $\Theta_n = \Theta n^\alpha$, *where* $\alpha < \frac{\delta}{2(1+2\delta)}$, $\delta = \min(\Delta, 1)$, $0 < \Theta < \infty$.

Theorem 7.3.1 has been proved in Boldin [**14**].

PROOF. Methodologically the proof is similar to that of Theorem 6.4.1 for $q = 1$. This allows us to skip some details.

It is easy to show (see Kreiss [**59**], Section 2) that for any $\boldsymbol{\beta}$ in some neighbourhood of $\boldsymbol{\beta}_0$

$$|\delta_t(\boldsymbol{\beta}) - \delta_t(\boldsymbol{\beta}_0)| \le c\,|\boldsymbol{\beta} - \boldsymbol{\beta}_0|\,b^t \tag{7.3.3}$$

for some $c > 0$ and $0 < b < 1$ independent of t and $\boldsymbol{\beta}$. Now (7.3.3) and (7.1.4) imply that for $n > n_0$, $|\boldsymbol{\theta}| \le \Theta_n$, and $t = 0, 1, 2, \dots$

$$|\delta_t(\boldsymbol{\beta}_n)| \le c_1 b^t, \qquad |\delta_t(\boldsymbol{\beta}_n + n^{-1/2}\boldsymbol{\theta})| \le c_1 b^t, \tag{7.3.4}$$

and

$$|\delta_t(\boldsymbol{\beta}_n + n^{-1/2}\boldsymbol{\theta}) - \delta_t(\boldsymbol{\beta}_n)| \le c_2\Theta_n n^{-1/2} b^t, \tag{7.3.5}$$

where the constants $c_1, c_2, \dots$ depend only on $\boldsymbol{\beta}_0$.

Recall that in the one-parameter autoregression, $\delta_t(\beta) = \beta^t$, $t = 0, 1, 2, \dots$, and, moreover, for some $0 < b < b_1 < 1$ and $|\theta| \le \Theta_n$

$$|(\beta_n + \theta n^{-1/2})^t - \beta_n^t| \le c_3\Theta_n n^{-1/2} t b^{t-1} \le c_4\Theta_n n^{-1/2} b_1^t$$

(see the proof of Lemma 6.9.1). Thus the inequalities (7.3.4) and (7.3.5) establish similar properties of the sequence $\{\delta_t,\ t = 0, 1, 2, \dots\}$ for an arbitrary q.

Let $\mathbf{k}_j^T(\boldsymbol{\beta}) = \big(k_{j1}(\boldsymbol{\beta}), \ldots, k_{jq}(\boldsymbol{\beta})\big)$ be the jth row of the matrix $\mathbf{K}(\boldsymbol{\beta})$. For the proof of the theorem it suffices to show

$$\sup_{|\boldsymbol{\theta}|\le\Theta_n} \Big| n^{-1/2} l_{jn}^S(\boldsymbol{\beta}_n + n^{-1/2}\boldsymbol{\theta}) - n^{-1/2} l_{jn}^S(\boldsymbol{\beta}_n) + 2g(0)\,\mathbf{E}|\varepsilon_1|\,\mathbf{k}_j^T(\boldsymbol{\beta})\boldsymbol{\theta}\Big| = o_p(1), \qquad j = 1, \ldots, q. \tag{7.3.6}$$

For notational convenience we will prove (7.3.6) for $j = 1$.

By the definition (7.2.12) one has

$$n^{-1/2} l_{1n}^S(\boldsymbol{\beta}_n + n^{-1/2}\boldsymbol{\theta}) - n^{-1/2} l_{1n}^S(\boldsymbol{\beta}_n)$$

$$= n^{-1/2} \sum_{t=1}^{n-1} \delta_{t-1}(\boldsymbol{\beta}_n + n^{-1/2}\boldsymbol{\theta}) \times \big(\Gamma_{tn}(\boldsymbol{\beta}_n + n^{-1/2}\boldsymbol{\theta}) - \Gamma_{tn}(\boldsymbol{\beta}_n)\big) \tag{7.3.7}$$

$$+ n^{-1/2} \sum_{t=1}^{n-1} \big(\delta_{t-1}(\boldsymbol{\beta}_n + n^{-1/2}\boldsymbol{\theta}) - \delta_{t-1}(\boldsymbol{\beta}_n)\big)\Gamma_{tn}(\boldsymbol{\beta}_n). \tag{7.3.8}$$

LEMMA 7.3.1. *Let Condition* 6.1(i) *be satisfied and let* $H_{1n}(\mathbf{a})$ *hold. Then the supremum of* (7.3.8) *over* $|\boldsymbol{\theta}| \le \Theta n^r$ *with* $0 < \Theta < \infty$, $r < 1/2$ *is* $o_p(1)$.

This lemma is proved similarly to Lemma 6.9.1 making use of (7.3.5) and (7.2.14).

Consider now (7.3.7). Arguing as in the proof of Theorem 6.4.1 (see (6.9.4) and subsequent relations), it can be rewritten as

$$\sum_{t=1}^{n-1} \delta_{t-1}(\boldsymbol{\beta}_n + \boldsymbol{\eta}_n)\big(4x_{tn}(\boldsymbol{\theta}) + 4y_{tn}(\boldsymbol{\theta}) - 2u_{tn}(\boldsymbol{\theta})\big),$$

where $\widetilde{\mathbf{u}}_t = (u_t, u_{t-1}, \ldots, u_{t-q+1})^T$, $\Delta_k(x)$ is defined by (6.9.3), $\boldsymbol{\eta}_n = n^{-1/2}\boldsymbol{\theta}$,

$$\begin{aligned} x_{tn}(\boldsymbol{\theta}) = n^{-1/2} \sum_{k=t+1}^{n} \big(&\Delta_{k-t}(\boldsymbol{\eta}_n^T\widetilde{\mathbf{u}}_{k-t-1})\Delta_k(\boldsymbol{\eta}_n^T\widetilde{\mathbf{u}}_{k-1}) \\ &- \Delta_{k-t}(\boldsymbol{\eta}_n^T\widetilde{\mathbf{u}}_{k-t-1})G(\boldsymbol{\eta}_n^T\widetilde{\mathbf{u}}_{k-1}) \\ &- \Delta_{k-t}(0)\Delta_k(0) + \Delta_{k-t}(0)G(0)\big), \end{aligned}$$

$$y_{tn}(\boldsymbol{\theta}) = n^{-1/2} \sum_{k=t+1}^{n} \big(\Delta_{k-t}(\boldsymbol{\eta}_n^T\widetilde{\mathbf{u}}_{k-t-1})G(\boldsymbol{\eta}_n^T\widetilde{\mathbf{u}}_{k-1}) - \Delta_{k-t}(0)G(0)\big),$$

$$\begin{aligned} u_{tn}(\boldsymbol{\theta}) &= n^{-1/2} \sum_{k=t+1}^{n} \big(\Delta_{k-t}(\boldsymbol{\eta}_n^T\widetilde{\mathbf{u}}_{k-t-1}) - \Delta_{k-t}(0)\big) \\ &\quad + n^{-1/2} \sum_{k=t+1}^{n} \big(\Delta_k(\boldsymbol{\eta}_n^T\widetilde{\mathbf{u}}_{k-1}) - \Delta_k(0)\big). \end{aligned}$$

LEMMA 7.3.2. *Let Conditions* 6.1(ii, iii) *be satisfied and let* $H_{1n}(\mathbf{a})$ *hold. Put* $\Theta_n = \Theta n^{-r}$ *with* $r > 1/4$, $0 < \Theta < \infty$. *Then*:

1°. $$\sup_{|\boldsymbol{\eta}_n| \le \Theta_n} \Big| \sum_{t=1}^{n-1} \delta_{t-1}(\boldsymbol{\beta}_n + \boldsymbol{\eta}_n) x_{tn}(\boldsymbol{\theta}) \Big| = o_p(1),$$

2°. $$\sup_{|\boldsymbol{\eta}_n| \le \Theta_n} \Big| \sum_{t=1}^{n-1} \delta_{t-1}(\boldsymbol{\beta}_n + \boldsymbol{\eta}_n) u_{tn}(\boldsymbol{\theta}) \Big| = o_p(1).$$

This lemma is proved similarly to Lemma 6.9.4 making use of (7.3.4).

LEMMA 7.3.3. *Let Conditions* 6.1(ii, iii) *be satisfied and let* $H_{1n}(\mathbf{a})$ *hold. Put* $\delta = \min(\Delta, 1)$ *and let* $\Theta_n = \Theta n^{-r}$ *with* $r > \frac{1+\delta}{2(1+2\delta)}$, $0 < \Theta < \infty$. *Then*

$$\sup_{|\boldsymbol{\eta}_n| \le \Theta_n} \left| \sum_{t=1}^{n-1} \delta_{t-1}(\boldsymbol{\beta}_n + \boldsymbol{\eta}_n) y_{tn}(\boldsymbol{\theta}) + \frac{1}{2} g(0) \mathbf{E}|\varepsilon_1| \mathbf{k}_1^T(\boldsymbol{\beta}_0)\boldsymbol{\theta} \right| = o_p(1).$$

The proof is similar to that of Lemma 6.9.6. As a counterpart of (6.9.35) we have the expression

$$g(0)\boldsymbol{\eta}^T n^{-1/2} \sum_{t=1}^{N_n} \delta_{t-1}(\boldsymbol{\beta}_n + \boldsymbol{\eta}_n) \Big(n^{-1} \sum_{k=t+1}^{n} \Delta_{k-t}(0) \widetilde{\mathbf{u}}_{k-1} \Big).$$

In view of (7.1.5) this expression by the law of large numbers becomes

$$-\frac{1}{2} g(0)\, \mathbf{E}|\varepsilon_1|\, \boldsymbol{\theta}^T \sum_{t=1}^{\infty} \delta_{t-1}(\boldsymbol{\beta}_0) \big(\delta_{t-1}(\boldsymbol{\beta}_0), \delta_{t-2}(\boldsymbol{\beta}_0), \dots, \delta_{t-q}(\boldsymbol{\beta}_0)\big)^T + o_p(1)$$
$$= -\frac{1}{2} g(0) \mathbf{E}|\varepsilon_1| \mathbf{k}_1^T(\boldsymbol{\beta}_0)\boldsymbol{\theta} + o_p(1),$$

which yields the required assertion.

Now Lemmas 7.3.1–7.3.3 along with (7.3.7), (7.3.8) and the inequality $\frac{\delta}{2(1+2\delta)} < 1/4$ imply (7.3.5) for $j = 1$, which completes the proof of the theorem. □

The following corollary, to be used in the sequel, is an analogue of Corollary 6.4.1.

COROLLARY 7.3.1. *Let Conditions* 6.1(i–iii) *be satisfied and let* $H_{1n}(\mathbf{a})$ *as in* (7.1.12) *hold. Then for any sequence* $\widehat{\boldsymbol{\beta}}_n$ *such that* $n^{1/2-\alpha}(\widehat{\boldsymbol{\beta}}_n - \boldsymbol{\beta}_n) = O_p(1)$ *with* $\alpha < \frac{\delta}{2(1+2\delta)}$, $\delta = \min(\Delta, 1)$,

$$n^{-1/2}\mathbf{l}_n^S(\widehat{\boldsymbol{\beta}}_n) = n^{-1/2}\mathbf{l}_n^S(\boldsymbol{\beta}_n) + \boldsymbol{\lambda}_S(\boldsymbol{\beta}_0)\sqrt{n}(\widehat{\boldsymbol{\beta}}_n - \boldsymbol{\beta}_n) + o_p(1). \tag{7.3.9}$$

It is easy to see that, subject to Condition 6.1(i), the asymptotic distribution of $n^{-1/2}\mathbf{l}_n^S(\boldsymbol{\beta}_n)$ under $H_{1n}(\mathbf{a})$ is the same as that of $n^{-1/2}\mathbf{l}_n^S(\boldsymbol{\beta}_0)$ under H_0. Hence Theorem 7.2.3 and Corollary 7.3.1 imply

THEOREM 7.3.2. *Let Conditions* 6.1(i–iii) *be satisfied and let* $H_{1n}(\mathbf{a})$ *as in* (7.1.12) *hold. Then*

$$n^{-1/2}\mathbf{l}_n^S(\boldsymbol{\beta}_0) = n^{-1/2}\mathbf{l}_n^S(\boldsymbol{\beta}_n) - \boldsymbol{\lambda}_S(\boldsymbol{\beta}_0)\,\mathbf{a} + o_p(1),$$

and therefore

$$n^{-1/2}\mathbf{l}_n^S(\boldsymbol{\beta}_0) \xrightarrow{d_{\beta_n}} N\big(-\boldsymbol{\lambda}_S(\boldsymbol{\beta}_0)\,\mathbf{a},\ \mathbf{K}(\boldsymbol{\beta}_0)\big).$$

Thus the asymptotic shift of $n^{-1/2}\mathbf{l}_n^S(\boldsymbol{\beta}_0)$ under $H_{1n}(\mathbf{a})$ equals

$$-\boldsymbol{\lambda}_S(\boldsymbol{\beta}_0)\mathbf{a} = 2g(0)\mathbf{E}|\varepsilon_1|\mathbf{K}(\boldsymbol{\beta}_0)\,\mathbf{a}.$$

Hence the asymptotic shift of $n^{-1/2}l_{jn}^S(\boldsymbol{\beta}_0)$, $j = 1, \ldots, q$ (recall that $l_{jn}^S(\boldsymbol{\beta}_0)$ is the test statistic for $H_0\colon \boldsymbol{\beta} = \boldsymbol{\beta}_0$ against H_j^+, H_j^-, H_{1j}, see (7.2.9), (7.2.10)) under $H_{1n}(\mathbf{a})$ with $\mathbf{a} = (0, \ldots, 0, a_j, 0, \ldots, 0)^T$ equals

$$2g(0)\mathbf{E}|\varepsilon_1|k_{jj}(\boldsymbol{\beta}_0)a_j = 2g(0)\mathbf{E}|\varepsilon_1|a_j \sum_{t=0}^{\infty} \delta_t^2(\boldsymbol{\beta}_0).$$

Let $\chi^2(q, \lambda^2)$ denote the noncentral chi-square distribution with q degrees of freedom and noncentrality parameter λ^2. Theorem 7.3.2 implies the following theorem for $L_{n,S}(\boldsymbol{\beta}_0)$ as in (7.2.24).

THEOREM 7.3.3. *Let Conditions* 6.1(i–iii) *be satisfied and let* $H_{1n}(\mathbf{a})$ *hold. Then*

$$L_{n,S}(\boldsymbol{\beta}_0) \xrightarrow{d_{\beta_n}} \chi^2(q, \lambda_S^2),$$

where the noncentrality parameter equals

$$\lambda_S^2 = \big(2g(0)\,\mathbf{E}|\varepsilon_1|\big)^2\mathbf{a}^T\mathbf{K}(\boldsymbol{\beta}_0)\,\mathbf{a}.$$

Note that by (7.2.22) $\lambda_S^2 \sim n^{-1}\mathbf{a}^T\mathbf{I}_n(\boldsymbol{\beta}_0, \boldsymbol{\beta}_0)\,\mathbf{a}$, where $\mathbf{I}_n(\boldsymbol{\beta}, \boldsymbol{\theta})$ is the Fisher information about β contained in $\mathbf{S}^n(\boldsymbol{\theta})$.

Theorem 7.3.3 enables us to evaluate the ARE's of our sign test with statistic $L_{n,S}(\boldsymbol{\beta}_0)$ with respect to known tests. In particular, if $\mathbf{E}\varepsilon_1 = 0$, $0 < \mathbf{E}\varepsilon_1^2 < \infty$, then under $H_{1n}(\mathbf{a})$

$$\sqrt{n}(\widehat{\boldsymbol{\beta}}_{n,LS} - \boldsymbol{\beta}_0) \xrightarrow{d_{\beta_n}} N\big(\mathbf{a}, \mathbf{K}^{-1}(\boldsymbol{\beta}_0)\big), \tag{7.3.10}$$

which entails that

$$L_{n,LS}(\boldsymbol{\beta}_0) := n(\widehat{\boldsymbol{\beta}}_{n,LS} - \boldsymbol{\beta}_0)^T\mathbf{K}(\boldsymbol{\beta}_0)(\widehat{\boldsymbol{\beta}}_{n,LS} - \boldsymbol{\beta}_0) \xrightarrow{d_{\beta_n}} \chi^2(q, \lambda_{LS}^2),$$

where $\lambda_{LS}^2 = \mathbf{a}^T\mathbf{K}(\boldsymbol{\beta}_0)\,\mathbf{a}$. The proof of (7.3.10) is similar to that of Theorem 5.3.2.

Hence the ARE of the sign test based on $L_{n,S}(\boldsymbol{\beta}_0)$ with respect to the widely used test based on $L_{n,LS}(\boldsymbol{\beta}_0)$ equals

$$e_{S,LS} = \big(2g(0)\,\mathbf{E}|\varepsilon_1|\big)^2,$$

which coincides with the corresponding ARE in the one-parameter case (see §6.5).

7.4. Testing linear hypotheses

In this section we construct tests for linear hypotheses about $\boldsymbol{\beta}$ and study their asymptotic power against local alternatives.

Namely, let $\boldsymbol{\beta}$ consist of two subvectors, $\boldsymbol{\beta}^T = (\boldsymbol{\beta}^{(1)T}, \boldsymbol{\beta}^{(2)T})$, of dimension m and $q-m$ respectively, $1 \le m < q$. For a given vector $\boldsymbol{\beta}_0^{(2)}$ we test the hypothesis

$$H_0'\colon \boldsymbol{\beta}^{(2)} = \boldsymbol{\beta}_0^{(2)}, \tag{7.4.1}$$

with $\boldsymbol{\beta}^{(1)}$ being a nuisance parameter. When $\boldsymbol{\beta}_0^{(2)} = \mathbf{0}$, the hypothesis (7.4.1) states that the equation (7.1.1) is actually of order $m < q$.

The power of the tests will be studied under local alternatives

$$H_{1n}(\mathbf{a})\colon \boldsymbol{\beta} = \boldsymbol{\beta}_n := \boldsymbol{\beta}_0 + \mathbf{a}n^{-1/2} + o(n^{-1/2}), \tag{7.4.2}$$

where $\boldsymbol{\beta}_0^T = (\boldsymbol{\beta}^{(1)T}, \boldsymbol{\beta}_0^{(2)T})$ and $\mathbf{a}^T = (\mathbf{a}^{(1)T}, \mathbf{a}^{(2)T})$ is a fixed q-vector (partitioned into subvectors of dimension m and $q-m$). Thus the alternatives $H_{1n}(\mathbf{a})$ allow the nuisance parameter $\boldsymbol{\beta}^{(1)}$ to vary in a range of order $O(n^{-1/2})$.

Now we proceed to the construction of test statistics. Let the matrix $\mathbf{K}(\boldsymbol{\beta})$ defined by (7.1.11) be partitioned as

$$\mathbf{K}(\boldsymbol{\beta}) = \begin{pmatrix} \mathbf{J}(\boldsymbol{\beta}) & \mathbf{B}^T(\boldsymbol{\beta}) \\ \mathbf{B}(\boldsymbol{\beta}) & \mathbf{M}(\boldsymbol{\beta}) \end{pmatrix},$$

with matrices $\mathbf{J}(\boldsymbol{\beta})$, $\mathbf{M}(\boldsymbol{\beta})$, and $\mathbf{B}^T(\boldsymbol{\beta})$ of order $m \times m$, $(q-m) \times (q-m)$, and $m \times (q-m)$, respectively. Put

$$\mathbf{V}(\boldsymbol{\beta}) = \begin{pmatrix} \mathbf{J}^{-1/2}(\boldsymbol{\beta}) & -\mathbf{J}^{-1}(\boldsymbol{\beta})\mathbf{B}^T(\boldsymbol{\beta})\mathbf{C}^{-1/2}(\boldsymbol{\beta}) \\ \mathbf{0} & \mathbf{C}^{-1/2}(\boldsymbol{\beta}) \end{pmatrix},$$

where $\mathbf{C}(\boldsymbol{\beta}) = \mathbf{M}(\boldsymbol{\beta}) - \mathbf{B}(\boldsymbol{\beta})\mathbf{J}^{-1}(\boldsymbol{\beta})\mathbf{B}^T(\boldsymbol{\beta})$. It is easy to check that

$$\mathbf{V}(\boldsymbol{\beta})\mathbf{V}^T(\boldsymbol{\beta}) = \mathbf{K}^{-1}(\boldsymbol{\beta}). \tag{7.4.3}$$

Let $\widehat{\boldsymbol{\beta}}_n$ be a sequence of q-vectors which may depend on the hypothesized vector $\boldsymbol{\beta}_0^{(2)}$ in (7.4.1) and on the observations $u_{1-q}, \dots, u_n$, such that

$$\sqrt{n}(\widehat{\boldsymbol{\beta}}_n - \boldsymbol{\beta}_n) = O_p(1) \qquad \text{under } H_{1n}(\mathbf{a}).$$

In particular, if $\mathbf{E}\varepsilon_1 = 0$ and $0 < \mathbf{E}\varepsilon_1^2 < \infty$ then $\sqrt{n}(\widehat{\boldsymbol{\beta}}_{n,LS} - \boldsymbol{\beta}_n) = O_p(1)$ under $H_{1n}(\mathbf{a})$ (see (7.3.10)). Similarly to $\boldsymbol{\beta}$ we partition $\widehat{\boldsymbol{\beta}}_n$ into subvectors, $\widehat{\boldsymbol{\beta}}_n^T = (\widehat{\boldsymbol{\beta}}_n^{(1)T}, \widehat{\boldsymbol{\beta}}_n^{(2)T})$ of dimension m and $q-m$. In particular, $\widehat{\boldsymbol{\beta}}_n^{(2)}$ can be taken to be $\boldsymbol{\beta}_0^{(2)}$. In this case we let

$$\widehat{\boldsymbol{\beta}}_{n0}^T := (\widehat{\boldsymbol{\beta}}_n^{(1)\,T}, \boldsymbol{\beta}_0^{(2)\,T}), \tag{7.4.4}$$

where $\widehat{\boldsymbol{\beta}}_n^{(1)}$ is a $\sqrt{n}$-consistent estimator of $\boldsymbol{\beta}^{(1)}$.

Let $\widehat{\mathbf{V}}_n$ be an estimate for $\mathbf{V}(\boldsymbol{\beta}_0)$ consistent under $H_{1n}(\mathbf{a})$. For example, one can take $\widehat{\mathbf{V}}_n = \mathbf{V}_n(\widehat{\boldsymbol{\beta}}_n)$.

Let $\mathbf{l}_n^S(\boldsymbol{\theta})$ be the vector defined by (7.2.15), (7.2.12). For testing H_0' as in (7.4.1) we will use the test statistic

$$L_{n,S}^{\pi}(\widehat{\boldsymbol{\beta}}_{n0}) := n^{-1}\left|\pi \circ \widehat{\mathbf{V}}_n^T \mathbf{l}_n^S(\widehat{\boldsymbol{\beta}}_{n0})\right|^2, \tag{7.4.5}$$

where π stands for the projection onto the subspace spanned by the last $q-m$ coordinate vectors.

The following theorem describes the limiting distribution of this statistic under the alternative $H_{1n}(\mathbf{a})$ specified by (7.4.2).

THEOREM 7.4.1. *Let Conditions* 6.1(i–iii) *be satisfied and let* $H_{1n}(\mathbf{a})$ *as in* (7.4.2) *hold. Then*

$$L^{\pi}_{n,S}(\widehat{\boldsymbol{\beta}}_{n0}) \xrightarrow{d_{\boldsymbol{\beta}_n}} \chi^2(q-m, \lambda^2_{S,\pi}),$$

where the noncentrality parameter equals

$$\lambda^2_{S,\pi} = \big(2g(0)\,\mathbf{E}|\varepsilon_1|\big)^2 \mathbf{a}^{(2)T} C(\boldsymbol{\beta}_0)\,\mathbf{a}^{(2)}.$$

In particular, under H_0' *as in* (7.4.1)

$$L^{\pi}_{n,S}(\widehat{\boldsymbol{\beta}}_{n0}) \xrightarrow{d_{\boldsymbol{\beta}_0}} \chi^2(q-m).$$

PROOF. Let $\widehat{\boldsymbol{\beta}}_{n0}$ have the form (7.4.4). By assumption, $\sqrt{n}(\widehat{\boldsymbol{\beta}}_n - \boldsymbol{\beta}_n) = O_p(1)$ under $H_{1n}(\mathbf{a})$, hence $\sqrt{n}(\widehat{\boldsymbol{\beta}}_{n0} - \boldsymbol{\beta}_n) = O_p(1)$ as well. Therefore, in view of Corollary 7.3.1, consistency of $\widehat{\mathbf{V}}_n$, and (7.4.3), we have

$$\begin{aligned} n^{-1/2}\widehat{\mathbf{V}}_n^T \mathbf{l}_n^S(\widehat{\boldsymbol{\beta}}_{n0}) &= n^{-1/2}\widehat{\mathbf{V}}_n^T \mathbf{l}_n^S(\boldsymbol{\beta}_n) \\ &\quad - 2g(0)\mathbf{E}|\varepsilon_1|\widehat{\mathbf{V}}_n^T \mathbf{K}(\boldsymbol{\beta}_0)\sqrt{n}(\widehat{\boldsymbol{\beta}}_{n0} - \boldsymbol{\beta}_n) + o_p(1) \\ &= n^{-1/2}\mathbf{V}^T(\boldsymbol{\beta}_0)\,\mathbf{l}_n^S(\boldsymbol{\beta}_n) \\ &\quad - 2g(0)\mathbf{E}|\varepsilon_1|\mathbf{V}^T(\boldsymbol{\beta}_0)\mathbf{K}(\boldsymbol{\beta}_0)\sqrt{n}(\widehat{\boldsymbol{\beta}}_{n0} - \boldsymbol{\beta}_n) + o_p(1) \\ &= n^{-1/2}\mathbf{V}^T(\boldsymbol{\beta}_0)\,\mathbf{l}_n^S(\boldsymbol{\beta}_n) - 2g(0)\mathbf{E}|\varepsilon_1|\mathbf{V}^{-1}(\boldsymbol{\beta}_0) \\ &\quad \times \big(\sqrt{n}(\widehat{\boldsymbol{\beta}}_n^{(1)} - \boldsymbol{\beta}_n^{(1)})^T, -\mathbf{a}^{(2)T}\big)^T + o_p(1). \end{aligned} \tag{7.4.6}$$

Since

$$\mathbf{V}^{-1}(\boldsymbol{\beta}_0) = \begin{pmatrix} \mathbf{J}^{1/2}(\boldsymbol{\beta}_0) & \mathbf{J}^{-1/2}(\boldsymbol{\beta}_0)\mathbf{B}^T(\boldsymbol{\beta}_0) \\ \mathbf{0} & \mathbf{C}^{1/2}(\boldsymbol{\beta}_0) \end{pmatrix},$$

(7.4.6) implies

$$\pi \circ \widehat{\mathbf{V}}^T \mathbf{l}_n^S(\widehat{\boldsymbol{\beta}}_{n0}) = \pi \circ \mathbf{V}^T(\boldsymbol{\beta}_0)\,\mathbf{l}_n^S(\boldsymbol{\beta}_n) + 2g(0)\mathbf{E}|\varepsilon_1|\mathbf{C}^{1/2}(\boldsymbol{\beta}_0)\mathbf{a}^{(2)} + o_p(1). \tag{7.4.7}$$

Now the theorem follows from (7.4.7) and the convergence

$$n^{-1/2}\pi \circ \mathbf{V}^T(\beta_0)\,\mathbf{l}_n^S(\boldsymbol{\beta}_n) \xrightarrow{d_{\boldsymbol{\beta}_n}} N(0, \mathbf{E}_{q-m}),$$

where $\mathbf{E}_{q-m}$ denotes the identity matrix of order $q-m$. □

In the model (7.1.1) satisfying the LAN condition, Kreiss [**58**] proposed, in particular, the following tests for H_0'. Let

$$\boldsymbol{\psi}_n(\boldsymbol{\theta}) = n^{-1/2}\sum_{k=1}^{n} \psi(\varepsilon_k(\boldsymbol{\theta}))\widetilde{\boldsymbol{u}}_{k-1},$$

where

$$\widetilde{\boldsymbol{u}}_{k-1} = (u_{k-1}, \dots, u_{k-q})^T, \qquad \varepsilon_k(\boldsymbol{\theta}) = u_k - \boldsymbol{\theta}^T\widetilde{\boldsymbol{u}}_{k-1},$$

$$\int \psi^2 dG < \infty, \qquad \int \psi \, dG = 0.$$

Let $\sigma^2 = \mathbf{E}\varepsilon_1^2$, $\mathbf{W} = \sigma^{-1}\mathbf{V}$ and let $\widehat{\mathbf{W}}_n$ be a consistent estimator for $\mathbf{W}$. The test statistic is taken to be

$$\Big(\int \psi^2 dG\Big)^{-1} \big|\pi \circ \widehat{\mathbf{W}}_n^T \boldsymbol{\psi}_n(\widehat{\boldsymbol{\beta}}_{n0})\big|^2. \tag{7.4.8}$$

Theorem 4.1 in Kreiss [**58**] states that under $H_{1n}(\mathbf{a})$ the statistic (7.4.8) is asymptotically distributed as $\chi^2(q-m, \delta^2(\psi))$, where

$$\delta^2(\psi) = \mathbf{a}^{(2)T}\mathbf{C}(\boldsymbol{\beta}_0)\,\mathbf{a}^{(2)}\Big(\int \psi\, g'/\,g\, dG\Big)^2\Big(\int \psi^2 dG\Big)^{-1}\sigma^2,$$

provided ψ and G are twice continuously differentiable with bounded derivatives and $\mathbf{E}\varepsilon_1^4 < \infty$. For $\psi(x) = x/\sigma^2$ the test (7.4.8) is asymptotically equivalent to the one based on the LSE. In this case $\delta^2(\psi) = \mathbf{a}^{(2)T}\mathbf{C}(\boldsymbol{\beta}_0)\,\mathbf{a}^{(2)}$, and the ARE of the test (7.4.5) with respect to (7.4.8) equals again

$$\frac{\lambda_{S,\pi}^2}{\delta^2(\psi)} = \big(2g(0)\mathbf{E}|\varepsilon_1|\big)^2 = e_{S,LS}.$$

For $\psi = g'/g$ the test (7.4.8) is an asymptotically locally maximin test (see Theorem 3.2 and definitions in Section 3 of Kreiss [**58**]). In this case

$$\delta^2(\psi) = \mathbf{a}^{(2)T}\mathbf{C}(\boldsymbol{\beta}_0)\,\mathbf{a}^{(2)}I(g)\sigma^2,$$

and the ARE of the sign test (7.4.5) with respect to this asymptotically optimal parametric test equals

$$(2g(0)\mathbf{E}|\varepsilon_1|)^2/(I(g)\sigma^2).$$

As we pointed out in §7.2 (after (7.2.23)), this quantity is always less than one. However, the tests in Kreiss [**58**] as well as other known tests for linear hypotheses (see, for example, Anderson [**2**], Hallin and Puri [**34**], Dzhaparidze [**25**], and the references therein) are constructed under the assumptions needed to ensure the LAN condition, which require, in particular, that $\mathbf{E}\varepsilon_1^2 < \infty$. In contrast to these tests the one based on the statistic (7.4.5) is applicable when this assumption fails.

7.5. Sign-based estimators

Our aim in this section is to estimate the vector $\boldsymbol{\beta}$ in the model (7.1.1) making use of the sign test statistics from §7.2.

In §6.6 we considered several approaches to the construction of sign estimators in the one-parameter autoregression model. Unfortunately, some approaches discussed in §6.6 do not allow a straightforward generalization to the multiparameter case. In particular, certain difficulties are met in extending to this case the estimator $\widehat{\beta}_{n,S}$ determined by equation (6.6.6). We begin with this estimator.

7.5.1. Sign estimator $\widehat{\boldsymbol{\beta}}_{n,S}$. We assume Conditions 6.1(i–iii) to hold. Consider the vector $\mathbf{l}_n^S(\boldsymbol{\theta})$ as in (7.2.15), whose components are defined by (7.2.12). By analogy with the one-parameter case we could determine an estimator for $\boldsymbol{\beta}$ as a solution of the equation

$$\mathbf{l}_n^S(\boldsymbol{\theta}) \div \mathbf{0}, \tag{7.5.1}$$

where the sign $\div$, as before, means zero crossing by the function $\mathbf{l}_n^S(\boldsymbol{\theta})$. However, in contrast to the univariate case, a vector function can "cross" a point in many directions, and there may be different ways to give a precise meaning to (7.5.1). The simplest of them is as follows.

The vector function $\mathbf{l}_n^S(\boldsymbol{\theta})$ is said to cross zero at the point $\boldsymbol{\theta}_0 = (\theta_{10}, \dots, \theta_{q0})^T$ if for any $j = 1, \dots, q$ the scalar function $l_{jn}^S(\theta_{10}, \dots, \theta_{j-1,0}, \theta_j, \theta_{j+1,0}, \dots, \theta_{q0})^T$ of one variable θ_j crosses zero at the point θ_{j0}.

Similarly to the one-parameter case (see the proof of Theorem 6.6.2) it can be shown that for any solution $\widehat{\boldsymbol{\beta}}_n$ of (7.5.1)

$$n^{-1/2}\mathbf{l}_n^S(\widehat{\boldsymbol{\beta}}_n) = o_p(1). \tag{7.5.2}$$

Therefore, if there exists a $\sqrt{n}$-consistent solution $\widehat{\boldsymbol{\beta}}_{n,S}$ of (7.5.1) (only such solutions are of interest for us), then in view of (7.5.2) and Corollary 7.3.1 (with $\boldsymbol{\beta}_n = \boldsymbol{\beta}_0 = \boldsymbol{\beta}$, $\widehat{\boldsymbol{\beta}}_n = \widehat{\boldsymbol{\beta}}_{n,S}$ and $\alpha = 0$)

$$\sqrt{n}(\widehat{\boldsymbol{\beta}}_{n,S} - \boldsymbol{\beta}) = -\boldsymbol{\lambda}_S^{-1}(\boldsymbol{\beta}) n^{-1/2}\mathbf{l}_n^S(\boldsymbol{\beta}) + o_p(1), \tag{7.5.3}$$

where the matrix $\boldsymbol{\lambda}_S(\boldsymbol{\beta})$ is defined by (7.3.1).

Theorem 7.2.3 and (7.5.3) immediately imply that

$$\sqrt{n}(\widehat{\boldsymbol{\beta}}_{n,S} - \boldsymbol{\beta}) \xrightarrow{d_{\boldsymbol{\beta}}} N\big(\mathbf{0}, \boldsymbol{\Sigma}_S(\boldsymbol{\beta})\big), \tag{7.5.4}$$

where

$$\boldsymbol{\Sigma}_S(\boldsymbol{\beta}) = \big(2g(0)\mathbf{E}|\varepsilon_1|\big)^{-2}\mathbf{K}^{-1}(\boldsymbol{\beta}),$$

with matrix $\mathbf{K}(\beta)$ defined by (7.1.11). The relations (7.5.3)–(7.5.4) extend Theorem 6.6.2 to the case of arbitrary q.

It is worth noticing that by (7.2.22) the limiting covariance matrix $\boldsymbol{\Sigma}_S(\boldsymbol{\beta})$ in (7.5.4) coincides with the limit of $n\mathbf{I}_n^{-1}(\boldsymbol{\beta}, \boldsymbol{\beta})$, where $\mathbf{I}_n(\boldsymbol{\beta}, \boldsymbol{\theta})$ is the Fisher information about $\boldsymbol{\beta}$ contained in $\mathbf{S}^n(\boldsymbol{\theta})$.

Unlike the one-parameter case (see Theorem 6.6.2), for an arbitrary q it is by no means a simple task to establish the existence of a $\sqrt{n}$-consistent solution of the equation (7.5.1), even with the aid of Theorem 7.3.1, and we will not pursue this matter here. One more difficulty connected with defining the estimator by (7.5.1) is the problem of selection of a suitable ($\sqrt{n}$-consistent) solution from among all solutions of equation (7.5.1).

Consider another possibility. A natural alternative to (7.5.1) is to define the estimator as a solution to the extremal problem

$$|\mathbf{l}_n^S(\boldsymbol{\theta})| \Longrightarrow \inf_{\boldsymbol{\theta}\in\mathbb{R}^q}. \tag{7.5.5}$$

It is easily shown that $\sqrt{n}$-consistent solutions of the problems (7.5.1) and (7.5.5) are asymptotically equivalent with regard to their asymptotic distributions.

Indeed, define $\widetilde{\boldsymbol{\beta}}_n$ by the equality

$$\sqrt{n}(\widetilde{\boldsymbol{\beta}}_n - \boldsymbol{\beta}) = -\boldsymbol{\lambda}_S^{-1}(\boldsymbol{\beta})n^{-1/2}\mathbf{l}_n^S(\boldsymbol{\beta}). \tag{7.5.6}$$

By Theorem 7.2.3 the vector $\sqrt{n}(\widetilde{\boldsymbol{\beta}}_n - \boldsymbol{\beta})$ has an asymptotically normal distribution, and hence $\sqrt{n}(\widetilde{\boldsymbol{\beta}}_n - \boldsymbol{\beta}) = O_p(1)$.

Substituting $\widetilde{\boldsymbol{\beta}}_n$ for $\widehat{\boldsymbol{\beta}}_n$ in (7.3.9) (with $\boldsymbol{\beta}_n = \boldsymbol{\beta}_0 = \boldsymbol{\beta}$ and $\alpha = 0$) we see that

$$n^{-1/2}\mathbf{l}_n^S(\widetilde{\boldsymbol{\beta}}_n) = o_p(1). \tag{7.5.7}$$

This implies that any solution $\widehat{\boldsymbol{\beta}}_n$ of the problem (7.5.5) *a fortiori* fulfills the condition $n^{-1/2}\mathbf{l}_n^S(\widehat{\boldsymbol{\beta}}_n) = o_p(1)$.

Moreover, by Corollary 7.3.1 (again with $\boldsymbol{\beta}_n = \boldsymbol{\beta}_0 = \boldsymbol{\beta}$ and $\alpha = 0$) relations (7.5.3)–(7.5.4) hold for any solution $\widehat{\boldsymbol{\beta}}_{n,S}$ of (7.5.5) which is $\sqrt{n}$-consistent.

Thus the problems (7.5.1) and (7.5.5) define asymptotically equivalent $\sqrt{n}$-consistent sign estimators (provided, of course, such solutions exist).

Clearly, the problems of existence of a $\sqrt{n}$-consistent solution of (7.5.5) and selecting it from among all solutions are as difficult as in the case of equation (7.5.1). These difficulties always arise when dealing with M-estimators. While these problems are tractable for smooth or at least convex objective functions, the function $\mathbf{l}_n^S(\boldsymbol{\theta})$ presents much more serious difficulties. However, there are some ways to circumvent these difficulties.

7.5.2. Sign estimator $\widetilde{\boldsymbol{\beta}}_{n,S}$. Instead of $\mathbf{l}_n^S(\boldsymbol{\theta})$, let us use a simpler piecewise-constant function

$$\mathbf{l}_n^S(\widehat{\boldsymbol{\beta}}_n, \boldsymbol{\theta}) = (l_{1n}^S(\widehat{\boldsymbol{\beta}}_n, \boldsymbol{\theta}), \dots, l_{qn}^S(\widehat{\boldsymbol{\beta}}_n, \boldsymbol{\theta}))^T,$$

where

$$l_{jn}^S(\widehat{\boldsymbol{\beta}}_n, \boldsymbol{\theta}) = \sum_{t=0}^{n-j-1} \delta_t(\widehat{\boldsymbol{\beta}}_n)\Gamma_{t+j,n}(\boldsymbol{\theta}), \qquad j = 1, \dots, q,$$

and $\widehat{\boldsymbol{\beta}}_n$ is a preliminary $\sqrt{n}$-consistent estimator for $\boldsymbol{\beta}$. Recall that $\Gamma_{tn}(\boldsymbol{\theta})$ are defined by (7.2.11), and the sequence $\{\delta_t(\boldsymbol{\beta}), t = 1-q, 2-q, \dots\}$ by (7.1.3).

We will take any random variable solving the problem

$$|\mathbf{l}_n^S(\widehat{\boldsymbol{\beta}}_n, \boldsymbol{\theta})| \Longrightarrow \inf_{\boldsymbol{\theta}\in Q(\widehat{\boldsymbol{\beta}}_n)}, \tag{7.5.8}$$

where $Q(\widehat{\boldsymbol{\beta}}_n) = \{\boldsymbol{\theta}\colon |\sqrt{n}(\boldsymbol{\theta} - \widehat{\boldsymbol{\beta}}_n)| \le \log n\}$, as a nonparametric sign estimator for $\boldsymbol{\beta}$ to be denoted by $\widetilde{\boldsymbol{\beta}}_{n,S}$. Obviously, there always exists a solution of (7.5.8). The following theorem describes the properties of $\widetilde{\boldsymbol{\beta}}_{n,S}$.

THEOREM 7.5.1. *If Conditions* 6.1(i–iii) *are satisfied and* $\sqrt{n}(\widehat{\boldsymbol{\beta}}_n - \boldsymbol{\beta}) = O_p(1)$, *then the solution* $\widetilde{\boldsymbol{\beta}}_{n,S}$ *of the problem* (7.5.8) *fulfills the following relations*:

$$\sqrt{n}(\widetilde{\boldsymbol{\beta}}_{n,S} - \boldsymbol{\beta}) = -\boldsymbol{\lambda}_S^{-1}(\boldsymbol{\beta})n^{-1/2}\mathbf{l}_n^S(\boldsymbol{\beta}) + o_p(1), \tag{7.5.9}$$

$$\sqrt{n}(\widetilde{\boldsymbol{\beta}}_{n,S} - \boldsymbol{\beta}) \xrightarrow{d_{\boldsymbol{\beta}}} N(\mathbf{0}, \boldsymbol{\Sigma}_S(\boldsymbol{\beta})), \qquad \boldsymbol{\Sigma}_S(\boldsymbol{\beta}) = \big(2g(0)\mathbf{E}|\varepsilon_1|\big)^{-2}\mathbf{K}^{-1}(\boldsymbol{\beta}). \tag{7.5.10}$$

PROOF. It can be shown (see Boldin [**14**], Section 5) that under the assumptions of the theorem

$$\sup_{|\boldsymbol{\theta}|\le\Theta\log n}\left|n^{-1/2}\mathbf{l}_n^S(\widehat{\boldsymbol{\beta}}_n,\boldsymbol{\beta}+n^{-1/2}\boldsymbol{\theta})-n^{-1/2}\mathbf{l}_n^S(\boldsymbol{\beta}+n^{-1/2}\boldsymbol{\theta})\right|=o_p(1),$$

which together with Theorem 7.3.1 (with $\boldsymbol{\beta}_n=\boldsymbol{\beta}_0=\boldsymbol{\beta}$) implies the stochastic expansion

$$\sup_{|\boldsymbol{\theta}|\le\Theta\log n}\left|n^{-1/2}\mathbf{l}_n^S(\widehat{\boldsymbol{\beta}}_n,\boldsymbol{\beta}+n^{-1/2}\boldsymbol{\theta})-n^{-1/2}\mathbf{l}_n^S(\boldsymbol{\beta})-\boldsymbol{\lambda}_S(\boldsymbol{\beta})\theta\right|=o_p(1). \tag{7.5.11}$$

Similarly to Corollary 7.3.1 we infer from (7.5.11) that

$$n^{-1/2}\mathbf{l}_n^S(\widehat{\boldsymbol{\beta}}_n,\boldsymbol{\beta}_n^*)=n^{-1/2}\mathbf{l}_n^S(\boldsymbol{\beta})+\boldsymbol{\lambda}_S(\boldsymbol{\beta})\sqrt{n}(\boldsymbol{\beta}_n^*-\boldsymbol{\beta})+o_p(1) \tag{7.5.12}$$

for any statistic $\boldsymbol{\beta}_n^*$ such that

$$\sqrt{n}(\boldsymbol{\beta}_n^*-\boldsymbol{\beta})/\log n=O_p(1).$$

Finally, $\widetilde{\boldsymbol{\beta}}_n$ defined by (7.5.6) belongs to the set $Q(\widehat{\boldsymbol{\beta}}_n)$ with probability tending to 1, and by (7.5.12) $n^{-1/2}\mathbf{l}_n^S(\widehat{\boldsymbol{\beta}}_n,\widetilde{\boldsymbol{\beta}}_n)=o_p(1)$. Hence, *a fortiori*,

$$n^{-1/2}\mathbf{l}_n^S(\widehat{\boldsymbol{\beta}}_n,\widetilde{\boldsymbol{\beta}}_{n,S})=o_p(1). \tag{7.5.13}$$

By definition,

$$\sqrt{n}(\widetilde{\boldsymbol{\beta}}_{n,S}-\boldsymbol{\beta})/\log n=O_p(1),$$

which together with (7.5.12) and (7.5.13) implies (7.5.9). The convergence (7.5.10) follows from (7.5.9) and Theorem 7.2.3. Thus the proof is completed. □

The numerical solution of the problem (7.5.8) should not be too complicated. For small q it can be solved by enumeration because $\mathbf{l}_n^S(\widehat{\boldsymbol{\beta}}_n,\boldsymbol{\theta})$ takes a finite number of values.

7.5.3. Sign estimator $\boldsymbol{\beta}_{n,S}^*$. As an alternative to $\widetilde{\boldsymbol{\beta}}_{n,S}$, we can consider another estimator, $\boldsymbol{\beta}_{n,S}^*$, which is equivalent to the solution of (7.5.8) with regard to the limiting distribution. Namely, let $\widehat{\mathbf{e}}_n$ be a consistent estimator for $\boldsymbol{\lambda}_S(\boldsymbol{\beta})$. Put

$$\boldsymbol{\beta}_{n,S}^*=\widehat{\boldsymbol{\beta}}_n-(n\widehat{\mathbf{e}}_n)^{-1}\mathbf{l}_n^S(\widehat{\boldsymbol{\beta}}_n). \tag{7.5.14}$$

Then Corollary 7.3.1 (with $\boldsymbol{\beta}_n=\boldsymbol{\beta}_0=\boldsymbol{\beta}$ and $\alpha=0$) and Theorem 7.2.3 imply the following theorem.

THEOREM 7.5.2. *Assume that Conditions* 6.1(i–iii) *are satisfied. Let* $\sqrt{n}(\widehat{\boldsymbol{\beta}}_n-\boldsymbol{\beta})=O_p(1)$ *and let* $\widehat{\mathbf{e}}_n$ *be a consistent estimator for the matrix* $\boldsymbol{\lambda}_S(\beta)$. *Then*

$$\sqrt{n}(\boldsymbol{\beta}_{n,S}^*-\boldsymbol{\beta})=-\boldsymbol{\lambda}_S(\beta)n^{-1/2}\mathbf{l}_n^S(\boldsymbol{\beta})+o_p(1),$$

$$\sqrt{n}(\boldsymbol{\beta}_{n,S}^*-\boldsymbol{\beta})\xrightarrow{d_{\boldsymbol{\beta}}}N\big(\mathbf{0},\boldsymbol{\Sigma}_S(\boldsymbol{\beta})\big),\qquad\boldsymbol{\Sigma}_S(\boldsymbol{\beta})=\big(2g(0)\mathbf{E}|\varepsilon_1|\big)^{-2}\mathbf{K}^{-1}(\boldsymbol{\beta}).$$

Note that $\beta^*_{n,S}$ does not solve (7.5.8), but it satisfies the condition

$$n^{-1/2}\mathbf{l}_n^S(\widehat{\beta}_n, \beta^*_{n,S}) = o_p(1).$$

For obtaining $\widehat{\mathbf{e}}_n$ it suffices to consistently estimate the constant $-2g(0)\mathbf{E}|\varepsilon_1|$ and the matrix $\mathbf{K}(\beta)$. If the vector $\widehat{\beta}^1_n = (\widehat{\beta}^1_{1n}, \widehat{\beta}_{2n}, \dots, \widehat{\beta}_{qn})^T$ differs from $\widehat{\beta}_n = (\widehat{\beta}_{1n}, \dots, \widehat{\beta}_{qn})^T$ only by the first component $\widehat{\beta}^1_{1n} = \widehat{\beta}_{1n} + hn^{-1/2}$, with a constant $h \neq 0$, then

$$(hn^{1/2})^{-1}\big(\Gamma_{1n}(\widehat{\beta}^1_n) - \Gamma_{1n}(\widehat{\beta}_n)\big)$$

is a consistent estimator for $-2g(0)\mathbf{E}|\varepsilon_1|$. Moreover, $\mathbf{K}(\widehat{\beta}_n)$ is a consistent estimator for $\mathbf{K}(\beta)$.

Another consistent estimator $\widehat{\mathbf{e}}_n$ is of the form

$$\widehat{\mathbf{e}}_n = -2\widehat{g}_n\widehat{m}_n\mathbf{K}(\widehat{\beta}_n),$$

where

$$\widehat{g}_n = (nh_n)^{-1}\sum_{i=1}^n \varphi\big(h_n^{-1}(u_i - \widehat{\beta}_n^T\widetilde{\mathbf{u}}_{i-1})\big)$$

is a Parzen–Rosenblatt type estimator with Gaussian kernel $\varphi(x)$, $h_n = n^{-\alpha}$, $0 < \alpha < 1/4$, $\widetilde{u}_{k-1} = (u_{k-1}, \dots, u_{k-q})^T$, and

$$\widehat{m}_n = n^{-1}\sum_{i=1}^n |u_i - \widehat{\beta}_n^T\widetilde{\mathbf{u}}_{i-1}|.$$

Next we notice that the covariance matrices of the estimators $\widetilde{\beta}_{n,S}$ (or $\beta^*_{n,S}$), $\widehat{\beta}_{n,LS}$, and $\widehat{\beta}_{n,LD}$ differ only by a scalar factor. Namely, if $\mathbf{E}\varepsilon_1 = 0$, $0 < \mathbf{E}\varepsilon_1^2 < \infty$,

$$\sqrt{n}(\widetilde{\beta}_{n,LS} - \beta) \xrightarrow{d_\beta} N\big(0, \mathbf{K}^{-1}(\beta)\big)$$

(see Anderson [**2**], Theorem 5.5.7), and under the assumptions: $\mathbf{E}\varepsilon_1 = 0$, $0 < \mathbf{E}\varepsilon_1^2 < \infty$, $G(0) = 1/2$, $g(0) > 0$ and $g(x)$ is continuous in a neighborhood of zero,

$$\sqrt{n}(\widehat{\beta}_{n,LD} - \beta) \xrightarrow{d_\beta} N\big(0, \big((2g(0))^2\mathbf{E}\varepsilon_1^2\big)^{-1}\mathbf{K}^{-1}(\beta)\big)$$

(see Pollard [**72**]). The ARE of $\widetilde{\beta}_{n,S}$ with respect to $\widehat{\beta}_{n,LS}$ or $\widehat{\beta}_{n,LD}$ is equal to the ratio of the respective coefficients of $\mathbf{K}^{-1}(\beta)$ in the limiting covariance matrices. Thus $e_{S,LS} = (2g(0)\mathbf{E}|\varepsilon_1|)^2$ and $e_{S,LD} = (\mathbf{E}|\varepsilon_1|)^2/\mathbf{E}\varepsilon_1^2$, which coincide with the respective ARE's in the one-parameter case (see §§6.6 and 6.5).

To conclude this section, we discuss the choice of the preliminary $\sqrt{n}$-consistent estimator involved in the definition of sign estimators. An immediate option is to take the LSE, which is $\sqrt{n}$-consistent provided $G(x)$ is symmetric and $\mathbf{E}\log^+|\varepsilon| < \infty$ (see Yohai and Maronna [**93**]). However, $\widehat{\beta}_{n,LS}$ has an unbounded influence functional, which is inherited by $\widetilde{\beta}_{n,S}$. This was shown in 5.6.1 and 6.7.2 for one-parameter autoregression; the general case will be considered in §7.6. In this respect the weighted estimators of least absolute deviations, which in the univariate case were treated in 5.5.3, may be of interest. So, we take the random variable $\widehat{\beta}_{n,LDW}$, solving the problem

$$\sum_{k=1}^n |\varphi(\widetilde{\mathbf{u}}_{k-1})(\mathbf{u}_k - \theta^T\widetilde{\mathbf{u}}_{k-1})| \Longrightarrow \inf_{\theta\in\mathbb{R}^q}, \tag{7.5.15}$$

as an estimator for β. The objective function in (7.5.15) is convex, so that there always exists a solution. We assume the following conditions.

CONDITION 7.5(i). $\mathbf{E}\varepsilon_1 = 0$, $\mathbf{E}_{\boldsymbol{\beta}}|\varphi(\widetilde{\mathbf{u}}_1)| < \infty$,

$$-\infty < \boldsymbol{\lambda}_{LDW}(\boldsymbol{\beta}) := -2g(0)\mathbf{E}_{\boldsymbol{\beta}}|\varphi(\widetilde{\mathbf{u}}_1)|\widetilde{\mathbf{u}}_1\widetilde{\mathbf{u}}_1^T < 0,$$

and

$$0 < \boldsymbol{K}_{LDW}(\boldsymbol{\beta}) := \mathbf{E}_{\boldsymbol{\beta}}\varphi^2(\widetilde{\mathbf{u}}_1)\widetilde{\mathbf{u}}_1\widetilde{\mathbf{u}}_1^T < \infty.$$

CONDITION 7.5(ii). $\max_{1 \le k \le n} n^{-1/2}|\varphi(\widetilde{\mathbf{u}}_k)\widetilde{\mathbf{u}}_k| = o_p(1)$.

CONDITION 7.5(iii). $G(0) = 1/2$, *$g(x)$ is continuous in a neighborhood of zero.*

Following the lines of the proof of Theorem 2 in Pollard [**72**] we obtain that under Conditions 7.5(i–iii)

$$\sqrt{n}(\widehat{\boldsymbol{\beta}}_{n,LDW} - \boldsymbol{\beta}) \xrightarrow{d_{\boldsymbol{\beta}}} N(0, \boldsymbol{\Sigma}_{LDW}(\boldsymbol{\beta})),$$

where

$$\boldsymbol{\Sigma}_{LDW}(\boldsymbol{\beta}) = \boldsymbol{\lambda}_{LDW}^{-1}(\boldsymbol{\beta})\mathbf{K}_{LDW}(\boldsymbol{\beta})\boldsymbol{\lambda}_{LDW}^{-1}(\boldsymbol{\beta}).$$

Notice that, unlike the one-parameter case where the extremal problem (5.5.8) for the weighted LAD estimators was reduced to equation (5.5.9), it does not make sense to pass from the convex problem (7.5.15) to the analogue of (5.5.9),

$$\sum_{k=1}^{n} \widetilde{\mathbf{u}}_{k-1}|\varphi(\widetilde{\mathbf{u}}_{k-1})| \operatorname{sign}(u_k - \theta^T\widetilde{\mathbf{u}}_{k-1}) \div 0,$$

because it would lead to the same difficulties which we faced when defining the sign estimator by the equation (7.5.1).

The results of this section are published in Boldin [**14**].

7.6. Influence functionals of estimators in the multiparameter autoregression

Suppose we have contaminated observations $\mathbf{Y}_n = (y_{1-q}, \dots, y_n)$, where

$$y_i = u_i + z_i^\gamma \xi_i, \qquad i \in \mathbb{Z}. \tag{7.6.1}$$

We assume that the variables $\{u_i\}$ in (7.6.1) satisfy (7.1.1), $\{z_i^\gamma\}$ form a Bernoulli sequence of i.i.d. random variables taking values 1 and 0 with probabilities γ (contamination level) and $1-\gamma$, $0 \le \gamma \le 1$, and $\{\xi_i\}$ are i.i.d. random variables with distribution μ_ξ from some class of distributions $\mathfrak{M}_\xi$; the sequences $\{u_i\}$, $\{z_i^\gamma\}$, $\{\xi_i\}$ are mutually independent.

Thus we consider the simplest model of independent outliers. This model for one-parameter autoregression was considered in §§5.6 and 6.7.

The definitions of the influence functional and the gross error sensitivity for estimators of the vector-valued parameter $\boldsymbol{\beta}$ are essentially the same as for the scalar case in §5.6. In order to characterize the performance of an estimator $\widehat{\boldsymbol{\beta}}_n$ for $\boldsymbol{\beta}$ based on observations $\mathbf{Y}_n$, assume that there exists the limit

$$\widehat{\boldsymbol{\beta}}_n \xrightarrow{\mathbf{P}_{\boldsymbol{\beta}}} \boldsymbol{\theta}_\gamma, \quad \text{and} \quad \boldsymbol{\theta}_0 = \boldsymbol{\beta}.$$

The infinitesimal characteristic of the estimator robustness against contamination is the derivative

$$\mathrm{IF}(\boldsymbol{\theta}_\gamma, \mu_\xi) = \lim_{\gamma \to 0} \frac{\boldsymbol{\theta}_\gamma - \boldsymbol{\theta}_0}{\gamma}, \tag{7.6.2}$$

which is called the influence functional of the estimator $\widehat{\boldsymbol{\beta}}_n$ (provided, of course, this derivative does exist). This quantity determines the main term of the asymptotic bias

$$\boldsymbol{\theta}_\gamma - \boldsymbol{\theta}_0 = \mathrm{IF}(\boldsymbol{\theta}_\gamma, \mu_\xi)\gamma + o(\gamma).$$

The gross error sensitivity of $\widehat{\boldsymbol{\beta}}_n$ is

$$\mathrm{GES}(\mathfrak{M}_\xi, \boldsymbol{\theta}_\gamma) := \sup_{\mu_\xi \in \mathfrak{M}_\xi} \big| \mathrm{IF}(\boldsymbol{\theta}_\gamma, \mu_\xi) \big|.$$

If this sensitivity is finite then the main term of the asymptotic bias is uniformly small over all possible contamination distributions and small γ. This means qualitatively that for small γ even very large outliers have little effect on the estimator.

Consider particular estimators.

7.6.1. Influence functional of the least squares estimator.

CONDITION 7.6(i). $\mathbf{E}\varepsilon_1 = 0,\ 0 < \mathbf{E}\varepsilon_1^2 < \infty,\ \mathbf{E}\xi_1^2 < \infty.$

For the contaminated sample the LSE $\widehat{\boldsymbol{\beta}}_{n,LS}$ is defined as a solution of the extremal problem

$$\sum_{i=1}^{n} (y_i - \boldsymbol{\theta}^T \widetilde{\boldsymbol{y}}_{i-1})^2 \Longrightarrow \inf_{\boldsymbol{\theta} \in \mathbb{R}^q},$$

where $\widetilde{\boldsymbol{y}}_{i-1} = (y_{i-1}, \dots, y_{i-q})^T$. This problem is equivalent to the equation

$$\sum_{i=1}^{n} \boldsymbol{\psi}_i^{LS}(\mathbf{Y}_n, \boldsymbol{\theta}) = \mathbf{0}, \tag{7.6.3}$$

where

$$\boldsymbol{\psi}_i^{LS}(\mathbf{Y}_n, \boldsymbol{\theta}) = \big(y_{i-1}(y_i - \boldsymbol{\theta}^T \widetilde{\boldsymbol{y}}_{i-1}), \dots, y_{i-q}(y_i - \boldsymbol{\theta}^T \widetilde{\boldsymbol{y}}_{i-1})\big)^T.$$

By the ergodic theorem, for $0 \le \gamma \le 1$ and $\boldsymbol{\theta} \in \mathbb{R}^q$

$$n^{-1} \sum_{i=1}^{n} \boldsymbol{\psi}_i^{LS}(\mathbf{Y}_n, \boldsymbol{\theta}) \xrightarrow{\mathbf{P}_{\boldsymbol{\beta}}} \boldsymbol{\Lambda}_{LS}(\gamma, \boldsymbol{\theta}),$$

where $\boldsymbol{\Lambda}_{LS}(\gamma, \boldsymbol{\theta}) = \big(\Lambda_1^{LS}(\gamma, \boldsymbol{\theta}), \dots, \Lambda_q^{LS}(\gamma, \boldsymbol{\theta})\big)^T$ with

$$\Lambda_j^{LS}(\gamma, \boldsymbol{\theta}) = \mathbf{E}_{\boldsymbol{\beta}} y_0 (y_j - \boldsymbol{\theta}^T \widetilde{\boldsymbol{y}}_{j-1}), \qquad j = 1, \dots, q. \tag{7.6.4}$$

It is easily seen that $\boldsymbol{\Lambda}_{LS}(0, \boldsymbol{\beta}) = \mathbf{0}$.

Substituting the expression (7.6.1) for y_j in (7.6.4) and invoking the mutual independence of $\{u_i\}$, $\{z_i^\gamma\}$, and $\{\xi_i\}$, as well as the independence between $\{z_i^\gamma\}$

and $\{\xi_i\}$, we obtain that the derivatives $\frac{\partial}{\partial\gamma}\mathbf{\Lambda}_{LS}(\gamma,\boldsymbol{\theta})$ and $\frac{\partial}{\partial\boldsymbol{\theta}}\mathbf{\Lambda}_{LS}(\gamma,\boldsymbol{\theta})$ exist and are continuous in $(\gamma,\boldsymbol{\theta})$ for $0\le\gamma\le 1$, $\boldsymbol{\theta}\in\mathbb{R}^q$, and

$$\frac{\partial}{\partial\gamma}\mathbf{\Lambda}_{LS}(0,\boldsymbol{\beta}) = -\mathbf{E}\xi_1^2\boldsymbol{\beta}, \tag{7.6.5}$$

$$\frac{\partial}{\partial\boldsymbol{\theta}}\mathbf{\Lambda}_{LS}(0,\boldsymbol{\beta}) = -\big(\mathbf{E}_{\boldsymbol{\beta}}u_0u_{|i-j|}\big)_{i,j=1,\dots,q} = -\mathbf{E}\varepsilon_1^2\mathbf{K}(\boldsymbol{\beta}). \tag{7.6.6}$$

Therefore the matrix $\frac{\partial}{\partial\boldsymbol{\theta}}\mathbf{\Lambda}_{LS}(0,\boldsymbol{\beta})$ is nondegenerate. By the implicit function theorem, the equation $\mathbf{\Lambda}_{LS}(\gamma,\boldsymbol{\theta})=\mathbf{0}$ has in a neighborhood of $(0,\boldsymbol{\beta})$ a unique solution $\boldsymbol{\theta}_\gamma^{LS}$ with $\boldsymbol{\theta}_0^{LS}=\boldsymbol{\beta}$, which is differentiable with respect to γ, and by (7.6.5), (7.6.6)

$$\left.\frac{d\boldsymbol{\theta}_\gamma^{LS}}{d\gamma}\right|_{\gamma=0} = -(\mathbf{E}\xi_1^2/\mathbf{E}\varepsilon_1^2)\mathbf{K}^{-1}(\boldsymbol{\beta})\,\boldsymbol{\beta}. \tag{7.6.7}$$

For small γ the root $\boldsymbol{\theta}_\gamma^{LS}$ is unique in the entire $\mathbb{R}^q$, and it can be written down explicitly. With probability tending to 1 for small γ there exists a unique solution of the equation (7.6.3) (i.e., the LSE $\widehat{\boldsymbol{\beta}}_{n,LS}$), which can also be written down explicitly. One easily obtains from these explicit expressions that $\widehat{\boldsymbol{\beta}}_{n,LS}\xrightarrow{\mathbf{P}_\beta}\boldsymbol{\theta}_\gamma^{LS}$. Along with (7.6.7) and the definition (7.6.2), this implies

$$\mathrm{IF}(\boldsymbol{\theta}_\gamma^{LS},\mu_\xi) = -(\mathbf{E}\xi_1^2/\mathbf{E}\varepsilon_1^2)\mathbf{K}^{-1}(\boldsymbol{\beta})\,\boldsymbol{\beta}.$$

For $q=1$ this is the equality (5.6.9). Denoting by $\mathfrak{M}_i$, $i=1,2$, the class of distributions μ_ξ with finite ith moment, we see that

$$\mathrm{GES}(\mathfrak{M}_2,\theta_\gamma)=\infty,\ \boldsymbol{\beta}\neq 0.$$

7.6.2. Influence functional of the least absolute deviations estimator. Here we assume Condition 7.6(i) and

CONDITION 7.6(ii). *$G(x)$ has a continuous density $g(x)$ such that* $\sup_x g(x)<\infty$ *and $g(0)>0$, and $G(0)=1/2$.*

The LAD estimator $\widehat{\boldsymbol{\beta}}_{n,LD}$ is defined as a solution of the problem

$$\sum_{k=1}^n |y_k-\boldsymbol{\theta}^T\widetilde{\mathbf{y}}_{k-1}| \Longrightarrow \inf_{\boldsymbol{\theta}\in\mathbb{R}^q}. \tag{7.6.8}$$

The objective function in (7.6.8) is convex, hence this problem has a solution, which is a.s. unique since the observations have a continuous distribution. By the ergodic theorem

$$n^{-1}\sum_{k=1}^n |y_k-\boldsymbol{\theta}^T\widetilde{\mathbf{y}}_{k-1}| \xrightarrow{\mathbf{P}_\beta} \mathbf{E}_\beta|y_1-\boldsymbol{\theta}^T\widetilde{\mathbf{y}}_0|.$$

As a limit of convex functions, the function $\mathbf{E}_\beta|y_1-\boldsymbol{\theta}^T\widetilde{\mathbf{y}}_0|$ is convex. Since it is smooth, its minimum is attained at a point satisfying the equation

$$\frac{\partial}{\partial\boldsymbol{\theta}}\mathbf{E}_\beta|y_1-\boldsymbol{\theta}^T\widetilde{\mathbf{y}}_0| = -\mathbf{E}_\beta\widetilde{\mathbf{y}}_0\,\mathrm{sign}(y_1-\boldsymbol{\theta}^T\widetilde{\mathbf{y}}_0)=\mathbf{0}. \tag{7.6.9}$$

Put

$$\mathbf{\Lambda}_{LD}(\gamma, \boldsymbol{\theta}) = \mathbf{E}_\beta \widetilde{\mathbf{y}}_0 \operatorname{sign}(y_1 - \boldsymbol{\theta}^T \widetilde{\mathbf{y}}_0),$$

then $\mathbf{\Lambda}_{LD}(0, \boldsymbol{\beta}) = \mathbf{0}$. By the formula for total probability rewrite $\mathbf{\Lambda}_{LD}(\gamma, \boldsymbol{\theta})$ in the form

$$\begin{aligned} \mathbf{\Lambda}_{LD}(\gamma, \boldsymbol{\theta}) &= \mathbf{E}_\beta \widetilde{\mathbf{y}}_0 \Big[1 - 2G\Big((\boldsymbol{\theta} - \boldsymbol{\beta})^T \widetilde{\mathbf{u}}_0 + \sum_{j=1}^{q} \theta_j z_{1-j}^\gamma \xi_{1-j} + z_1^\gamma \xi_1 \Big) \Big] \\ &= \sum_{i=o}^{q} \mathbf{E}_\beta \Big[\widetilde{\mathbf{y}}_0 \Big(1 - 2G\Big((\boldsymbol{\theta} - \boldsymbol{\beta})^T \widetilde{\mathbf{u}}_0 + \sum_{j=1}^{q} \theta_j z_{1-j}^\gamma \xi_{1-j} + z_1^\gamma \xi_1 \Big) \Big) \Big| H_i \Big] \mathbf{P}(H_i), \end{aligned}$$

where H_i occurs when precisely i variables among $z_{-q+1}^\gamma, \ldots, z_1^\gamma$ are different from zero. The conditional expectations here do not depend on γ, while the probabilities of H_i are polynomials in γ. Hence one easily infers with the help of Conditions 7.6(i, ii) that:

- the matrix $\frac{\partial \mathbf{\Lambda}_{LD}(\gamma, \boldsymbol{\theta})}{\partial \boldsymbol{\theta}}$ exists and is continuous in $(\gamma, \boldsymbol{\theta})$ for $\boldsymbol{\theta} \in \mathbb{R}^q$ and $0 \le \gamma \le 1$, in particular,

$$\frac{\partial \mathbf{\Lambda}_{LD}(0, \boldsymbol{\beta})}{\partial \boldsymbol{\theta}} = -2g(0) \mathbf{E} \varepsilon_1^2 \mathbf{K}(\boldsymbol{\beta}); \tag{7.6.10}$$

- the derivative $\frac{\partial \mathbf{\Lambda}_{LD}(\gamma, \boldsymbol{\theta})}{\partial \gamma}$ exists and is continuous, in particular,

$$\mathbf{\Delta}_{LD}(\boldsymbol{\beta}) := \frac{\partial \mathbf{\Lambda}_{LD}(0, \boldsymbol{\beta})}{\partial \gamma} = \left(\Delta_1^{LD}(\boldsymbol{\beta}), \ldots, \Delta_q^{LD}(\boldsymbol{\beta}) \right)^T, \tag{7.6.11}$$

where $\Delta_j^{LD}(\boldsymbol{\beta}) = \mathbf{E}_\beta \xi_1 \big(1 - 2G(\beta_j \xi_1)\big)$, $j = 1, \ldots, q$.

Therefore, the equation (7.6.9) in a neighborhood of $(0, \boldsymbol{\beta})$ has a unique solution $\boldsymbol{\theta}_\gamma^{LD}$ with $\boldsymbol{\theta}_0^{LD} = \boldsymbol{\beta}$, which is differentiable w.r.t. γ and by (7.6.10), (7.6.11)

$$\left. \frac{d\boldsymbol{\theta}_\gamma^{LD}}{d\gamma} \right|_{\gamma=0} = \left(2g(0) \mathbf{E} \varepsilon_1^2 \mathbf{K}(\boldsymbol{\beta}) \right)^{-1} \mathbf{\Delta}_{LD}(\boldsymbol{\beta}). \tag{7.6.12}$$

Since the matrix $-\frac{\partial \mathbf{\Lambda}_{LD}(0, \boldsymbol{\beta})}{\partial \boldsymbol{\theta}}$ is positive definite, so is $-\frac{\partial \mathbf{\Lambda}_{LD}(\gamma, \boldsymbol{\theta})}{\partial \boldsymbol{\theta}}$ in some neighborhood of $(0, \boldsymbol{\beta})$. Therefore, at the point $\boldsymbol{\theta}_\gamma^{LD}$ the function $\mathbf{E}_\beta |y_1 - \boldsymbol{\theta}^T \widetilde{\mathbf{y}}_0|$ attains a strict minimum, which is unique since the function is convex. This fact suffices to establish the convergence of $\widehat{\boldsymbol{\beta}}_{n,LD}$ to $\boldsymbol{\theta}_\gamma^{LD}$ in probability following the lines of the proof of Theorem 2.2.1 in Bloomfield and Steiger [**7**].

By (7.6.12) and (7.6.2) we obtain

$$\mathrm{IF}(\boldsymbol{\theta}_\gamma^{LD}, \mu_\xi) = (2g(0) \mathbf{E} \varepsilon_1^2 \mathbf{K}(\boldsymbol{\beta}))^{-1} \mathbf{\Delta}_{LD}(\boldsymbol{\beta}). \tag{7.6.13}$$

For $q = 1$ this is the equation (5.6.12). If $\xi_i = \xi$ with a constant ξ then for $\boldsymbol{\beta} \neq 0$ $|\,\mathrm{IF}(\boldsymbol{\theta}_\gamma^{LD}, \mu_\xi)| \to \infty$ as $\xi \to \infty$. Therefore

$$\mathrm{GES}(\mathfrak{M}_2, \theta_\xi) = \infty \qquad \text{if} \quad \boldsymbol{\beta} \neq 0.$$

7.6.3. Influence functionals of weighted LAD estimators. We assume here Condition 7.6(ii) and the following conditions:

CONDITION 7.6(iii). $\mathbf{E}\varepsilon_1 = 0$, $\mathbf{E}|\xi_1| < \infty$.

CONDITION 7.6(iv). $\mathbf{E}_\beta|\varphi(\widetilde{\mathbf{y}}_1)| < \infty$, $\mathbf{E}_\beta|\varphi(\widetilde{\mathbf{y}}_1)\widetilde{\mathbf{y}}_1\widetilde{\mathbf{y}}_1^T| < \infty$ *for small* γ.

CONDITION 7.6(v). $-\infty < \boldsymbol{\lambda}_{LDW}(\boldsymbol{\beta}) := -2g(0)\mathbf{E}_\beta|\varphi(\widetilde{\mathbf{u}}_1)|\widetilde{\mathbf{u}}_1\widetilde{\mathbf{u}}_1^T < 0$.

Similarly to the LAD estimator one can show that any sequence of random variables $\widehat{\boldsymbol{\beta}}_{n,LDW}$ solving the problem

$$\sum_{k=1}^{n} |\varphi(\widetilde{\mathbf{y}}_{k-1})(y_k - \boldsymbol{\theta}^T\widetilde{\mathbf{y}}_{k-1})| \Longrightarrow \inf_{\boldsymbol{\theta}\in\mathbb{R}^q}, \tag{7.6.14}$$

converges in probability to $\boldsymbol{\theta}_\gamma^{LDW}$. The influence functional of $\widehat{\boldsymbol{\beta}}_{n,LDW}$ equals

$$\mathrm{IF}(\boldsymbol{\theta}_\gamma^{LDW}, \mu_\xi) = -\boldsymbol{\lambda}_{LDW}^{-1}(\boldsymbol{\beta})\boldsymbol{\Delta}_{LDW}(\boldsymbol{\beta}), \tag{7.6.15}$$

where $\boldsymbol{\Delta}_{LDW}(\boldsymbol{\beta}) = (\Delta_1^{LDW}(\boldsymbol{\beta}), \dots, \Delta_q^{LDW}(\boldsymbol{\beta}))^T$,

$$\Delta_j^{LDW}(\boldsymbol{\beta}) = \mathbf{E}_\beta\Big\{u_{1-j}|\varphi(\widetilde{\mathbf{u}}_0)|\big(1 - 2G(-\xi_1)\big) + \sum_{k=1}^{q}(u_{1-j} + \xi_1\delta_{kj})\,|\varphi(\widetilde{\mathbf{u}}_{k,\xi})|\big(1 - 2G(\beta_k\xi_1)\big)\Big\},$$

$\widetilde{\mathbf{u}}_{k,\xi} = (u_0, \dots, u_{2-k}, u_{1-k} + \xi_1, u_{-k}, \dots, u_{-q+1})^T$, and δ_{ij} is Kronecker's delta. If

$$\sup_{\mathbf{x}\in\mathbb{R}^q} |\varphi(\mathbf{x})\mathbf{x}| < \infty, \tag{7.6.16}$$

then

$$\mathrm{GES}(\mathfrak{M}_\xi, \theta_\gamma^{LDW}) < \infty.$$

Here $\mathfrak{M}_\xi$ stands for the class of distributions μ_ξ satisfying the conditions $\mathbf{E}|\xi_1| < \infty$ and 7.6(iv). Note that in 5.6.3 we dealt with the equation (5.6.14), which is equivalent to the extremal problem (7.6.14) for $q = 1$. This enabled us to obtain the influence functional $\widehat{\beta}_{n,LDW}$ under Conditions 5.6(v, vi), weaker than those imposed here.

Of course, for $q = 1$ (7.6.15) implies the relations in 5.6.3, and (7.6.15) coincides with (7.6.13) for $\varphi(\widetilde{\mathbf{y}}_1) = 1$ and an arbitrary q.

7.6.4. Influence functional of the sign estimator $\widetilde{\boldsymbol{\beta}}_{n,S}$. Consider now the sign estimator $\widetilde{\boldsymbol{\beta}}_{n,S}$, which was defined in the no-contamination case by (7.5.8). For contaminated observations (7.6.1), $\widetilde{\boldsymbol{\beta}}_{n,S}$ is defined as the solution of the problem

$$|\mathbf{l}_n^S(\widehat{\boldsymbol{\beta}}_n, \boldsymbol{\theta})| \Longrightarrow \inf_{\theta\in Q(\widehat{\boldsymbol{\beta}}_n)},$$

where $Q(\widehat{\boldsymbol{\beta}}_n) = \{\boldsymbol{\theta}\colon |\sqrt{n}(\boldsymbol{\theta} - \widehat{\boldsymbol{\beta}}_n)| \le \log n\}$ and the preliminary estimator $\widehat{\boldsymbol{\beta}}_n$ as well as $\mathbf{l}_n^S(\widehat{\boldsymbol{\beta}}_n, \boldsymbol{\theta})$ are based on $\mathbf{Y}_n$.

CONDITION 7.6(vi). *$\widehat{\boldsymbol\beta}_n \xrightarrow{\mathbf{P}_{\boldsymbol\beta}} \boldsymbol\theta_\gamma$, and its influence functional* $\mathrm{IF}(\boldsymbol\theta_\gamma, \mu_\xi)$ *exists.*

The definitions of $\widetilde{\boldsymbol\beta}_{n,S}$ and Condition 7.6(vi) directly imply that

$$\widetilde{\boldsymbol\beta}_{n,S} \xrightarrow{\mathbf{P}_{\boldsymbol\beta}} \widetilde{\boldsymbol\theta}_\gamma^S = \boldsymbol\theta_\gamma,$$

and therefore $\mathrm{IF}(\widetilde{\boldsymbol\theta}_\gamma^S, \mu_\xi) = \mathrm{IF}(\boldsymbol\theta_\gamma, \mu_\xi)$. Thus the sign estimator $\widetilde{\boldsymbol\beta}_{n,S}$ inherits the influence functional of the preliminary estimator $\widehat{\boldsymbol\beta}_n$. The latter can be taken to be a weighted LAD estimator satisfying conditions (7.6.16), then $\widetilde{\boldsymbol\beta}_{n,S}$ will have a finite gross error sensitivity.

7.6.5. Influence functional of the sign estimator $\boldsymbol\beta^*_{n,S}$. By (7.5.14) this estimator for a contaminated sample is defined by

$$\boldsymbol\beta^*_{n,S} = \widehat{\boldsymbol\beta}_n - (n\widehat{\mathbf{e}}_n)^{-1}\mathbf{l}_n^S(\widehat{\boldsymbol\beta}_n),$$

where the preliminary estimator $\widehat{\boldsymbol\beta}_n$ as well as $\widehat{\mathbf{e}}_n$ and $\mathbf{l}_n^S$ are based on $\mathbf{Y}_n$.

We assume here Conditions 7.6(ii, iii, vi) and the following

CONDITION 7.6(vii). *$\widehat{\mathbf{e}}_n \xrightarrow{\mathbf{P}_{\boldsymbol\beta}} \boldsymbol\theta_\gamma^e$, $\boldsymbol\theta_0^e = \boldsymbol\lambda_S(\boldsymbol\beta)$ and the function $\boldsymbol\theta_\gamma^e$ is continuous at $\gamma = 0$.*

Recall that $\boldsymbol\lambda_S(\boldsymbol\beta)$ is defined by (7.3.1).

If $\sqrt{n}(\widehat{\boldsymbol\beta}_n - \boldsymbol\theta_\gamma) = O_p(1)$, then Condition 7.6(vii) is met by

$$\widehat{\mathbf{e}}_n = -2\widehat{g}_n\widehat{m}_n\mathbf{K}(\widehat{\boldsymbol\beta}_n),$$

where

$$\widehat{g}_n = (nh_n)^{-1}\sum_{i=1}^n \varphi(h_n^{-1}(y_i - \widehat{\boldsymbol\beta}_n^T\widetilde{\mathbf{y}}_{i-1}))$$

is a Parzen–Rosenblatt type estimator with Gaussian kernel $\varphi(x)$, $h_n = n^{-\alpha}$, $0 < \alpha < 1/4$, and

$$\widehat{m}_n = n^{-1}\sum_{i=1}^n \left|y_i - \widehat{\boldsymbol\beta}_n^T\widetilde{\mathbf{y}}_{i-1}\right|.$$

For small γ under Conditions 7.6(ii, iii, vi, vii)

$$\boldsymbol\beta^*_{n,S} \xrightarrow{\mathbf{P}_{\boldsymbol\beta}} \boldsymbol\theta_\gamma^{*S} := \boldsymbol\theta_\gamma - (\boldsymbol\theta_\gamma^e)^{-1}\boldsymbol\Lambda_S(\gamma, \boldsymbol\theta_\gamma), \tag{7.6.17}$$

where

$$\boldsymbol\Lambda_S(\gamma, \boldsymbol\theta) = \left(\Lambda_1^S(\gamma, \boldsymbol\theta), \dots, \Lambda_q^S(\gamma, \boldsymbol\theta)\right)^T$$

with

$$\Lambda_q^S(\gamma, \boldsymbol\theta) = \sum_{t=0}^\infty \delta_t(\boldsymbol\theta)\mathbf{E}_{\boldsymbol\beta}\operatorname{sign}\left[(y_0 - \boldsymbol\theta^T\widetilde{\mathbf{y}}_{-1})(y_{t+j} - \boldsymbol\theta^T\widetilde{\mathbf{y}}_{t+j-1})\right] \tag{7.6.18}$$

for $j = 1, \dots, q$.

The function $\mathbf{\Lambda}_S(\gamma, \boldsymbol{\theta})$ is well defined for any $0 \leq \gamma \leq 1$ and for $\boldsymbol{\theta}$ at least in some neighborhood $|\boldsymbol{\theta} - \boldsymbol{\beta}| \leq \varepsilon$. For small enough ε the quantities $\delta_t(\boldsymbol{\theta})$ decrease at an exponential rate quite similarly to $\delta_t(\boldsymbol{\beta})$ (see (7.1.4) and (7.3.3)). For an arbitrary q relation (7.6.17) is proved along the same lines as the analogous relation (6.7.19) for $q = 1$. A detailed proof of (6.7.19) was given in 6.7.3, which allows us to omit the justification of (7.6.17).

It follows from (7.6.17) and (7.6.18) that

$$\mathbf{\Lambda}_S(0, \boldsymbol{\beta}) = \mathbf{0}, \qquad \boldsymbol{\theta}_0^{*S} = \boldsymbol{\beta}. \tag{7.6.19}$$

Similarly to the proof of Lemma 6.7.1, one shows that under Conditions 7.6(ii, iii) the partial derivatives of $\mathbf{\Lambda}_S(\gamma, \boldsymbol{\theta})$ exist and are continuous in $(\gamma, \boldsymbol{\theta})$ for $|\boldsymbol{\theta} - \boldsymbol{\beta}| \leq \varepsilon$, $0 \leq \gamma \leq 1$. In particular,

$$\frac{\partial \mathbf{\Lambda}_S(0, \boldsymbol{\beta})}{\partial \boldsymbol{\theta}} = \boldsymbol{\lambda}_S(\boldsymbol{\beta}), \tag{7.6.20}$$

$$\frac{\partial \mathbf{\Lambda}_S(0, \boldsymbol{\beta})}{\partial \gamma} = \mathbf{\Delta}_S(\boldsymbol{\beta}), \tag{7.6.21}$$

where $\mathbf{\Delta}_S(\boldsymbol{\beta}) = \left(\Delta_1^S(\boldsymbol{\beta}), \dots, \Delta_q^S(\boldsymbol{\beta})\right)^T$ and, with $\beta_0 = 1$

$$\Delta_j^S(\boldsymbol{\beta}) = \sum_{t=0}^{q-j} \delta_t(\boldsymbol{\beta}) \sum_{k=0}^{q-j-t} \mathbf{E}_{\boldsymbol{\beta}}\big(1 - 2G(\beta_k \xi_1)\big)\big(1 - 2G(\beta_{k+j+t}\xi_1)\big), \quad j = 1, \dots, q.$$

For $q = 1$, (7.6.20) and (7.6.21) coincide with (6.7.6) and (6.7.7), respectively.

Since $\mathbf{\Lambda}(\gamma, \boldsymbol{\theta})$ is continuously differentiable, it follows from (7.6.17) and Conditions 7.6(vi, vii) that

$$\begin{aligned} \boldsymbol{\theta}_\gamma^{*S} &= \boldsymbol{\theta}_\gamma - \left[\boldsymbol{\lambda}_S(\boldsymbol{\beta}) + o(1)\right]^{-1} \\ &\quad \times \left[\mathbf{\Lambda}_S(0, \boldsymbol{\beta}) + \frac{\partial \mathbf{\Lambda}_S(0, \boldsymbol{\beta})}{\partial \gamma}\gamma + \frac{\partial \mathbf{\Lambda}_S(0, \boldsymbol{\beta})}{\partial \boldsymbol{\theta}}(\boldsymbol{\theta}_\gamma - \boldsymbol{\beta}) + o(\gamma)\right]. \end{aligned}$$

Along with (7.6.19)–(7.6.21), this implies

$$\mathrm{IF}(\boldsymbol{\theta}_\gamma^{*S}, \mu_\xi) := \lim_{\gamma \to +0} \frac{\boldsymbol{\theta}_\gamma^{*S} - \boldsymbol{\beta}}{\gamma} = \boldsymbol{\lambda}_S^{-1}(\boldsymbol{\beta}) \mathbf{\Delta}_S(\boldsymbol{\beta}). \tag{7.6.22}$$

For $q = 1$ (7.6.22) coincides with the influence functional of the nonparametric sign estimator defined by (6.7.13). It is easily seen that $\mathrm{GES}(\mathfrak{M}_1, \theta_\gamma^{*S}) < \infty$.

7.7. Empirical distribution function of residuals and related empirical processes

Consider the autoregression model (7.1.1) assuming that Conditions 6.5(ii, iii) are satisfied. For convenience, we recall them:

$$\mathbf{E}\varepsilon_1 = 0, \qquad \mathbf{E}|\varepsilon_1|^{1+\Delta} < \infty \quad \text{for some} \quad \Delta > 0;$$

$G(x)$ has a density $g(x)$ such that $\sup_x g(x) < \infty$ and $g(x)$ satisfies the Lipschitz condition.

Let $u_{1-q}, \dots, u_n$ be observations from the strictly stationary solution (7.1.5) of the equation (7.1.1). In this section we construct Kolmogorov and omega-square type tests for hypotheses about the distribution function $G(x)$ based on these observations. We consider the simple hypothesis

$$H_0\colon G(x) = G_0(x) \tag{7.7.1}$$

for a completely specified $G_0(x)$ as well as the composite hypothesis that $G(x)$ belongs to a parametric family,

$$H_G\colon G(x) \in \{G(x, \lambda), \lambda = (\lambda_1, \dots, \lambda_r)^T \in \Lambda\}, \tag{7.7.2}$$

where $\Lambda \subset \mathbb{R}^r$.

The tests will be based on the empirical distribution function $\widehat{G}_n(x)$ of estimated variables $\varepsilon_1, \dots, \varepsilon_n$. Our main task in this section is to study the asymptotic properties of $\widehat{G}_n(x)$.

Let, for $\boldsymbol{\theta} = (\theta_1, \dots, \theta_q)^T \in \mathbb{R}^q$,

$$\varepsilon_k(\boldsymbol{\theta}) = u_k - \theta_1 u_{k-1} - \cdots - \theta_q u_{k-q}, \qquad k = 1, \dots, n,$$
$$G_n(x, \boldsymbol{\theta}) = n^{-1} \sum_{k=1}^{n} I(\varepsilon_k(\boldsymbol{\theta}) < x).$$

The random function $G_n(x, \boldsymbol{\theta})$ is the empirical distribution function of (dependent unless $\boldsymbol{\theta} = \boldsymbol{\beta}$) random variables $\varepsilon_1(\boldsymbol{\theta}), \dots, \varepsilon_n(\boldsymbol{\theta})$. Since $\varepsilon_k(\boldsymbol{\beta}) = \varepsilon_k$, $G_n(x, \boldsymbol{\beta})$ is the true empirical distribution fuction of $\varepsilon_1, \dots, \varepsilon_n$. However $\boldsymbol{\beta}$ is unknown, and so is $G_n(x, \boldsymbol{\beta})$. We will show that $G_n(x, \boldsymbol{\beta})$ are close to $G_n(x, \boldsymbol{\theta})$ uniformly in $x \in \mathbb{R}$ for $|\boldsymbol{\theta} - \boldsymbol{\beta}| \le \Theta n^{-1/2}$, $0 < \Theta < \infty$. Hence, we will derive properties of $\widehat{G}_n(x) := G_n(x, \widehat{\boldsymbol{\beta}}_n)$, where $\widehat{\boldsymbol{\beta}}_n$ is a $\sqrt{n}$-consistent estimator of $\boldsymbol{\beta}$, from well-known properties of $G_n(x, \boldsymbol{\beta})$.

The main result of this section is the following

THEOREM 7.7.1. *Assume that Conditions* 6.5(ii, iii) *are satisfied. Then for any* $0 < \Theta < \infty$

$$\sup_x \sup_{|\boldsymbol{\theta}| \le \Theta} \sqrt{n} \big| G_n(x, \boldsymbol{\beta} + n^{-1/2}\boldsymbol{\theta}) - G_n(x, \boldsymbol{\beta}) \big| = o_p(1).$$

The proof is given in §7.8. For the case of finite variance $\mathbf{E}\varepsilon_1^2 < \infty$ this theorem has been published in Boldin [**8**].

Consider some applications of Theorem 7.7.1. Let, as in §5.7, $D[0,1]$ denote the Skorokhod metric space of functions on [0,1] without discontinuities of the second kind, and let $w(t)$, $t \in [0,1]$, be a Brownian bridge. Let $G^{-1}(t)$ denote the inverse function to $G(x)$,

$$G^{-1}(t) = \sup\{x\colon G(x) \le t\}.$$

For brevity, write $G_n(x) = G_n(x, \boldsymbol{\beta})$.

COROLLARY 7.7.1. *If Conditions* 6.5(ii, iii) *are satisfied and* $\sqrt{n}(\widehat{\boldsymbol{\beta}}_n - \boldsymbol{\beta}) = O_p(1)$, *then*

$$\sup_x \sqrt{n} |\widehat{G}_n(x) - G_n(x)| = o_p(1). \tag{7.7.3}$$

PROOF. By assumption $\sqrt{n}(\widehat{\boldsymbol{\beta}}_n - \boldsymbol{\beta}) = O_p(1)$, given $\delta > 0$ one can find (sufficiently large) $0 < \Theta < \infty$ such that $\mathbf{P}_{\boldsymbol{\beta}}\{|\sqrt{n}(\widehat{\boldsymbol{\beta}}_n - \boldsymbol{\beta}| > \Theta\} < \delta/2$ for all n. Then by Theorem 7.7.1 for any $\varepsilon > 0$, $\delta > 0$

$$\begin{aligned}
&\mathbf{P}_{\boldsymbol{\beta}}\{\sup_x \sqrt{n}|\widehat{G}_n(x) - G_n(x)| > \varepsilon\} \\
&\quad\le \mathbf{P}_{\boldsymbol{\beta}}\{\sup_x \sqrt{n}|\widehat{G}_n(x) - G_n(x)| > \varepsilon, |\sqrt{n}(\widehat{\boldsymbol{\beta}}_n - \boldsymbol{\beta})| \le \Theta\} \\
&\qquad\qquad + \mathbf{P}_{\boldsymbol{\beta}}\{|\sqrt{n}(\widehat{\boldsymbol{\beta}}_n - \boldsymbol{\beta})| > \Theta\} \\
&\quad\le \mathbf{P}_{\boldsymbol{\beta}}\{\sup_x \sup_{|\boldsymbol{\theta}|\le\Theta} \sqrt{n}|G_n(x, \boldsymbol{\beta} + n^{-1/2}\boldsymbol{\theta}) - G_n(x, \boldsymbol{\beta})| > \varepsilon\} + \delta/2 < \delta
\end{aligned}$$

for all sufficiently large n. □

Corollary 7.7.1 enables us to construct asymptotic tests for hypotheses about the distribution of the random errors in the autoregression model, similar to the tests in the case of an i.i.d. sample.

We begin with hypothesis H_0 as in (7.7.1). Define the random processes

$$\begin{aligned}
\widehat{w}_n(t) &= \sqrt{n}[\widehat{G}_n(G^{-1}(t)) - t], \\
w_n(t) &= \sqrt{n}[G_n(G^{-1}(t)) - t], \ t \in [0, 1].
\end{aligned}$$

It is well known that $w_n(t)$ weakly converges to $w(t)$ in $D[0, 1]$. By (7.7.3)

$$\sup_t \sqrt{n}|\widehat{w}_n(t) - w_n(t)| = o_p(1). \tag{7.7.4}$$

Now (7.7.4) and the convergence of $w_n(t)$ to $w(t)$ imply

COROLLARY 7.7.2. *Let Conditions* 6.5(ii, iii) *be satisfied and let* $\sqrt{n}(\widehat{\boldsymbol{\beta}}_n - \boldsymbol{\beta}) = O_p(1)$. *Then* $\widehat{w}_n(t)$ *weakly converges to* $w(t)$ *in* $D[0, 1]$.

Put

$$\widehat{D}_n = \sup_x |\sqrt{n}[\widehat{G}_n(x) - G_0(x)]|$$

and

$$\widehat{\omega}_n^2 = \int_{-\infty}^{\infty} \{\sqrt{n}[\widehat{G}_n(x) - G_0(x)]\}^2 dG_0(x).$$

By Corollary 7.7.2 under the hypothesis H_0 as in (7.7.1)

$$\begin{aligned}
\mathbf{P}_{\boldsymbol{\beta}}\{\widehat{D}_n < \lambda\} &= \mathbf{P}_{\boldsymbol{\beta}}\{\sup_t |\widehat{w}_n(t)| < \lambda\} \to \mathbf{P}\{\sup_t |w(t)| < \lambda\} = \boldsymbol{K}(\lambda), \\
\mathbf{P}_{\boldsymbol{\beta}}\{\widehat{\omega}_n^2 < \lambda\} &= \mathbf{P}_{\boldsymbol{\beta}}\Big\{\int_0^1 \widehat{w}_n^2(t)dt < \lambda\Big\} \to \mathbf{P}\Big\{\int_0^1 w^2(t)dt < \lambda\Big\} = \boldsymbol{S}(\lambda).
\end{aligned}$$

Here $\boldsymbol{K}(\lambda)$ and $\boldsymbol{S}(\lambda)$ are well-known limiting distribution functions of Kolmogorov and Cramér–von Mises statistics, which do not depend on G_0. Their tables can be found in Bol′shev and Smirnov [**17**]. Thus the statistics $\widehat{D}_n$ and $\widehat{\omega}_n^2$ can be used for hypothesis testing about $G(x)$.

Moreover, let $k_{1-\alpha}$ denote the $(1-\alpha)$-quantile of $\boldsymbol{K}(\lambda)$. Then by Corollary 7.7.2

$$\mathbf{P}_{\beta}\{\sup_x \sqrt{n}|\widehat{G}_n(x) - G(x)| < k_{1-\alpha}\} \to 1 - \alpha.$$

Therefore

$$\widehat{G}_n(x) - n^{-1/2}k_{1-\alpha} < G(x) < \widehat{G}_n(x) + n^{-1/2}k_{1-\alpha}$$

is a uniform in x confidence "belt" for $G(x)$ of asymptotic level $1 - \alpha$.

Consider now the testing problem of the parametric hypothesis H_G as in (7.7.2). For i.i.d. observations this problem has been the subject of numerous studies, so that it is difficult to mention all contributions. Without attempting this we restrict ourselves to some selected results.

Denote

$$f(x, \lambda) = \left(\frac{\partial G(x,\lambda)}{\partial \lambda_1}, \dots, \frac{\partial G(x,\lambda)}{\partial \lambda_r}\right)^T.$$

Suppose that the hypothesis H_G holds, and let λ_0 be the true parameter value, so that $G(x) = G(x, \lambda_0)$. Let $\widehat{\lambda}_n$ be an estimate of λ_0. Denote by $g(x, \lambda)$ the density function of $G(x, \lambda)$.

Suppose the following conditions to hold under hypothesis H_G as in (7.7.2).

CONDITION 7.7(i). *For any $\lambda \in \Lambda$, $\sup_x g(x, \lambda) < \infty$, $g(x, \lambda)$ as a function of x satisfies the Lipschitz condition,*

$$\int_{\infty}^{\infty} xg(x, \lambda)dx = 0, \qquad \int_{-\infty}^{\infty} |x|^{1+\Delta} g(x, \lambda)dx < \infty \quad \text{for some} \quad \Delta > 0.$$

CONDITION 7.7(ii). *The function $f(G^{-1}(t, \lambda), \lambda)$ is continuous in (t, λ) for $t \in [0, 1]$, $\lambda \in \Lambda$.*

CONDITION 7.7(iii). *The estimator $\widehat{\lambda}_n$ is representable in the form*

$$\sqrt{n}(\widehat{\lambda}_n - \lambda_0) = n^{-1/2} \sum_{k=1}^{n} l(\varepsilon_k, \lambda_0) + o_p(1), \tag{7.7.5}$$

where

$$\int_{-\infty}^{\infty} l(x, \lambda)g(x, \lambda)dx = 0, \quad L(\lambda) := \int_{-\infty}^{\infty} l(x, \lambda)l^T(x, \lambda)g(x, \lambda)dx < \infty.$$

A typical structure of an estimator fulfilling Condition 7.7(iii) is as follows. Let $\widetilde{\lambda}_n = \widetilde{\lambda}_n(\varepsilon_1, \dots, \varepsilon_n)$ be an estimator, which would be used to estimate λ_0 from i.i.d. observations $\varepsilon_1, \dots, \varepsilon_n$. Usually a "good" estimator can be represented as

$$\sqrt{n}(\widetilde{\lambda}_n - \lambda_0) = n^{-1/2} \sum_{k=1}^{n} l(\varepsilon_k, \lambda_0) + o_p(1), \tag{7.7.6}$$

where the terms in (7.7.6) have zero mean and a finite covariance matrix. For instance, if $\widetilde{\lambda}_n$ is the maximum likelihood estimator, then under usual regularity conditions (7.7.6) holds with

$$l(x, \lambda) = I^{-1}(\lambda)\frac{\partial \log g(x, \lambda)}{\partial \lambda}, \tag{7.7.7}$$

where $I(\lambda)$ is the Fisher information about λ contained in $\varepsilon_1, \dots, \varepsilon_n$. It is natural to take for $\widehat{\lambda}_n$ an estimator of the form $\widehat{\lambda}_n := \widetilde{\lambda}_n(\varepsilon_1(\widehat{\boldsymbol{\beta}}_n), \dots, \varepsilon_n(\widehat{\boldsymbol{\beta}}_n))$, which handles the residuals $\varepsilon_1(\widehat{\boldsymbol{\beta}}_n), \dots, \varepsilon_n(\widehat{\boldsymbol{\beta}}_n)$ as if they were the unobservable errors $\varepsilon_1, \dots, \varepsilon_n$ themselves. If

$$\sqrt{n}\Big(\widetilde{\lambda}_n(\varepsilon_1, \dots, \varepsilon_n) - \widetilde{\lambda}_n\big(\varepsilon_1(\widehat{\boldsymbol{\beta}}_n), \dots, \varepsilon_n(\widehat{\boldsymbol{\beta}}_n)\big)\Big) = o_p(1), \tag{7.7.8}$$

then (7.7.6) and (7.7.8) imply Condition 7.7(iii). Later on we will give an example of such an estimator.

Denote for brevity $L = L(\lambda_0)$, $f(t) = f(G^{-1}(t), \lambda_0)$, $l(t) = l(G^{-1}(t, \lambda_0), \lambda_0)$, $e(t) = \int_0^t l(s)ds$, and define the random process

$$\widehat{u}_n[G(x, \widehat{\lambda}_n)] = \sqrt{n}[\widehat{G}_n(x) - G(x, \widehat{\lambda}_n)].$$

Making the substitution $G(x, \widehat{\lambda}_n) = t$ we obtain the process $\widehat{u}_n(t)$ on [0, 1].

COROLLARY 7.7.3. *Let Conditions* 7.7(i–iii) *be satisfied and* $\sqrt{n}(\widehat{\boldsymbol{\beta}}_n - \boldsymbol{\beta}) = O_p(1)$. *Then under the hypothesis* H_G, $\widehat{u}_n(t)$ *weakly converges in* $D[0, 1]$ *to a Gaussian process* $u(t)$ *with zero mean and covariance function*

$$\min(t, s) - ts - f^T(t)e(s) - f^T(s)e(t) + f^T(t)Lf(s).$$

PROOF. By (7.7.3)

$$\sup_x \Big| \widehat{u}_n[G(x, \widehat{\lambda}_n)] - \sqrt{n}[G_n(x) - G(x, \widehat{\lambda}_n)] \Big| = o_p(1). \tag{7.7.9}$$

Define

$$u_n[G(x, \widehat{\lambda}_n)] = \sqrt{n}[G_n(x) - G(x, \widehat{\lambda}_n)].$$

It is well known that $u_n(t)$ weakly converges in $D[0, 1]$ to $u(t)$ (see Durbin [**24**], in the particular case (7.7.7) this convergence was established by Tyurin [**84**]). Together with (7.7.9) this fact implies the corollary. □

As a particular case, consider the normality hypothesis

$$H_\Phi \colon G(x) \in \{\Phi(x/\sigma), 0 < \sigma < \infty\}$$

(which is equivalent to the property that the process $\{u_i\}$ in (7.1.5) is Gaussian).

Estimate the variance σ_0^2 by

$$\widehat{s}_n^2 = n^{-1} \sum_{k=1}^{n} \varepsilon_k^2(\widehat{\boldsymbol{\beta}}_n).$$

It is easy to verify that under H_Φ, (7.7.6) and (7.7.8) hold with $l(x, \sigma) = x^2 - \sigma^2$, hence Condition 7.7(iii) is fulfilled. Obviously, Conditions 7.7(i–ii) are also satisfied. Put

$$\widehat{u}_{1n}(t) = \sqrt{n}[\widehat{G}_n(\widehat{s}_n \Phi^{-1}(t)) - t].$$

By a simple calculation we infer from Corollary 7.7.3 that $\widehat{u}_{1n}(t)$ under H_Φ weakly converges in $D[0,1]$ to a Gaussian process $u_1(t)$ with zero mean and covariance function

$$\min(t,s) - ts - \tfrac{1}{2}\Phi^{-1}(t)\varphi(\Phi^{-1}(t))\Phi^{-1}(s)\varphi(\Phi^{-1}(s)),$$

where $\varphi(x)$ is the standard normal density function. The distribution of $\int_0^1 u_1^2(t)dt$ has been tabulated (see, for example, Martynov [**67, 68**] or Pearson and Hartley [**70**] for tables). Hence for testing H_Φ one can apply the ω^2-type test with test statistic

$$\widehat{\omega}_{1n}^2 = \int_{-\infty}^{\infty} \widehat{u}_{1n}^2[\Phi(x/\widehat{s}_n)]d\Phi(x/\widehat{s}_n) = \int_0^1 \widehat{u}_{1n}^2(t)dt.$$

It is worth making the following comment concerning this example. It turns out that for testing H_Φ one can construct a more powerful test on a natural class of alternatives. To do this, one should use the integral test statistic involving the process

$$\widehat{u}_{2n}(t) = \sqrt{n}[\widehat{G}_n(\widetilde{\varepsilon}_n + \widehat{s}_n\Phi^{-1}(t)) - t],$$

where

$$\widetilde{\varepsilon}_n = n^{-1}\sum_{k=1}^{n} \varepsilon_k(\widehat{\boldsymbol{\beta}}_n),$$

i.e., to proceed as if the mean value of ε_i were also unknown. We refer the reader to the paper of Tyulyagin [**83**], where a similar phenomenon is discussed in the case of i.i.d. sample. By Corollary 7.7.1 our test statistics behave asymptotically as in the i.i.d. case. On the other hand, in practical applications it is natural to assume the mean value to be unknown and treat it as one of the parameters to be estimated. The application of the Kolmogorov and omega-square tests in this setting is described by Boldin [**9**].

In general, the process $u(t)$ depends on the hypothesized family and (except for location-scale families) on the parameter λ_0. Hence so are the asymptotic distributions of Kolmogorov or omega-square type functionals of $\widehat{u}_n(t)$, which could be used as test statistics. For example, the above process $u_1(t)$ and the asymptotic distribution of $\widehat{\omega}_{1n}^2$ are specific for the family of normal distributions with a scale parameter. To circumvent this difficulty, one can transform the process $u(t)$ (and, similarly, $\widehat{u}_n(t)$) into a standard process with distribution independent of the family or parameter value. Such transformations were, in particular, proposed by Tyurin [**84**] and Khmaladze [**50, 51**]. Then test statistics based on the transformed process $\widehat{u}_n(t)$ will be asymptotically distribution free.

Sometimes another approach may be useful. Namely, the limiting distributions of Kolmogorov–Smirnov test statistics based on $\varepsilon_1, \dots, \varepsilon_n$ under the hypothesis H_G have tail probabilities fulfilling the following asymptotic relation

$$\lim_{n\to\infty} \mathbf{P}\{\sup_x |u_n(G(x,\widetilde{\lambda}_n))| > z\} = ae^{-bz^2}(1+o(1)) \qquad \text{as} \quad z\to\infty$$

with some constants $a > 0$, $b > 0$.

This relation provides approximate quantiles of the statistic $\sup_x |u_n(G(x,\widetilde{\lambda}_n))|$ for large n and small significance levels. Results of this kind can be found, for example, in Tyurin [**86**] and Piterbarg [**71**]. The statistic $\sup_x |\widehat{u}_n(G(x, \widehat{\lambda}_n))|$ has the same asymptotic quantiles by Corollary 7.7.3. For example, under H_Φ

$$\lim_{n\to\infty} \mathbf{P}_\beta\{ \sup_x |\widehat{u}_{1n}[\Phi(x, \widehat{s}_n)]| > z\} = \frac{2\sqrt{6}}{3} \exp(-2z^2)(1 + o(1)) \qquad \text{as} \quad z \to \infty.$$

Note that results similar to Theorem 7.7.1 and Corollaries 7.7.1–7.7.3 have been obtained also for models of observations different from the ones treated in this book. We can point out, in particular, the papers by Boldin [**9**], where the linear regression model with autocorrelated errors is considered; Boldin [**10**] dealing with the moving average model; Kreiss [**59**] dealing with autoregressive–moving average model and linear processes; Koul and Levental [**54**] who treat the first order autoregression with $|\beta| > 1$.

7.8. Proof of Theorem 7.7.1

We present the proof for $q = 1$. In the general case the proof is more cumbersome, though follows essentially the same lines. The proof to be given here is a modification of the one in Boldin [**8**].

So, let henceforth β and θ be real valued. Denote

$$\Delta_k(x) = \begin{cases} 1, & \varepsilon_k < x, \\ 0, & \varepsilon_k \geq x. \end{cases}$$

Then $G_n(x, \beta) = n^{-1} \sum_{k=1}^n \Delta_k(x)$, and since $\varepsilon_k(\theta) = \varepsilon_k - (\theta - \beta)u_{k-1}$, it can be written as

$$G_n(x, \theta) = n^{-1} \sum_{K=1}^n \Delta_k\big(x + (\theta - \beta)u_{k-1}\big).$$

Therefore

$$\sqrt{n}\big[G_n(x, \beta + n^{-1/2}\theta) - G_n(x, \beta)\big] = z_{1n}(x, \theta) + z_{2n}(x, \theta),$$

where

$$z_{1n}(x, \theta) = n^{-1/2} \sum_{k=1}^n \big[\Delta_k(x + n^{-1/2}\theta u_{k-1}) - G(x + n^{-1/2}\theta u_{k-1}) - \Delta_k(x) + G(x)\big],$$

$$z_{2n}(x, \theta) = n^{-1/2} \sum_{k=1}^n \big[G(x + n^{-1/2}\theta u_{k-1}) - G(x)\big].$$

The density function $g(x)$ is bounded and satisfies the Lipschitz condition, hence it satisfies Hölder's condition of any order $0 < \delta \leq 1$ with some constant L. In what follows we let $\delta = \min(\Delta, 1)$, where Δ is the constant appearing in Condition 6.5(ii).

On applying the Taylor formula to the terms in $z_{2n}(x, \theta)$ we obtain

$$z_{2n}(x, \theta) = \theta g(x) n^{-1} \sum_{k=1}^n u_{k-1} + \varepsilon_n(x, \theta),$$

where

$$\sup_x \sup_{|\theta|\le\Theta} |\varepsilon_n(x,\theta)| \le \Theta^{1+\delta} L n^{-(1+\delta/2)} \sum_{k=1}^n |u_{k-1}|^{1+\delta} = o_p(1).$$

By (7.1.1) and the law of large numbers

$$n^{-1}\sum_{k=1}^n u_{k-1} = (1-\beta)^{-1} n^{-1} \sum_{k=1}^n \varepsilon_k + o_p(1) = o_p(1).$$

Therefore

$$\sup_x \sup_{|\theta|\le\Theta} |z_{2n}(x,\theta)| = o_p(1). \tag{7.8.1}$$

Next we show that

$$\sup_x \sup_{|\theta|\le\Theta} |z_{1n}(x,\theta)| = o_p(1), \tag{7.8.2}$$

which together with (7.8.1) implies the theorem. For the proof we reduce the supremum in (7.8.2) to the supremum over finite sets of x and θ. To that end, divide the interval $[-\Theta n^{-1/2}, \Theta n^{-1/2}]$ into 3^{m_n} subintervals by points $\eta_{sn} = -\Theta n^{-1/2} + 2\Theta n^{-1/2} 3^{-m_n} s$, $s = 0, 1, \dots, 3^{m_n}$, where m_n is such that $3^{m_n} \sim \log n$. Moreover, divide the real line into N_n intervals by points $-\infty = x_0 < x_1 < \cdots < x_{N_n-1} < x_{N_n} = +\infty$, where $G(x_i) = iN_n^{-1}$, and let $N_n \sim \sqrt{n}\log n$. Define the variables

$$\begin{aligned} \widetilde{u}_{ks} &= u_k\bigl[1 - 2\Theta n^{-1/2} 3^{-m_n} \eta_{sn}^{-1} I(u_k \ge 0)\bigr], \\ \widehat{u}_{ks} &= u_k\bigl[1 - 2\Theta n^{-1/2} 3^{-m_n} \eta_{sn}^{-1} I(u_k \le 0)\bigr]. \end{aligned} \tag{7.8.3}$$

Then

$$\max\bigl(|\widetilde{u}_{ks}|, |\widehat{u}_{ks}|\bigr) \le 3|u_k|. \tag{7.8.4}$$

For η_{jn} such that

$$0 \le \eta_{jn} - \theta n^{-1/2} \le 2\Theta n^{-1/2} 3^{-m_n},$$

the following inequalities hold:

$$\eta_{jn}\widetilde{u}_{k-1,j} \le \theta n^{-1/2} u_{k-1} \le \eta_{jn}\widehat{u}_{k-1,j}, \qquad k = 1, \dots, n.$$

Hence for x_r and x_{r+1} such that $x_r \le x \le x_{r+1}$,

$$x_r + \eta_{jn}\widetilde{u}_{k-1,j} \le x + n^{-1/2}\theta u_{k-1} \le x_{r+1} + \eta_{jn}\widehat{u}_{k-1,j}, \qquad k = 1, \dots, n. \tag{7.8.5}$$

Put $\widehat{U}_s = (\widehat{u}_{0s}, \dots, \widehat{u}_{n-1,s})$, let

$$z_n(x, \eta_{sn}, \widehat{U}_s) = n^{-1/2} \sum_{k=1}^n \bigl[\Delta_k(x + \eta_{sn}\widehat{u}_{k-1,s}) - G(x + \eta_{sn}\widehat{u}_{k-1,s}) - \Delta_k(x) + G(x)\bigr],$$

and define $\widetilde{U}_s$ and $z_n(x, \eta_{sn}, \widetilde{U}_s)$ through $\widetilde{u}_{ks}$ in a similar way. By monotonicity of $\Delta_k(x)$ and $G(x)$ we obtain from (7.8.5) (cf. a similar argument in the proof of Lemma 5.8.1)

$$\sup_x \sup_{|\theta|\le\Theta} |z_{1n}(x,\theta)|$$

$$\le \sup_{i\le N_n-1} \sup_{s\le 3^{m_n}} |z_n(x_{i+1}, \eta_{sn}, \widehat{U}_s)| \tag{7.8.6}$$

$$+ \sup_{i\le N_n} \sup_{s\le 3^{m_n}} |z_n(x_i, \eta_{sn}, \widetilde{U}_s)| \tag{7.8.7}$$

$$+ \sup_{|t_1-t_2|\le N_n^{-1}} \Big| n^{-1/2} \sum_{k=1}^{n} \big[\Delta_k\big(G^{-1}(t_1)\big) - t_1 - \Delta_k\big(G^{-1}(t_2)\big) + t_2\big] \Big| \tag{7.8.8}$$

$$+ \sup_{i\le N_n} \sup_{s\le 3^{m_n}} n^{-1/2} \sum_{k=1}^{n} \big[G(x_{i+1} + \eta_{sn}\widehat{u}_{k-1,s}) - G(x_i + \eta_{sn}\widetilde{u}_{k-1,s})\big]. \tag{7.8.9}$$

We will show that (7.8.6)–(7.8.9) are $o_p(1)$. This conclusion for (7.8.8) is obtained for any sequence $N_n \to \infty$ as in the proof of Theorem 13.1 in Billingsley [**6**]. Consider (7.8.9). By the Taylor formula rewrite the expression under the sup sign as

$$\sqrt{n}\big[G(x_{i+1}) - G(x_i)\big] + g(x_{i+1})\eta_{sn} n^{-1/2} \sum_{k=1}^{n} \widehat{u}_{k-1,s} - g(x_i)\eta_{sn} n^{-1/2} \sum_{k=1}^{n} \widetilde{u}_{k-1,s} + \varepsilon_n, \tag{7.8.10}$$

where

$$|\varepsilon_n| \le 2\,L\,\Theta^{1+\delta} n^{-(1+\delta/2)} \sum_{k=1}^{n} |\widehat{u}_{k-1,s}|^{1+\delta} = o_p(1)$$

by (7.8.4). The third term in (7.8.10) is no greater than

$$\Theta \sup_x g(x) \left[\Big| n^{-1} \sum_{k=1}^{n} u_{k-1} \Big| + 2\,3^{-m_n} n^{-1} \sum_{k=1}^{n} |u_{k-1}| \right] = o_p(1),$$

since $3^{-m_n} = o(1)$ and $n^{-1}\sum_{k-1}^{n} u_{k-1} = o_p(1)$. In a similar way the second term in (7.8.10) is shown to be $o_p(1)$. The first one is $\sqrt{n}N_n^{-1} = o(1)$ by the choice of N_n. Thus (7.8.9) is $o_p(1)$.

Consider now (7.8.7). The arguments here are rather laborious. First we approximate the variables

$$u_{k-1} = \sum_{j=0}^{\infty} \beta^j \varepsilon_{k-1-j}$$

by

$$u^*_{k-1} = \sum_{j=0}^{l_n-1} \beta^j \varepsilon_{k-1-j},$$

which depend on a finite number of ε_j's. Take $l_n \sim -r\log_b n$ with b such that $|\beta| < b < 1$ and an integer $r > 2(1+\delta)/\delta$. Similarly to $\widetilde{u}_{ks}$ in (7.8.3) define

$$\widetilde{u}^*_{ks} = u^*_k\big[1 - 2\Theta n^{-1/2}3^{-m_n}\eta_{sn}^{-1}I(u^*_k \geq 0)\big].$$

Then we can rewrite $z_n(x_i, \eta_{sn}, \widetilde{U}_s)$ as

$$z_n(x_i, \eta_{sn}, \widetilde{U}_s) = n^{-1/2}\sum_{k=1}^n \nu_k(i,s) + n^{-1/2}\sum_{k=1}^n \xi_k(i,s),$$

where

$$\begin{aligned}\nu_k(i,s) &= \Delta_k(x_i + \eta_{sn}\widetilde{u}_{k-1,s}) - G(x_i + \eta_{sn}\widetilde{u}_{k-1,s}) \\ &\qquad - \Delta_k(x_i + \eta_{sn}\widetilde{u}^*_{k-1,s}) + G(x_i + \eta_{sn}\widetilde{u}^*_{k-1,s}), \\ \xi_k(i,s) &= \Delta_k(x_i + \eta_{sn}\widetilde{u}^*_{k-1,s}) - G(x_i + \eta_{sn}\widetilde{u}^*_{k-1,s}) - \Delta_k(x_i) + G(x_i).\end{aligned}$$

Thus (7.8.7) is no greater than

$$\sup_{i\leq N_n}\sup_{s\leq 3^{m_n}}\left(\Big|n^{-1/2}\sum_{k=1}^n \nu_k(i,s)\Big| + \Big|n^{-1/2}\sum_{k=1}^n \xi_k(i,s)\Big|\right).$$

Hence for any $\varepsilon > 0$ one has by Chebyshev's inequality

$$\begin{aligned}&\mathbf{P}_\beta\{\sup_{i\leq N_n}\sup_{s\leq 3^{m_n}}|z_n(x_i, \eta_{sn}, \widetilde{U}_s)| > 2\,\varepsilon\} \\ &\quad\leq \sum_{i=0}^{N_n}\sum_{s=0}^{3^{m_n}}\Big[\mathbf{P}_\beta\Big\{\Big|n^{-1/2}\sum_{k=1}^n \nu_k(i,s)\Big| > \varepsilon\Big\} \\ &\qquad\qquad + \mathbf{P}_\beta\Big\{\Big|n^{-1/2}\sum_{k=1}^n \xi_k(i,s)\Big| > \varepsilon\Big\}\Big] \\ &\quad\leq \sum_{i=0}^{N_n}\sum_{s=0}^{3^{m_n}}\Big[\varepsilon^{-2}\mathbf{E}_\beta\Big|n^{-1/2}\sum_{k=1}^n \nu_k(i,s)\Big|^2 \\ &\qquad\qquad + \varepsilon^{-4}\mathbf{E}_\beta\Big|n^{-1/2}\sum_{k=1}^n \xi_k(i,s)\Big|^4\Big].\end{aligned} \tag{7.8.11}$$

The next two lemmas establish the rate of convergence for the expected values in (7.8.11).

LEMMA 7.8.1.

$$\sup_{i\leq N_n}\sup_{s\leq 3^{m_n}}\mathbf{E}_\beta\Big|n^{-1/2}\sum_{k=1}^n \nu_k(i,s)\Big|^2 = O(n^{-3/2}).$$

PROOF. First we show that

$$\sup_{k,s} \mathbf{E}_\beta |\widetilde{u}_{ks} - \widetilde{u}^*_{ks}| = O(n^{-1}). \tag{7.8.12}$$

By the definition of u^*_{k-1} and the choice of l_n one has

$$\sup_k \mathbf{E}_\beta |u_{ks} - u^*_{ks}| \le \sup_k \mathbf{E}_\beta \sum_{j \ge l_n} b^j |\varepsilon_{k-j}| = O(b^{l_n}) = O(n^{-r}). \tag{7.8.13}$$

In view of the definitions of $\widetilde{u}_{ks}$ and $\widetilde{u}^*_{ks}$ and the inequality $|2\,\Theta n^{-1/2} 3^{-m_n} \eta_{sn}^{-1}| \le 2$,

$$\sup_{k,s} \mathbf{E}_\beta |\widetilde{u}_{ks} - \widetilde{u}^*_{ks}| \le 3\,\mathbf{E}_\beta |u_1 - u^*_1| + 2\,\mathbf{E}_\beta \big|u_1 \big[I(u^*_1 \ge 0) - I(u_1 \ge 0)\big]\big|. \tag{7.8.14}$$

By Hölder's inequality the second term in the right-hand side of (7.8.14) is no greater than

$$\begin{aligned} &2\big(\mathbf{E}_\beta |u_1|^{1+\delta}\big)^{1/(1+\delta)} \big(\mathbf{E}_\beta |I(u^*_1 \ge 0) - I(u_1 \ge 0)|^{(1+\delta)/\delta}\big)^{\delta/(1+\delta)} \\ &\quad \le c\big(\mathbf{E}_\beta |I(u^*_1 \ge 0) - I(u_1 \ge 0)|\big)^{\delta/(1+\delta)} \\ &\quad = c\big[\mathbf{P}_\beta\{u^*_1 \ge 0,\, u_1 < 0\} + \mathbf{P}_\beta\{u^*_1 < 0,\, u_1 \ge 0\}\big]^{\delta/(1+\delta)}. \end{aligned}$$

By Chebyshev's inequality we get

$$\begin{aligned} \mathbf{P}_\beta\{u^*_1 \ge 0,\, u_1 < 0\} &\le \mathbf{P}_\beta\{|u^*_1 - u_1| \ge n^{-(1+\delta)/\delta}\} + \mathbf{P}_\beta\{-n^{(1+\delta)/\delta} < u_1 < 0\} \\ &\le n^{(1+\delta)/\delta} \mathbf{E}_\beta |u^*_1 - u_1| + n^{-(1+\delta)/\delta} \sup_x g(x) \le c_1 n^{-(1+\delta)/\delta} \end{aligned}$$

where the constants $c_1, c_2, \dots$ are independent of n, i, k, s. In a similar way one estimates $\mathbf{P}_\beta\{u^*_1 < 0,\, u_1 \ge 0\}$. Thus the second term in the right-hand side of (7.8.14) is $O(n^{-1})$. Together with (7.8.14), (7.8.13) this implies (7.8.12).

It remains to note that the variables $\{\nu_k(i,s),\ k = 1, 2, \dots, n\}$ are centered, uncorrelated, and

$$\mathbf{E}_\beta \nu_k^2(i,s) \le \mathbf{E}_\beta |\eta_{sn} \widetilde{u}_{k-1,s} - \eta_{sn} \widetilde{u}^*_{k-1,s}| \le c_2 n^{-3/2}$$

by (7.8.12) (a similar argument is given in more detail in the proof of Lemma 5.8.1). Hence the lemma follows. □

LEMMA 7.8.2.

$$\sup_{i,s} \mathbf{E}_\beta \Big| n^{-1/2} \sum_{k=1}^{n} \xi_k(i,s) \Big|^4 = O(n^{-1} \log^3 n).$$

PROOF. Observe that $\{\xi_k(i,s),\ k=1,\dots,n\}$ are centered and uncorrelated p-dependent random variables with $p = l_n \sim -r\log_b n$. Moreover,

$$|\xi_k(i,s)| \le 2 \quad \text{and} \quad \mathbf{E}_\beta \xi_k^2(i,s) \le \mathbf{E}_\beta |\eta_{sn} \widetilde{u}_{k-1}^*| \le 3\,\Theta n^{-1/2} \mathbf{E}_\beta |u_{k-1}|$$

by (7.8.4). Hence one easily infers that

$$\mathbf{E}_\beta |n^{-1/2} \sum_{k=1}^n \xi_k(i,s)|^4 \le c_3 n^{-1} l_n^3 + \sup_{k\le n} \left(\mathbf{E}_\beta \xi_k^2(i,s)\right)^2 = O(n^{-1}\log^3 n)$$

uniformly in i, s. This proves the lemma. □

Lemmas 7.8.1 and 7.8.2 imply that (7.8.11) is

$$O(N_n\, 3^{m_n}\, n^{-1} \log^3 n) = O(n^{-1/2} \log^5 n) = o(1)$$

by the choice of the sequences N_n and 3^{m_n}. Therefore (7.8.7) is $o_p(1)$. In a similar way one shows that (7.8.6) is also $o_p(1)$. Thus (7.8.2) is established, which together with (7.8.1) proves the theorem. □

Bibliography

1. S. V. Al′tshuler, *Parameter estimation methods for autoregressive-moving average processes*, Avtomat. i Telemekh. **8** (1982), 5–18; English transl., Remote Control **43** (1982), no. 8, 979–990.
2. T. W. Anderson, *The statistical analysis of time series*, Wiley, New York, 1971.
3. B. von Bahr and C. G. Esséen, *Inequalities for the rth absolute moment of a sum of random variables,* $1 \leqslant r \leqslant 2$, Ann. Math. Statist. **36** (1965), 299–303.
4. G. Basset and R. Koenker, *Asymptotic theory of least absolute error regression*, J. Amer. Statist. Assoc. **73** (1978), 618–622.
5. P. J. Bickel and K. A. Doksum, *Mathematical statistics. Basic ideas and selected topics*, Holden-Day, San Francisco, 1977.
6. P. Billingsley, *Convergence of probability measures*, Wiley, New York, 1968.
7. P. Bloomfield and W. L. Steiger, *Least absolute deviations. Theory, applications, and algorithms*, Birkhäuser, Boston, 1983.
8. M. V. Boldin, *Estimation of the distribution of noise in an autoregression scheme*, Teor. Veroyatnost. i Primenen. **27** (1982), no. 4, 905–910; English transl., Theory Probab. Appl. **27** (1982), no. 4, 866–871.
9. ______, *Testing of hypotheses in autoregressive schemes by Kolmogorov and* ω^2 *tests*, Dokl. Akad. Nauk SSSR **273** (1983), no. 1, 19–22; English transl., Soviet Math. Dokl. **28** (1983).
10. ______, *On hypothesis testing in a moving average scheme by the Kolmogorov–Smirnov and the* ω^2 *tests*, Teor. Veroyatnost. i Primenen. **34** (1989), no. 4, 758–764; English transl., Theory Probab. Appl. **34** (1990), no. 4, 699–704.
11. ______, *On median estimates and tests in autoregressive models*, Math. Methods Statist. **3** (1994), no. 2, 114–129.
12. ______, *On nonparametric sign tests in multiparameter autoregression*, Math. Methods Statist. **4** (1995), no. 4, 435–448.
13. ______, *On the least absolute deviations estimator in a nonstationary autoregression and testing for stationarity*, Teor. Veroyatnost. i Primenen. **41** (1996), no. 2, 409–417; English transl. in Theory Probab. Appl. **34** (1996), no. 2.
14. ______, *On nonparametric sign estimators in multiparameter autoregression*, Math. Methods Statist. **5** (1996), no. 2, 154–172.
15. ______, *On the empirical distribution function of residuals and rank estimators in autoregression*, Math. Methods Statist. **6** (1997), no. 1 (to appear).
16. M. V. Boldin and Yu. N. Tyurin, *On nonparametric sign procedures in autoregression models*, Math. Methods Statist. **3** (1994), no. 4, 279–305.
17. L. N. Bol′shev and N. V. Smirnov, *Tables of mathematical statistics*, "Nauka", Moscow, 1983. (Russian)
18. G. E. P. Box and G. M. Jenkins, *Time series analysis, forecasting, and control*, Holden-Day, San Francisco, 1970.
19. O. Bustos and V. J. Yohai, *Robust estimates for ARMA models*, J. Amer. Statist. Assoc. **81** (1986), 155–168.
20. D. R. Cox and D. V. Hinkley, *Theoretical statistics*, Chapman and Hall, London, 1974.
21. E. Denoël and J. P. Solvay, *Linear prediction of speech with a least absolute error criterion*, IEEE Trans. on ASSP **33** (1985), 1337–1403.
22. B. A. Dubrovin, S. P. Novikov, and A. T. Fomenko, *Modern Geometry – methods and applications. Part I. The Geometry of surfaces, transformation groups, and fields*, 2nd ed., Graduate Texts in Mathematics, vol. 93, Springer–Verlag, New York, 1992.
23. J. M. Dufour, *Rank tests for serial dependence*, J. Time Ser. Anal. **2** (1981), 117–128.

24. J. Durbin, *Weak convergence of the sample distribution function when parameters are estimated*, Ann. Statist. **1** (1973), 279–290.
25. K. O. Dzhaparidze, *Tests of composite hypotheses for random variables and stochastic processes*, Teor. Veroyatnost. i Primenen. **22** (1977), 104–118; English transl., Theory Probab. Appl. **22** (1977), 104–118.
26. F. Eiker, *Asymptotic normality and consistency of the least squares estimators for families of linear regressions*, Ann. Math. Statist. **34** (1963), 447–456.
27. G. B. Evans and N. E. Savin, *The calculation of the least square estimator of the parameter in a random walk model*, Ann. Statist. **9** (1981), 1114–1118.
28. N. E. Ferretti, D. M. Kelmansky, and V. J. Yohai, *Estimators based on ranks for ARMA models*, Commun. Statist. A – Theory Meth. **20** (1991), no. 12, 3879–3907.
29. W. L. Freedman, B. F. Madore, J. R. Mould, R. Hill, L. Ferrarese, R. C. Kennicutt, Jr., A. Saha, P. B. Stetson, J. A. Graham, H. Ford, J. G. Hoessel, J. Huchra, S. M. Hughes, and G. D. Illingworth, *Distance to the virgo cluster galaxy M100 from Hubble Space Telescope observations of Cepheids*, Nature **371** (1994), 757–762.
30. J. Hájek and Z. Šidák, *Theory of rank tests*, Academia, Prague, 1967.
31. M. Hallin, J. Fr. Ingenbleek, and M. L. Puri, *Linear serial rank tests for randomness against ARMA alternatives*, Ann. Statist. **13** (1985), no. 3, 1156–1181.
32. ———, *Linear and quadratic rank tests for randomness against serial dependence*, J. Time Ser. Anal. **8** (1987), 409–424.
33. M. Hallin and M. L. Puri, *Optimal rank-based procedure for time-series analysis: Testing an ARMA model against other ARMA models*, Ann. Statist. **16** (1988), 402–432.
34. ———, *Aligned rank test for linear models with autocorrelated error terms*, J. Multivariate Anal. **50** (1994), no. 2, 175–237.
35. F. R. Hampel, *The influence curve and its role in robust estimation*, J. Amer. Statist. Assoc. **69** (1974), 383–393.
36. F. R. Hampel, E. M. Ronchetti, R. J. Rousseeuw, and W. A. Stahel, *Robust statistics. The approach based on influence functions*, Wiley, New York, 1986.
37. E. J. Hannan, *Multiple time series*, Wiley, New York, 1970.
38. S. Heiler and R. Willers, *Asymptotic normality of R-estimates in the linear model*, Statistics **19** (1988), 173–184.
39. Th. P. Hettmansperger, *Statistical inference based on ranks*, Wiley, New York, 1984.
40. J. L. Hodges, Jr., and E. L. Lehmann, *Rank methods for combinations of independent experiments in analysis of variance*, Ann. Math. Statist. **33** (1962), 482–497.
41. ———, *Estimates of location based on rank tests*, Ann. Math. Statist. **34** (1963), 598–611.
42. M. Hollander and D. A. Wolfe, *Nonparametric statistical methods*, Wiley, New York, 1973.
43. E. Hubble, *A relation between distance and radial velocity among extra-galactic nebulae*, Astronomy (Proc. N.A.S.) **15** (1929), 168–173.
44. P. J. Huber, *Robust statistics*, Wiley, New York, 1981.
45. I. A. Ibragimov and Yu. V. Linnik, *Independent and stationary sequences of random variables*, "Nauka", Moscow, 1965; English transl., Wolters–Noordhoff, Groningen, 1971.
46. I. A. Ibragimov and R. Z. Khas′minskii, *Statistical estimation: asymptotic theory*, "Nauka", Moscow, 1979; English transl., Springer–Verlag, Berlin–New York, 1981.
47. J. Jacod and A. N. Shiryaev, *Stochastic processes*, Springer–Verlag, New York, 1987.
48. G. M. Jenkins and D. G. Watts, *Spectral analysis and its applications*, Holden Day, San Francisco, 1969.
49. R. L. Kashyap and A. R. Rao, *Dynamic stochastic models from empirical data*, Academic Press, New York, 1976.
50. E. V. Khmaladze, *The use of ω^2 tests for testing parametric hypotheses*, Teor. Veroyatnost. i Primenen. **24** (1979), no. 2, 280–297; English transl., Theory Probab. Appl. **24** (1979), no. 2, 283–301.
51. ———, *Martingale approach in the theory of goodness-of-fit tests*, Teor. Veroyatnost. i Primenen. **26** (1981), 240–257; English transl., Theory Probab. Appl. **26** (1981), 240–257.
52. H. L. Koul, *Asymptotic behavior of Wilcoxon type confidence regions in multiple linear regression*, Ann. Math. Statist. **40** (1969), 1950–1979.
53. ———, *A class of ADF tests for subhypothesis in the multiple linear regression*, Ann. Math. Statist. **41** (1970), 1273–1281.

54. H. L. Koul and S. Levental, *Weak convergence of the residual empirical process in explosive autoregression*, Ann. Statist. **17** (1989), no. 4, 1784–1794.
55. H. L. Koul and J. C. Pflug, *Weakly adaptive estimators in explosive autoregression*, Ann. Statist. **18** (1990), no. 2, 939–960.
56. W. S. Krasker and R. E. Welsch, *Efficient bounded-influence regression estimation*, J. Amer. Statist. Assoc. **77** (1982), 595–604.
57. J. P. Kreiss, *On adaptive estimation in stationary ARMA processes*, Ann. Statist. **15** (1987), 112–133.
58. ———, *Testing linear hypotheses in autoregressions*, Ann. Statist. **18** (1990), 1470–1482.
59. ———, *Estimation of the distribution function of noise in stationary processes*, Metrica **38** (1991), 285–297.
60. E. L. Lehmann, *Nonparametrics: statistical methods based on ranks*, Holden-Day, San-Francisco, 1975.
61. ———, *Theory of point estimation*, Wiley, New York, 1983.
62. B. Maldelbrot, *The variation of certain speculative prices*, J. Business **36** (1963), 394–419.
63. ———, *The variation of some other speculative prices*, J. Business **40** (1967), 383–413.
64. R. D. Martin, *Robust estimation for time series autoregressions*, in Robustness in Statistics (R L. Launer and G. N. Wilkinson, eds.), Academic Press, New York, 1979, pp. 147–176.
65. ———, *The Cramér–Rao bound and robust M-estimates for autoregressions*, Biometrika **69** (1982), 437–442.
66. R. D. Martin and V. J. Yohai, *Influence functionals for time series*, Ann. Statist. **14** (1986), 781–818.
67. G. V. Martynov, *Computation of limit distributions of statistics for normality test of type* ω^2, Teor. Veroyatnost. i Primenen. **21** (1976), no. 1, 3–15; English transl., Theory Probab. Appl. **21** (1976), no. 1, 1–13.
68. ———, *Omega-squared tests*, "Nauka", Moscow, 1978. (Russian)
69. A. Mokkadem, *Mixing properties of ARMA processes*, Stochastic Processes Appl. **29** (1988), 309–315.
70. E. S. Pearson and H. O. Hartley (eds.), *Biometrika tables for statisticians,* II, Cambridge Univ. Press, Cambridge, 1972.
71. V. I. Piterbarg, *Asymptotic methods in the theory of gaussian processes and fields*, Transl. Math. Monographs, vol. 148, Amer. Math. Soc., Providence, RI, 1995.
72. D. Pollard, *Asymptotics for least absolute deviation regression estimators*, Econometrics Theory **7** (1991), 186–199.
73. B. T. Polyak and Ya. Z. Tsypkin, *Optimal methods of estimating autoregression coefficients in the case of incomplete information*, Izv. Akad. Nauk SSSR Tekhn. Kibernet. **6** (1983), no. 1, 118–126; English transl., Soviet J. Comput. Systems Sci. **21** (1983), no. 1, 100–109.
74. M. L. Puri and P. K. Sen, *Nonparametric methods in general linear models*, Wiley, New York, 1985.
75. M. M. Rao, *Asymptotic distribution of an estimator of the boundary parameter of an unstable process*, Ann. Statist. **6** (1978), 185–190.
76. S. N. Roy, *Some aspects of multivariate analysis*, Wiley, New York, 1957.
77. A. S. Sharov and I. D. Novikov, *Edwin Hubble, the discoverer of the big bang universe*, Cambridge University Press, Cambridge, 1993.
78. A. N. Shiryaev, *Probability*, "Nauka", Moscow, 1980; English transl, Springer–Verlag, New York, 1984.
79. A. N. Shiryaev and V. G. Spokoiny, *Statistical experiments and decisions: asymptotic theory*, unpublished manuscript.
80. G. Simonova, *The application of software SIGN in teaching new methods of statistical data analysis*, International Conference "Statistical Education in Modern World: Ideas, Orientations, Technologies," 3–5 July, 1996, Abstracts (1996), St.-Petersburg University of Economics and Finance, St.-Petersburg, 224–225.
81. S. M. Stigler, *The history of statistics. The measurements of uncertainty before 1900*, The Beeknap Press of Harvard University Press, Cambridge, 1986.
82. J. W. Tukey, *Exploratory data analysis. Preliminary Edition*, Addison-Wesley, Reading, MA, 1977.
83. A. N. Tyulyagin, *On asymptotic admissibility of goodness-of-fit tests*, Izv. Akad. Nauk SSSR Ser. Mat. **46** (1982), 569–616; English transl., Math. USSR-Izv. **20** (1983), no. 3, 535–575.

84. Yu. N. Tyurin, *On parametric hypotheses testing with non-parametric tests*, Teor. Veroyatnost. i Primenen. **15** (1970), no. 4, 745–749; English transl., Theory Probab. Appl. **15** (1970), no. 4, 722–725.
85. ———, *Linear model in multidimensional nonparametric statistics*, in Multidimensional Statistical Analysis in Social and Economics Research, "Nauka", Moscow, 1974, pp. 7–24. (Russian)
86. ———, *The limit distribution of the Kolmogorov–Smirnov statistics for a composite hypothesis*, Izv. Akad. Nauk SSSR Ser. Mat. **48** (1984), no. 6, 1314–1343; English transl., Math. USSR-Izv. **25** (1985), no. 3, 619–646.
87. ———, *Statistical inference for data with group structure*, in Data Analysis, Estimation of Parameters and Social Choice. Collections of Papers, All-Union Institute of System Studies, Moscow, 1984, pp. 3–28. (Russian)
88. ———, *Sign linear statistical analysis*, in Data Analysis, Estimation of Parameters and Social Choice in System Studies. Collections of Papers, All-Union Institute of System Studies, Moscow, 1986, pp. 4–16. (Russian)
89. ———, *Nonparametric methods for linear models of time series*, in System Research in Computer Sciences and Statistics, The Institute for System Analysis of Russian Academy, Moscow, 1992, pp. 54–61. (Russian)
90. Yu. N. Tyurin and G. L. Simonova, *Sign analysis of linear models*, Surveys in Applied and Industrial Mathematics, Probability and Statistics **1** (1994), no. 2, 214–278. (Russian)
91. G. Vaucouleurs, A. Vaucouleurs, H. G. Corwin, R. J. Buta, G. Paturel, and R. Fouque, *Third reference catalog of bright galaxies* (*RC3*), Springer–Verlag, New York, 1991.
92. A. T. Walden, *Non-Gaussian reflectivity, entropy and deconvolution*, Geophysics **50** (1985), 2862–2888.
93. V. J. Yohai and R. A. Maronna, *Asymptotic behaviour of least-squares estimates for autoregressive processes with infinite variances*, Ann. Statist. **5** (1977), no. 3, 554–560.

Selected Titles in This Series

(Continued from the front of this publication)

(See the AMS catalog for earlier titles)